The Story of the Herefords: Part Two
The Origin and Development of the Herfordshire Breed of Cattle

by Alvin H. Sanders

with an introduction by Jackson Chambers

This work contains material that was originally published in 1914.

This publication is within the Public Domain.

This edition is reprinted for educational purposes
and in accordance with all applicable Federal Laws.

Introduction Copyright 2018 by Jackson Chambers

Self Reliance Books

Get more historic titles on animal and stock breeding, gardening and old fashioned skills by visiting us at:

http://selfreliancebooks.blogspot.com/

Introduction

I am pleased to present another title in the "Cattle" series.

The work is in the Public Domain and is re-printed here in accordance with Federal Laws.

As with all reprinted books of this age that are intended to perfectly reproduce the original edition, considerable pains and effort had to be undertaken to correct fading and sometimes outright damage to existing proofs of this title. At times, this task is quite monumental, requiring an almost total "rebuilding" of some pages from digital proofs of multiple copies. Despite this, imperfections still sometimes exist in the final proof and may detract from the visual appearance of the text.

I hope you enjoy reading this book as much as I enjoyed making it available to readers again.

Jackson Chambers

CHAPTER XII.

THE SHOCK OF SHOWYARD WAR.

The very highest degree of excellence, indeed approximate perfection, is required to win blue and purple ribbons in great open competitions. This is the reason why breeders always have a pride in tracing the descent of their cattle, if possible, from ancestors whose titles to superiority have been made clear by notable showyard victories. It may therefore serve a useful purpose to take up in some detail the roster of great names developed by some of the more memorable contests that occurred during the years immediately following this great influx of Herefordshire's best blood into the western states, for out of this flood of importations emerged the herds, distributed far and wide, that have since made the blood practically available to all parts of the country upon a legitimate commercial basis.

As we have already pointed out, Sir Charles, Success, Hero, and the Anxieties led the early line of Herefords in the great battle of the breeds that constituted so marked a feature of the western fairs from 1876 to the later '80's. They were followed by many British showyard stars of the first magnitude; but within a comparatively short space of time American breeders were producing by a judicious blending of the various bloods now in their possession show cattle equaling, if not indeed excelling in some respects, the best showyard types evolved in England.

The Great Fight of 1882.—Let us start with a reference to certain old-time exhibitions at important fairs of the cornbelt states which gave a zest to those events previously unknown—those foolish old days of "picked-up" committees and breed competitions, when the shrewdest "wire-puller" or the most successful "bull-dozer" among the exhibitors stood a good chance of obtaining results which in these later times would be impossible.

There was a time when the real battle of the breeds began at two important district shows in the state of Indiana, one at Lafayette and the other at Crawfordsville. Liberal money prizes were provided for open classes at both places, and this chance to line up alongside the Shorthorns was eagerly accepted by the zealous new champions of the Hereford.

At that date no American newspaper was making any effort to report the contests at these events in any detail, and noting the necessity for such a service the newly established "Breeder's Gazette" undertook the task, not however without serious misgivings. The story of the Lafayette show of 1882, told in the issue of that paper for Sept. 14 of that year, would probably bring a broad smile to the seasoned reader of such reviews in these later days, but it nevertheless suffices to reflect clearly a decidedly tense situation.

The Shorthorn colors were borne by the following: J. H. Potts & Son, with their Duke of Richmond blood; "Uncle Harvey" Sodowsky, as sly an old showman as ever set foot in a showyard, and owner of that great show cow Jessie Hopewell; Thomas Wilhoit, one of the best breeders and feeders in Indiana, with his famous Knight of Athelstane; Launcelot Palmer of Missouri, with old Loudon Duke of Greenwood; besides Stevenson & Son and Croft Bros., both of Illinois.

The Hereford cause now no longer depended upon Mr. Miller, for Messrs. Culbertson, Earl & Stuart, Fowler & VanNatta, Thomas Clark, William Constable, C. K. Parmelee, and Benjamin Hershey had got their heads together during the summer months and resolved to go after the enemy en masse. Each of these fitted and sent into the Lafayette, Crawfordsville, Peoria and St. Louis shows of that year the best of their top cattle—some 70 head in all, and a mighty phalanx it was.

It does not seem to matter much just now as to what happened that week at Lafayette, but it certainly had meaning enough at the time, not only to those who participated as exhibitors, but to the thousands of farmers and cattle growers in all parts of the country who were interested in the warfare being waged at these exhibitions of breeding animals, as well as at the Chicago Fat Stock Show.

The class judging came first. Constable won with Hero over Clark's Anxiety 3d and Culbertson's Sir Garnet by The Grove 3d; Earl & Stuart had first on two-year-old bull with the massive imported Royal 16th, over Romeo and Tregrehan; Shadeland scored again on yearling bulls with Prince Edward, against Anxiety 4th and Anxiety 5th and others. Culbertson won first in a strong class for cows with imp. Downton Rose, second going to Shadeland's imp. Lady 3d*; Clark won in the two-year-old heifers with Peerless, as against Venus; a rare lot of yearling heifers were headed by Hershey's Miss Broadgauge 2d by Horatius, with Earl & Stuart's Wilton heifer Delight 2d as the runner-up. Prettymaid and Duchess 2d were unplaced in this extraordinary company. It will be noted that the entire outfit of winners was imported.

There was no end of trouble before a committee was finally secured to judge the open classes. With the Hereford men it was "anything to beat the Shorthorns." It mattered not at all which Hereford bull, cow or herd might be chosen. The only question was how to prevent the "Shorthorn crowd from putting up a job" to beat them. There were nightly councils of war at the old Hotel Lahr. Scouts were appointed to give warning of attempted unfair tactics. Culbertson was in command of a coterie of comrades under perfect discipline. Man after man tendered by the society to serve as a juror was challenged and rejected, sometimes because he was a cattle feeder instead of a breeder, and sometimes vice versa. Worn out at length by

---

*While this cow, Lady 3d by Horace 2d, had been shown regularly since a calf, she had dropped three calves by the time she was five years old.

the irreconcilable differences between the spokesmen for the rival hosts, the superintendent finally named a committee without reference to the wishes of either side, and the big fight was then waged with the following result:

In the graded herd competition Potts was first, Culbertson second and Sodowsky third. In young herds Fowler & VanNatta were first, Potts second and Parmelee third. Constable took the bull championship with Hero and Sodowsky the cow championship with Jessie Hopewell.

It might interest some of the younger generation of cattle breeders to know that of the 30 Shorthorns shown in these championship contests there were but two roans, and of course no whites.

"The Breeder's Gazette" commented on the esprit du corps in the Hereford camp upon this memorable occasion, and stated that the Hereford aggregation, including so many English Royal winners, would be seen all the way through the central circuit, with their campaign practically directed by one man, working not for himself but for the glory and honor of the breed. This drew out from Hon. J. H. Pickrell, the old-time Shorthorn showman, of Baron Booth of Lancaster fame, the following comment:

"Why should 70 cattle travel in one body, under one major general, to a fair that gives as the highest prize (in class) but $20, unless it is to awe the whole country with its grandeur? It must be a weak cause that needs such bolstering. Let me ask, why make such a war at all? Time settles all things and will in the end tell us what will best shorten a Texas horn, what breed will thrive best and make the greatest improvement generally."

At Crawfordsville the week following the excitement was again intense, and when the gates closed it was found that while some re-distribution of honors had been made the break was so even as to demonstrate that it was a case of Greek versus Greek. Potts had first in the big herd competition, Sodowsky second and Culbertson third. In young herds Earl & Stuart were first, Fowler & VanNatta second and Parmelee third. Tregrehan was champion bull and Jessie Hopewell champion female.

And so the battle raged that year—with new converts clearly gained for the "white faces," and with Shorthorn breeders turning more and more towards the Scotch blood that was bearing so well the brunt of this unprecedented attack upon their position.

"**Rudolph's Year.**"—The event of the year 1883 was the appearance of Rudolph, probably the greatest son of The Grove 3d, flashed upon the American public at Des Moines. G. S. Burleigh had his imp. Anxiety 2d on the western circuit, then three years old and shown at a weight of 2,535 pounds. Gudgell & Simpson had a strong herd of Aberdeen-Angus on the road, headed by old Knight of St. Patrick, so that the open beef-herd championship was hotly contested. Potts and Robert Miller won first and second respectively, and the Wyoming Herefords were third.

W. C. McGavock, who had started in with Herefords at Franklin, Mo., sent a show herd out this year headed by Gypsy Boy, a 1,900-pound three-year-old that took first at the Minnesota State Fair and at other shows.

J. S. Hawes of Colony, Kans., and W. E. Campbell of Caldwell, Kans., were conspicuous exhibitors this season in the west; at Topeka the latter beat his Shorthorn opponents in the bull championship, and the former gained an open prize for bull with five of his calves. It was here that Campbell first exhibited his famous yearling grade heifer, Texas Jane.*

In Michigan, Phelps of Pontiac, William Hamilton of Flint, and Thomas Foster flew the Hereford flag. Unfortunately we are without the names of their prize-winners.

The big show of 1883 was at the Illinois State Fair, held that year at Chicago. Rudolph, Tregrehan, Anxiety 2d, Anxiety 3d, and Sir Garnet were there, and Mr. Burleigh's bull had the honor of standing next to Rudolph. Fowler & VanNatta gained the grand beef herd championship over the Potts Shorthorns with a lot that included Tregrehan, Ethel, Viola, Fancy, and Jewel 3d. Tom

---

*Over Texas Jane's stall was this placard:
"I was born on W. E. Campbell's ranch Aug. 19, 1882, and was at once christened
Texas Jane.
My father was a Hereford thoroughbred;
My mother a wild Texas scrub;
The cross makes me easily fed,
And I am able to rustle for grub.

Don't stare at the meat on my back,
Or be surprised at my snowy-white face;
For 'tis all the work of Pa Pa
That gives me this Hereford grace."

Clark was already becoming famous for the quality of his heifers and drew first in both the three- and two-year-old classes, and second on yearlings.

At St. Louis the Wyoming herd, with Rudolph at the head, gained the lion's share of the honors, opposition being offered by J. S. Hawes and W. C. McGavock.

The Scare of 1884.—The discovery of what the official veterinarians called contagious pleuro-pneumonia among certain dairy and distillery-fed cattle in Illinois frightened most of those who had fitted herds for show in 1884 into keeping their cattle at home, so that the exhibits were more or less meager all around.

Tom Clark and George Leigh ventured out, however, at Chicago. Clark showed Hero this year, a bull that would have stood a lot of competition. He had first in his class, and ribbons decorated the halters of Peerless, Duchess 12th, Silver, Jessie 2d, and Flossie.

While the show season was practically a failure, there was a big trade in 1884 in Herefords all over the west, although prices had begun to sag off under excessive importations which were somewhat below the levels reached a few years previous. At a combination sale at Kansas City in September Gudgell & Simpson, George Leigh & Co., and Frank Crane sold 70 head at an average of $514.57. It was here that E. S. Shockey bought the Anxiety bulls Beau Monde, at $1,000, and Beau Real, as a calf, at $300.

George W. Henry bought imp. Royal 16th at the $1,000 figure. W. E. Campbell paid $1,230 for imp. Miss Archibald, and George Morgan gave $1,500 for Primrose 2d.

The Shows of 1885.—At the Iowa State Fair the Iowa Hereford Cattle Co.'s Washington was the first-prize aged bull, Sergeant Major heading the two-year-old class and the famous cow Melody 10th gaining first among the aged females.

In the competition for the grand beef herd championship, however, all breeds competing, the S. S. Brown Shorthorns were first, T. W. Harvey's Turlington Aberdeen-Angus were second, and the Herefords third.

At Minnesota J. O. Curry and J. C. Bertram of Aurora, Ill., made exhibits, Mr. Bertram showing his stylish yearling Wilton bull Sir Wilfred and Curry gaining first in the cow class with Delight (the dam of Peerless), then ten years old. Mr. Culbertson was also an exhibitor and had second prize here on Helen by Anxiety.

At the old Western National Show at Bismarck Grove, Lawrence, Kans., in those days one of the leading agricultural fairs of the west, J. H. Hawes was first with his famous Fortune, son of Sir Richard 2d. Other exhibitors were Lucien Scott, G. A. Fowler of Kansas City, and Shockey & Gibb of Lawrence.

At the Illinois State Fair, held in Chicago, the Herefords were unquestionably the strongest class on the grounds. Washington was the first-prize aged bull and Sergeant Major was again winner over a great string of two-year-olds. The Iowa company's Melody was adjudged the best cow and Fowler & VanNatta were second with Lark, one of the most notable members of their herd at that time.

At St. Louis Hawes, Crane, and F. H. Jackson of South Bend, Ind., competed; Fortune, shown at a weight of 2,550 pounds, won first, and Crane had second on the good bull Grimley 9443.

The Business Overdone.—During the year 1885, notwithstanding the activity on the surface, signs began to multiply which indicated that the market for purebred Herefords had for the time being been overstocked. Stimulated by the widespread display of interest in the "white faces", importation had followed importation until more cattle were on the market than could be taken care of at prices approaching those prevailing during the years immediately preceding. By this time many herds had sprung into existence in nearly all the leading agricultural states, and various speculators were bringing over cattle which were being offered both at public and private sale. Those who had borne the heat and burden of the day in introducing the breed in the west, and who had already made large investments in the very best of the British show and breeding stock, began to feel the desirability of in some way restricting the number of cattle being offered.

One Hundred Dollar Tax on Imported Cattle.—At the annual meeting of the American Hereford Breeders' Association in Chicago on Nov. 13, 1885, membership in the association was restricted to American citizens, and a new rule was adopted providing that all imported cattle thereafter offered for entry in the American herd book should pay a recording fee of $100. The object of this was obvious—the checking of free importations of the inferior or low-grade cattle which were being bought in England in the hope of reselling in America at a profit.

At this same meeting a rule was adopted providing that every animal imported before Nov. 13, 1885, and every animal calved in America prior to Jan. 1, 1886, should not be eligible to record in the American herd book after July 1, 1886; that application for entry must be made within six months; and that out of every ten bulls dropped as the property of any member after Jan. 1, 1886, only nine should be eligible to record, the object of this being to insure the discarding of 10 per cent of the bull calf crop. A resolution was also adopted providing that cattle imported prior to March 1, 1886, by breeders unaware of these new rules should be admitted on submission of proper evidence to that effect. This meeting was presided over by Dr. O. Bush of Sheldon, Ill., at that time president of the association. Needless to say, the radical action aroused some bitterness and provoked acrimonious debate.

The results of the trade both at public and private sale in England and America during the year 1885 indicated clearly that "the bloom was off the rye," so far as fancy prices for anything except the very best cattle were concerned. The fact is, that the pace had been too fast, the business was being overdone. While quite a number of importers and dealers were financially crippled during the slump in values that materialized about this date, there was no doubt but that in the long run the period of liquidation which now set in was beneficial so far as the ultimate best interest of the Herefords in the United States was concerned. Excellent cattle were now being produced in the United States from past importations, and it was no longer essential that the herds of Herefordshire be heavily drawn upon to supply home demands.

The Famous Invasion of Kentucky.—The year 1886 is memorable in the annals of American beef cattle breeding for the vigorous effort made by the Illinois and Indiana champions of the Herefords to storm the one great citadel of Shorthorn power—the blue grass region of Kentucky. For the span of two generations Kentucky had been wedded to the "red, white and roan." Throughout all the years that the "white faces" had been steadily gaining ground north of the Ohio River they had received no encouraging word from "the Blue Grass." The Alexanders, Renicks, Vanmeters, Cunninghams, Bedfords, Goffs, Clays, Warfields, Hamiltons, and their contemporaries had for years reigned supreme in their capacity as purveyors-in-chief to the farmers of the Mississippi Valley, and latterly to the ranchmen of Texas and the southwest in general, of all that was deemed best in the way of good cattle. Accustomed for two generations to the patronage of the leading cattle growers of the west they scoffed at the pretensions of the Herefords, and were slow to admit what their colleagues in the north had already conceded—that the newcomers from Herefordshire were destined henceforth to divide the honors with their favorites.

The two leading shows of 1886 in "the Blue Grass" were scheduled for Shelbyville and Lexington. At these two points the charge of Pickett's heroes at Gettysburg was in a bovine sense duplicated, and with like results. At the risk of wearying somewhat the readers of this volume, the author ventures to incorporate at this point liberal excerpts from his own attempt at telling the story of this undertaking—practically his first reportorial effort in the way of a detailed account of an event of such character. We quote from "The Breeder's Gazette" of Sept. 2, 1886:

"THE HEREFORDS IN KENTUCKY.

"REPELLED AT SHELBYVILLE BY BLUE GRASS SHORTHORNS.

" 'And darest thou then
To beard the lion in his den, the Douglas in his hall?'

"The sensation of the showyard season just inaugurated, so far as the beef breeds of cattle are concerned, is the 'nervy' attempt of leading breeders of Herefords to force a hearing in the most 'solid' of all American Shorthorn breeding districts, the Blue Grass country of Kentucky.

"For weeks past the threatened invasion by the 'white faces' has been the one theme of conversation among the Shorthorn people of the locality named, and as the clans began to gather for the fray last week the good citizens of Shelbyville suddenly found their quiet little city transformed into a bustling camp of warring factions with but one name upon every lip: 'The Herefords!' Verily the Shorthorn citadel was shaken from center to circumference, and as the long line of deep-fleshed wanderers from Herefordshire wound its way through their gates, with Fowler, Sir Bartle Frere, Bowdoin, Prince Edward, and Caractacus as their chiefs, those who had been born and raised with the supremacy of the 'red, white and roan,' undisputed and unchallenged were treated to a sight such as the eyes of many who gazed with eager interest had never before regaled themselves. Hundreds of those who came to visit the show had never seen a Hereford, and it is but simple justice to the breed to state that some, at least, who had apparently come to scoff remained to admire; and while the visitors were unable to snatch a victory from out the jaws of what they had all along expected would prove a defeat, they feel that a missionary work has been accomplished that will some day return a reasonable profit.

"It goes without saying that the invading column was a strong one, representing, as it did, the first-class herds of Messrs. Adams Earl, of Lafayette, Ind.; Fowler & VanNatta, Fowler, Ind.; C. M. Culbertson and G. W. Henry, both of Chicago, Ill.; and, while not so large a combination as the memorable white-faced array of 1882, and while by no means including all the best show beasts of the breed in the north, it was nevertheless a formidable force for any one breed of cattle to encounter single-handed, even upon its own soil. The charge was successfully, and we might say, good-humoredly repelled, however, with the herds of Col. T. S. Moberley, of Richmond, and T. S. Grundy & Sons of Springfield (Ky.), bearing the brunt of the fight. Mr. A. J. Alexander sent a small contingent of young things under the lead of the 37th Duke of Airdrie from Woodburn, and Messrs. J. G. Robbins & Sons reinforced the army of defense with a herd from Horace, Ind., while Shelby County breeders contributed their mite in aid of the general cause.

"The story of the placing of the prizes in the Hereford class by a local committee follows:

"Five animals eligible to the ring for Hereford bulls three years old and over were on the ground, three of which, it may be observed, were sons of the celebrated Lord Wilton, but one of their number, Mr. Henry's well known Stocktonbury bull imp. Prince Edward, was off his feed and was not led out. The quartette that did enter the amphitheatre, however, was one of extraordinary merit, and it would have been no easy task for a jury even of expert Hereford breeders to pass upon the bulls without considerable delay. The judges were, we understand, none of them familiar with the kind of cattle upon which they were required to pass, and we believe we are correct in saying that they did not 'handle' a single entry in this ring. Mr. Earl brought forward his $3,000 English Royal winner Sir Bartle Frere (6682), by Lord Wilton (4740), out of Tiny (4467) by Longhorns (4711), with a weight of nearly 2,500 pounds, the first appearance of the bull, we believe, in any American showyard. As stated in 'The Gazette' for Aug. 19, this bull has achieved a national reputation as the sire of the first-, second- and third-prize grade yearling steers at the Chicago Fat Stock Show, and in his present form, with his lovely yearling daughters by his side, must be a source of the highest pride to his owner as he is of admiration to the public. He has a head and horn of unusual beauty, full of strong character, and yet as handsome as could be wished. His ribs spring out with a noble arch, showing a back and loin of most extraordinary strength; and on this account, if for no other reason, he is a dangerous competitor in any field. The flesh carried by this bull on his top is certainly sufficient to justify at least some of the talk of our Hereford friends about the wealth of meat carried by their cattle in the most valuable parts, for if abundance of choice broiling and roasting beef is any desideratum Sir Bartle Frere is a type of beast that must delight the hand and eye of any man used to handling beeves of fine quality. He carried more flesh than any bull of any breed on exhibition, and carried it all with such smoothness and great show of fine breeding as to stamp him an animal of rare quality.

"Next to this son of Wilton stood Mr. G. W. Henry's imp. Caractacus (7470), showing rather more white than is deemed desirable in the perfectly marked Hereford, but a bull of great substance and much flesh withal. He is on the short, compact order, and if he had some of Royal 16th's great length would have probably pleased Kentucky cattlemen as well as any bull in the ring. He was bred by John W. Smith, Thinghill Court, Hereford, and was got by Rosarian (6139) (son of Marechal Neil 4485), out of Curly 7th 12250 by The Emperor 12257, he by Mercury (3967), the sire of Longhorns (4711), Tredegar (5077), Thoughtful (5063), etc.

"Fowler 12099, Messrs. Fowler & VanNatta's celebrated son of old Tregrehan 6203, made friends from the beginning; and the encomiums he received at the hands of both the public and members of the awarding committee were the spontaneous tribute paid by unbiased men to an animal that possesses merit of an uncommon kind. Fowler will be remembered by many of our readers as one of the sensational two-year-olds of the Illinois State Fair of 1885, and at that time was regarded by some good judges as the best Hereford bull of his age in the west. He carried 1,850 pounds at Chicago last September, if we remember aright, and while a slight indisposition robbed him of some of his flesh some weeks ago, he has been on the up-grade since, and now lifts the beam at about 2,300 pounds. When it is remembered that he is but a trifle past three years old, and is not in the highest flesh, it will be seen that he is of great scale. He has good length, good rib, good back and loin, covers well on nearly all his points and carries the meat low down on his carcass. He has marvelous width in front (as has his famous sire), giving great room for heart and lungs, and altogether shows much genuine Hereford character. His dam, Princess 1990, was by Seventy-Six 1093, son of that great old sire imp. Sir Richard 2d 4984.

"Mr. C. M. Culbertson's Bowdoin 8579 was one of Fowler's competitors at Chicago last year, and stood next to Sergeant Major in that interesting competition. He came forward at Shelbyville in his three-year-old form as showy as ever, and with twelve months' development adding to the charms that rendered him a successful bull in 1885. He has done extremely well since returned to Newman from Michigan, and with his great smoothness and style is a dangerous antagonist. Like Sir Bartle Frere, he might fill a little better in his flank, and a little more flesh to round him out a trifle behind the shoulder would not be amiss, but he is so neatly turned and so pleasing is the general effect produced upon the eye as one beholds him, that he leaves a most favorable impression. He is a son of the old hero of Stocktonbury (Lord Wilton), out of a Remus cow, and was bred by Mr. S. Goode, of Ivingtonbury.

"There was but one ribbon to be awarded and, by a vote of two to one, it was given to Fowler, the other vote being cast for Mr. Henry's Caractacus.

"No two-year-olds were entered and but a few yearlings, Mr. VanNatta securing the ribbon in the latter ring with Randolph, the son of Tregrehan that won first prize at the Illinois State Fair last fall as bull calf. In bull calves Mr. J. A. Pickett, of Shelby Co., Ky., who is quietly testing the 'white faces' in the 'penny-royal country,' came to the front with a son of Brant 12314, a Canada-bred sire, in competition with seven head from Fowler & VanNatta, Adams Earl and C. M. Culbertson.

"Before taking up the cows and heifers the prize for best bull of any age was awarded and developed quite a surprise to the knowing ones, Fowler, Bowdoin, Sir Bartle Frere, Caractacus and all giving way to Mr. Earl's yearling Earl of Shadeland 9th, a worthy youngster unnoticed by the committee on yearling bulls. This decision seemed to rather 'paralyze' our Hereford friends, but it must be admitted that the recipient of the ribbon is not only a choice individual, but richly bred. He is a son of that finely-fleshed and impressive sire Garfield (6975), dam Bramble 2d (6948) by Lord Wilton, and while a very elegant young bull the wisdom of rating him higher than the first-class matured animals pitted against him may be called in question.

"Hereford cows were an extra show, and considerable difficulty was experienced in arriving at a decision. With such animals in the fight as Mr. Earl's Ada 2d 7006 by Lord Wilton, and Duchess 21st 7551 by Commander (4453), Fowler & VanNatta's Lark, Ethel, Viola 2d, Miss Fawley, and Mr. Henry's Edwina by Prince Edward, it is not strange that our Kentucky friends split badly on first ballot. The first prize ultimately fell to Mr. Earl's daughter of Wilton, the second going to Mr. VanNatta's Sir Richard cow Viola 2d. We should rather have preferred the Commander cow to Ada, on account of her superior smoothness and quality, and in the same respect Miss Fawley, Ethel and Edwina would have probably beaten Viola; but if substance and weight were the objects sought the decision might have been worse. Both of the prize cows are uneven in their flesh, but they are animals of great depth and constitution and most worthy specimens of the breed. In neatness and finish Mr. Henry's Edwina surpassed all of her competitors, and if we mistake not by another year will give the cows, at our northern shows at least, a lively race.

"In two-year-old heifers a full chorus of objections attended the sending of the blue to Mr. Earl's Garfield heifer Erica 5th (dam the Rodney cow Cammilla 8478), and the red to the same exhibitor's Sparkle by Tom Clark's Anxiety 3d. Although both are good heifers there were certainly several to be preferred to Erica 5th. Mr. VanNatta's grand Anxiety heifer Peeress and Tregrehan's daughter Miss Fowler are gems in their way, but neither was fortunate in getting a place. The decision for first place at least was palpably wrong and was so regarded by the successful exhibitor.

"In yearling heifers Fowler & VanNatta's Violet (dam Pretty Maid), an animal of rare ripeness and carrying a grand lot of flesh, was selected to wear the only ribbon offered in this ring. She is another of the get of old Tregrehan, and is a worthy representative of the bull that has done so much for the great herd at Hickory Grove. Heifer calves were an admirable show, and while probably most people would have preferred Mr. VanNatta's wonderfully ripe heifer by Fowler, she had to give way to Mr. Earl's Elena, a promising daughter of Elton 1st 11245, the young Sir Richard bull that is the rising star at Shadeland. Elena is out of the Carwardine cow Flirt 6985 by Rodney, and will improve, but Mr. VanNatta's extraordinary calf is so wealthy in her flesh, so grand in her crops, so deep of rib and short of leg, that it is difficult to satisfactorily get by her. We can account for it in no other way than that the committee feared her back might not hold up to the required level.

"The best female in the Hereford class was adjudged to be Viola 2d, the Fowler & VanNatta cow alluded to above as receiving second in the cow ring. In spite of some unevenness in her back, she is yet a grand strong animal, abounding in flesh, with great show of substance, and being a daughter of the famous "old Dick" must be admitted to be a very valuable cow."

These prizes placed, the herd contest followed:

"The work in the classes only served to whet the curiosity of the great throng of visitors for the breed competition to follow, and as the ring for best bull of any age or breed was called excitement reached fever heat. The position of the exhibitors of the

Hereford was peculiar. They were strangers in a strange land. To three out of five people on the fair grounds their cattle were an utter novelty, and, while it was believed that honest decision would be rendered, it was scarcely anticipated that people born and bred to another ideal in cattle would discard at first sight all they had been led to admire in a beef animal and award the palm to beasts differing so widely in essential characteristics from the Blue Grass Shorthorn. One man would object that cattle built like these Herefords 'couldn't get through mud at all,' while others condemned for what they pleased to term their 'lack of style.' The 'white face' failed to carry his head high enough to suit the average spectator, while the fact that they were so superior in front, heart, crops, rib, back and loin, and so well let down in the twist, could not atone apparently for any weakness about the rump. They were called small, too, by many who had never seen a deep-fleshed, short-legged Hereford weighed, and, while the Shorthorn section abounded in animals deficient in more vital points than those which were objected to in the Herefords, there were few who could admit that the latter were the equal of the old-time favorites. There were some notable exceptions, however, and more than one farmer was heard to express a desire to try the Hereford on Kentucky soil; so that, while it cannot be said that the visitors did more than insert an entering wedge, they have 'broken the ice' in such a manner as to lead them to expect a more encouraging reception another year.

"A dozen animals filed into the arena in competition for the male championships of the yard, and as they fell into the semi-circular line formed by the amphitheatre with a Hereford for a base at either end, the keystone of the solid arch of bulls ominously enough was seen to be a Shorthorn and he the champion of his class. Sir Bartle Frere held the right and the Garfield yearling from Shadeland the left, with the Shorthorns massed from the centre to the right, while Bowdoin, Fowler and Caractacus, and indeed the entire Hereford strength (save Bartle Frere) was from the start on the side to which the verdict ultimately assigned them—left. As one scanned the field the one special impression gathered was that the better quality and the deepest flesh lay with the Herefords, the heaviest weights and the finer style with their adversaries. The one were of the low-down, thick-set, kindly feeding sort, the other possessing the greater scale and range with their flesh carried higher from the ground. Col. Grundy's Red Chief was shown with consummate skill, and this in itself, in a region where style and animation count so heavily as at Kentucky shows, was half the battle. The level top and neat well turned quarters of the red Bates-topped Phyllis, coupled with airy style, proved irresistible, and as he donned the blue the great crowd broke forth with tumultuous applause. The nervous tension of weeks found sudden relaxation. The strain was over. Hannibal had been thundering at their gates and in the first pitched battle was sent reeling back upon his base in discouragement if not dismay. The Hereford exhibitors had prided themselves upon the fine quality of their bulls above almost all other features of their exhibit, and believed their opponents to be weakest in their males, so that while not specially surprised at the result they felt that their severest attack had been successfully repelled.

"For best herd of 'thoroughbred' cattle, any breed, for beef purposes, to consist of one bull and four females, the females to consist of one cow three years old and over, one two years old and under three, one one year old and under two, and one under one year old, the following imposing array competed: Col. T. S. Moberley's Wild Eyes, Prince, Rosalina, Barrington Blanche (Roan Duchess), Juanita 16th (Desdemona), and Desdemona calf; Col. T. S. Grundy's Red Chief, Grundy's Young Marys 52d, 63d, 84th, and Mary calf; J. G. Robbins & Son's Royal Best, Kitty Wells (Amelia), Nora (Adelaide), Majesty and calf; Mr. Adams Earl's (Herefords) Sir Bartle Frere, Ada 2d, Sparkle (Anxiety heifer), Lady Wilton 8th (by Bartle Frere), and Elena by Elton 1st; Fowler & Van-Natta's Fowler, Viola 2d, Peeress (by Anxiety 5th), Violet (by Tregrehan), and Lassie by Fowler. These five herds, three Shorthorns and two Herefords, constituted the most interesting exhibit of the day, and the strength of some of the female Shorthorns made it morally certain that victory would perch upon Kentucky banners. The main contest lay between Grundy and Moberley. The Springfield exhibitor wisely enough, perhaps, headed his herd with the two-year-old that had become so popular with the committees, and again left the ring a winner. Had his older bull been at the head of his lot we should have considered it a stronger herd, but as it was there were many who thought the Moberley cattle entitled to the prize. Again did the welkin ring with Shorthorn exultations, and again did the Hereford clans retire under the shadow of defeat.

"Three entries came forward in competition for the best herd of 'thoroughbred' cattle of any breed, for beef purposes, to consist of one bull and four females one year old and under two, two being

Shorthorns and one Hereford, Messrs. Moberley and Alexander representing the former and Mr. Earl the latter. The visitors again thought their chances good, as their string was of extraordinary strength, including Earl of Shadeland 9th, Lady Wiltons 6th, 8th and 10th, and Elenora 9th. Mr. Alexander sent 37th Duke of Airdrie, 30th Duchess of Airdrie, 28th Duchess of Airdrie, Wild Eyes Lady 5th, Rosewood 5th and Miss Bates 20th, while Col. Moberley entered 6th Airdrie of Forest Grove, Forest Grove Duchess, one Mary, a Harriet, and a Desdemona. A committee of five inspected the cattle, and balloted as follows: Moberley two, Earl two, Alexander one. Before the sixth man could be found to tie the ribbon a shower set in that allayed both dust and interest in the fight. An umpire was finally secured, however, and casting his ballot for the Richmond herd sent the ribbon to adorn the already large collection of trophies won by Col. Moberley with stock from Forest Grove. There were many who regretted that the Herefords had not been allowed this prize, not for sweet charity's sake, but because it was held by a large number of people that Mr. Earl had the best of it on the merits of his stock. But it was not a Hereford day, and Lewis, and the Woodburn people as well, had to put not only a wetting but a beating in their pipes and smoke it.

"The only remaining prize was that for bull with three of his get, and as the only Shorthorn entry was of a bull in breeding condition with calves right from the pasture, Messrs. Fowler & VanNatta (the only other competitors) carried it away with Fowler.

"'Chewing the cud of sweet and bitter fancy' the plucky breeders from Indiana and Illinois made the best of their Waterloo, and exercising a wise philosophy parted company with their successful rivals in the best of spirits. A miniature Hereford bull (intended as a souvenir watch charm) was presented to Mr. Earl as the representative of the visiting Hereford breeders with an address of thanks for their attendance, and the proprietor of Shadeland responded in a fitting manner, setting forth briefly what is claimed for the red-with-white-face cattle. And so the show ended.''

Royal Grove Excites Admiration.—While the main fight was being waged in Kentucky J. O. Curry was arousing the enthusiasm of all good cattlemen in the north by showing his imported two-year-old The Grove 3d bull Royal Grove 21500. At the Minnesota State Fair this richly-furnished, furry-coated, low-legged youngster was easily the best animal of the beef breeds in the ring. "Harry" Yeomans' Washington was there, but had begun to lose his bloom. His flesh had always been rather soft to the touch, and had now begun to slip; nevertheless, he managed to defeat the good breeding bull Wild Eyes by Lord Wilton, that for some years headed the herd maintained by the Cosgrove Live Stock Co., LeSueur, Minn., of which Mr. C. N. Cosgrove, long identified with the Minnesota State Fair management, was president.

It was at this fair that Mr. Curry presented the bull calf Archibald A., by imp. Archibald and out of the celebrated Coral. This calf was thick and shapely enough, but had that wiry hair and thick hide that seemed so strangely and so unfortunately persistent in Archibald's progeny. Yeomans had old Melody, by John Hill's Merry Monarch, out again, and although now seven years old and the mother of five calves she was placed first. She was a cow of great scale, weighing 1,900 pounds, and had wonderfully arched ribs. In the herd contest Royal Grove's superb bloom and character carried the day for the Curry cattle.

Beau Real Unfairly Beaten.—At Des Moines manager Yeomans of the Iowa Hereford Cattle Co. had the honor of meeting and beating Shockey & Gibb's two-year-old wonder, Beau Real 11055 by Anxiety 4th, with Washington. This was in the championships open to all breeds. He was lucky to do it, for the younger bull was "coming," and the other "going." Harry Loveland was feeding for the Early Dawn people and led Beau Real into the ring this fall weighing near 2,200 pounds at a few days short of three years old. It is not believed that a better backed bull than Beau Real* has ever been

---

*Writing to "The Breeder's Gazette" for Oct. 31, 1900, E. S. Shockey referred to Beau Real in the following terms:

"Beau Real, calved Sept. 22, 1883, was rather thin in flesh when we bought him, but the way he responded to good treatment was remarkable. We had many tempting offers for him, but would not part with him. With personal care and feeding we introduced him to the showring at Bismarck Grove in 1885 as a yearling, where he took first in class and sweepstakes Hereford bull any age, thus beginnning a four-year show record in which he defied his antagonists to the end. He was once placed second by an 'expert' judge on account of the 'tie' in his back, and once by another 'expert' who said he was 'too fat to breed.' He was never taken out of service to fit for the showring. He never had any special preparation because he always insisted on carrying a wealth of firm flesh, transmitting the same feeding and thickmeated quality to his offspring. He was both a bull and a heifer getter, but most of his sons went west to do duty on the range. Those few that had an opportunity, such as Wild Tom, of Sunny Slope fame, and Kansas Lad, with Mr. Armour, prove the breeding-on quality so essential in a sire. Among his many magnificent daughters I will mention a few that were called upon to

seen in America. In expanse of loin he has certainly never been excelled by any bull of any breed ever seen in the American showyard, and he was smooth, but for a dimple in his back, and heavily wrapped in deep mellow flesh. The decision which sent first at this Des Moines show to Yeomans' Blenheim over Beau Real in the class for two-year-old Hereford bulls was not generally approved, not even by Mr. Yeomans himself. Melody 16th again was female champion.

At Lawrence, Kans., Shockey & Gibb had the senior bull prize in Beau Real's half-brother, Beau Monde. He was a broad-backed bull full of good flesh and with excellent character and quality. He represented a cross of Anxiety 4th upon The Grove 3d blood, his dam being Beauty 2d 9901. The exhibition by Shockey & Gibb of this extraordinary pair of bulls drew marked attention to the great Gudgell & Simpson sire.

At Lincoln, Neb., C. M. Sears of Aurora had first in aged bulls with Prince of Wales 8912 over Hawes' Fortune and Sir Evelyn. Beau Real was first in two-year-olds and champion male. Two very grand heifers destined to fame, Lady Wilton and Miss Beau Real, were features of this show. The Beau Real herd won the grand championship over all breeds.

---

sustain their sire's showyard reputation, such as Miss Beau Real, Miss Beau Real Lad 3d, Curly Lady and Beau Real's Maid.

"If I were to criticise Beau Real I would say to change his horn a trifle, remove the 'dimple' in his back and make him a bit straighter in hind legs. I never saw such a thick broad loin. He was well flanked fore and aft and his quarters were well filled and beautifully finished."

Sir Bartle Frere Wins Over Washington.—One of the best cattle shows of 1886 was that at the Indiana State Fair. This was the first meeting between Washington, the ranking aged bull of the western circuit, and Sir Bartle Frere. The latter was in fresher condition, and deservedly won. In two-year-olds Blenheim was preferred to Tom Clark's Peerless Wilton.

The progeny of the imported cattle were by this time beginning to show "class." Mr. Earl's young bulls by Garfield, the daughters of Sir Bartle Frere, and Clark's Anxiety 3d heifers on one side the Mississippi and the Anxiety 4th bulls on the other were foreshadowing a bright future for home breeding. In a great ring of cows seen upon this occasion Peerless, and Mr. Earl's Duchess 21st and Ada 2d, had to step back in favor of Clark's Flossie. The yearling heifers were also a wonderful lot, worthily headed by another daughter of Anxiety 3d, Peerless 2d. The Shadeland lot won the grand championship herd prize over all breeds and also the young herd championship with Earl of Shadeland 9th by Garfield, three Lady Wiltons by Sir Bartle Frere, and Edwina 4th by Prince Edward. Mr. Lewis and his assistants had been unusually successful in the fitting of the Shadeland show stock of 1886.

The Michigan breeders made a good show this year at Jackson. Messrs. Phelps of Pontiac, Merrill & Fifield of Bay City, Hamilton of Flint, Hart of Lapere, Driggs of Palmyra and the newly organized partnership of Sotham & Stickneys participating. Merrill & Fifield's Tom Wilton, an own brother to Mr. Bertram's Sir Wilfred, headed the three-year-old bulls at Jackson.

Beau Real Defeats Fowler.—Beau Real was first, Blenheim second and Fowler third in the senior bull class at Des Moines in September, 1887. Fowler should probably have been second. His half-brother Ethelbert 16633, out of the Tudge-bred cow Ethel, drew the blue in two-year-olds, while in yearlings his own son Fowler Prince headed the class. Miss Fowler by Tregrehan won by superior bloom over the matronly Hebe 8th and old Melody of the Early Dawn and Indianola herds. The Tregrehans won again in two-year-old heifers with Violet 19441. The star yearling was Miss Beau Real, whose dam was Bertha by Rudolph, and in the heifer calves Early Dawn drew both first and second with Miss Belle Monde 4th, also out of Bertha, and Miss Belle Monde 5th, from a Grove 3d dam, both being sired by Beau Monde. Beau Real was champion bull, and Miss Fowler champion female.

At the Nebraska show Beau Real was first and champion over all breeds, and in the bull-with-get class defeated the Fortunes and Sir Evelyns, shown by Hawes, as well as "Harry" Yeld's Gift Wiltons. With the late Senator Harris of Kansas as referee Hawes had the female championship, beating Miss Beau Real with Nutbrowne, a daughter of Anxiety 4th. This was a big good show participated in by a number of local breeders, including E. E. Day,

Milliken Bros., A. S. Harrington, C. M. Sears, William Baker and C. M. Leighton. John Gosling was one of the judges and was thus early grounding an opinion of the Anxiety blood which he has ever since stoutly maintained.

At Topeka the main fight was again between Early Dawn and Mr. Hawes, Beau Real beating Fortune, now eight years old, in the class, but losing to the latter in the bull championship open to all breeds. This was a rare victory for the son of old Sir Richard 2d, as the opposition included such Shorthorn bulls as Cupbearer and The Baronet.

These old-time breed contests created a lot of excitement and aroused keen interest, but they of course settled no breed difference and sometimes developed amusing situations. At this Kansas show of 1887, for example, after a preliminary examination Col. True, the judge, informed the superintendent that he did not care to assume the responsibility of tying the ribbons alone in the herd competition, and asked that another judge be added. Prof. Shelton was accordingly called, and an examination and a ballot revealed Col. True's vote for Clay & Winn's Shorthorns and Prof. Shelton's vote for the Early Dawn Herefords. Consultation did not result in agreement, and ex-Governor Glick, a Shorthorn breeder, was directed to decide the tie. As his name was announced as referee, "I've got it" and "I've lost it" came simultaneously from "Newt" Winn and Harry Loveland, who was feeding for Shockey & Gibb. It was even so, for the Missouri

Shorthorns were given premier position and the Kansas "white faces" were placed second.

The Illinois show of 1887 was held at Olney. The Herefords were fittingly presented by the two veteran showmen Clark and VanNatta, J. O. Curry, George W. Henry, Tom Ponting and others. John Imboden was judge, and in aged bulls properly placed Fowler first. At full maturity he was a bull of the real old sturdy Herefordshire stamp. Masculine, massive, rugged and active, with a commanding presence, rare depth and spread of rib, and big well filled quarters, he was an outstanding specimen of the type that made the conquest of the range by the Herefords a certainty. Mr. Henry's Caractacus, thick-fleshed and compactly fashioned but rather light in color, was second. Ponting's Defiance, by Culbertson's imp. Lord Wilton out of an Anxiety dam, one of the good bulls of his day, was much admired upon this occasion. Curry's Horace-Regulus bull Harold was first in two-year-old, and the same owner's Archibald A. was first in yearlings.

The cows were headed by Clark's Flossie, seen here with calf at foot and still showing the traditional Anxiety smoothness and wealth of good flesh. Fowler & VanNatta's eleven-year-old Truth 2d stood next, and Ponting's low-legged broad-topped "little one," Gertrude 2d, granddaughter of Gay Lass and dam of Defiance and other good ones, was third. The VanNatta herd had both first and second in three-year-olds with Miss Fowler, by Fowler, and Peeress, by Anxiety 5th. In two-year-olds Clark was handily first with Peerless 2d and VanNatta second on Violet by Tregrehan. Peerless 3d drew the blue for Clark in the yearlings with Henry's Countess of Rossland next. Fowler & VanNatta won on herd. A new committee made Caractacus champion bull and Peerless 2d champion female of the class. Harking back to J. H. Pickrell's sarcastic reference to the Hereford "syndicate" of 1882, it was significant of the now widespread recognition of Hereford excellence that a member of this famous old-time Shorthorn breeder's own family, Mr. A. A. Pickrell, acting as referee at Olney, sent the grand beef herd championship to Fowler & VanNatta.

By this date the fame of St. Louis as the great agricultural show of the middle west had sadly faded. In 1886 there had not been a Hereford on the grounds, and in 1887 but three were seen, those of Tom Clark, G. W. Henry and Fielding W. Smith. Caractacus was first and champion, and Dictator 2d, son of the famous Dictator, was second. Flossie won as usual in the cow class, but Peerless 2d was made champion female.*

At a show held at Kansas City this year Dictator

---

*Old Peerless died in the fall of 1887, and her record was so remarkable that it should find full recognition. We quote Mr. Clark's statement of her career made shortly after her death, as follows:

"Peerless was exhibited in England at Leominster show in 1881, and was one of four yearling heifers winning the first prize, also one of a pair of heifers winning first prize at the Lord Tredegar show at Newport the same year. In America, in 1882, Peerless won first prize as the best two-year-old at Hoopeston, Ill., first at Lafayette, Ind., second at Springfield, Ill., second at Illinois State Fair at Peoria, and first in class and sweepstakes female of any age at St. Louis. In 1883 she was first at Illinois State Fair at Chicago, and sweepstakes female any age. She was then withdrawn from the showyard until

2d defeated Fortune and Sir Evelyn. In Minnesota Fowler & VanNatta and the Cosgrove Co. were the only contestants, Fowler beating Wild Eyes in the bull section but Cosgrove taking a majority of the prizes on females, including first in the cow class with Bonnyface, a 1,975-pound daughter of Rudolph.

In Michigan Merrill & Fifield, Sotham & Stickneys and Hon. James M. Turner showed under F. H. Johnson of South Bend as judge. Tom Wilton and Clarence Grove, both owned by Merrill & Fifield, were first and second in senior bulls. Sotham & Stickneys were first in two-year-olds with Stockfield's Wilton by Hall's Hotspur. Merrill & Fifield won on cows with Lovely 2d and Greenhorn 5th.

1886, when she was first at the Illinois State Fair at Chicago over hot competition.

"The following is a list of her produce and their winnings: Empress 2d 12771 was one of the four yearling heifers shown by me at the Illinois State Fair in 1884 that took first prize in young herd over all breeds. She is now owned by H. H. Clough, Elyria, O., and was shown by him this year at the Loraine County Fair, taking first prize as cow in strong competition; she was also in the herd that took the grand sweepstakes over all breeds. Her next calf was Peerless Wilton 12774, which I am now using in my herd on Anxiety 3d 4466 heifers, and for which I refused, when he was eleven months old, $1,500. He was the sire of that remarkable sixteen-month-old bull that took second premium for yearlings at the last Illinois State Fair against much older and larger animals. Her next calf, Peerless 2d 16240, took first prize as a yearling at the Illinois State Fair in 1886, and in 1887 at the Illinois State Fair was first as two-year-old, sweepstakes for the best female of any age, and sweepstakes over all breeds in the two-year-old competition; at St. Louis in 1887 she was first as two-year-old in her class and sweepstakes for best female of any age, also one of five in sweepstakes herd. Her next calf, Peerless 3d 26664, took second as calf at the Illinois State Fair in 1886, first as yearling at the Illinois State Fair in 1887, and was one of sweepstakes young herd at St. Louis this year. Her last calf (but not least) is a bull named Anxiety Wilton, which bids fair to equal any of her other produce.

"At the time of her death Peerless was within three weeks of dropping a heifer calf, which would have made her sixth calf. She was seven years old last May. With her show career and the calves she produced, I would class her as one of the most remarkable cows that ever lived. The cause of her death was an abscess on her kidney, from which she had been suffering most of the summer; although she suffered so much, she retained her beautiful form until the day of her death."

CHAPTER XIII.

SOME ROUSING DEMONSTRATIONS.

The state of Ohio had thus far been somewhat neglected by the leading exhibitors of Herefords. Local breeders had made creditable shows each year, and new herds were founding, prominent among them being that of H. H. Clough, who restored in full measure the reputation which Elyria had many years before enjoyed as a Hereford headquarters.

The Ohio Show of 1888.—The "big chiefs" of the trade in those days decided to let the Ohio farmers see their best cattle at the state fair of September, 1888. Adams Earl, Fowler & VanNatta, George W. Henry, Tom Clark and C. M. Culbertson entered the state, coming, indeed, the week before and showing at the old tri-state fair at Toledo. They were reenforced at Columbus by the herds of Mr. Clough and F. C. Sayles of Berlin Heights.

Fowler at five years old came in at a weight of 2,800 pounds, still carrying himself like a two-year-old. However, he was side-tracked by the committee in the senior bull class in favor of Tom Clark's well-brought-out four-year-old, Prince Edward 2d 14117, by Mr. Henry's Prince Edward and out of Luna 4th by Horatius. This was a thick-fleshed bull particularly good in loin and twist. His sire, Prince

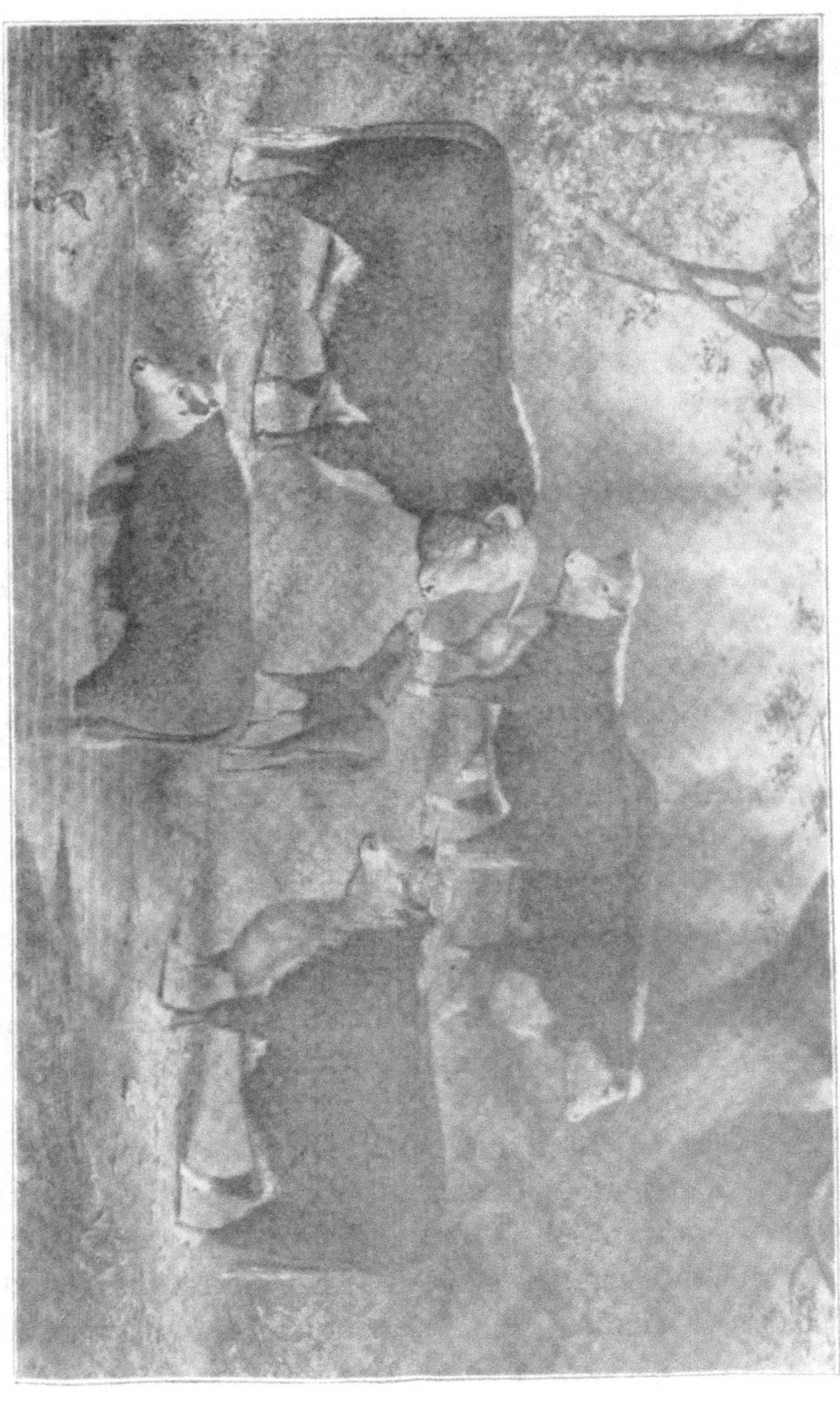

Imp. Prince Edward 7061.   Annie Laurie.   Marie Antoinette.   Countess of Rossland.   Edwina.
GEO. W. HENRY'S ROSSLAND PARK SHOW HERD—From the lithograph by Cecil Palmer.

Edward, with his attractive Lord Wilton head and his level quarters, and now nearly eight years old, was in this ring, but was unable to hold back the younger generation. Clough was showing a very good four-year-old bull called Sylvester 11123, bred at Gudgell & Simpson's from Anxiety 4th. Sayles' Cedric 8933 was also seen—an imported Turner-bred bull with the traditional Grove 3d-Spartan thickness of flesh, but lacking the scale of his competitors. Fowler was placed second, with Prince Edward and Cedric next in line.

Earl of Shadeland 22d.—Mr. Earl's Earl of Shadeland 22d 27147, by Garfield and out of Electra 2d by Sir Bartle Frere, second dam being that capital cow Anguilla by Sir Richard 2d, had been made champion at Toledo over bulls of all ages. He was easily enough first in the two-year-old division at Columbus and subsequently was declared champion male of the breed. A bull of rare balance and quite the sensation of the show season of 1888, Earl of Shadeland 22d had an illustrious career as a prize-winner. Fowler & VanNatta also were proving the merit of their breeding stock in convincing fashion; in this ring they supplied Cherry Boy 26495, son of Fowler and Cherry Pie 2d by Horatius, and destined to great fame later on. He was second here to Earl of Shadeland 22d. Mr. Culbertson's Star Grove bulls by The Grove 3d were full of flesh, but had not the size and stretch of their antagonists. Mr. Earl also headed the yearling class with another

Garfield bull, Earl of Shadeland 30th 30725, out of Snowdrop by Colorado 4252. It may be said in passing that the Shadeland herd contained several Colorado cows that proved most excellent producers. This Earl of Shadeland 30th was a youngster of fine promise, which he afterwards quite fulfilled.

The VanNatta stalls supplied the winner in the cow class in the nine-year-old Miss Mize 2015, by Sir Richard 2d, full of thick natural flesh and with a beatiful heifer calf at foot afterwards known to fame as May Fowler. Henry's Edwina, by Prince Edward and with her sire's loins and quarters, was second. Clark's Peerless 2d, suffering from a lame foot, was down in third place, while VanNatta's Flossie and Peeress were out entirely, a fact well illustrating the exceptional strength of this great show. The two-year-old heifers were a revelation to many of the spectators. Peerless 3d had first, Van-Natta's Polly Pink by Fowler was second, and Culbertson's Grove Maid 17th was third. The yearlings were also a royal lot headed by Earl's Lady Wilton 26th by Bartle Frere, with Henry's Lady Pitt 2d next, and Lady Wilton 28th third. In heifer calves May Fowler led. In the bull-with-get class Fowler won. Mr. Earl's fine yearling, Lady Wilton 26th, was female champion.

In the grand-beef-herd contest, open to all, Mr. Henry's Herefords were victorious. Mr. Earl won the young herd championship, as well as the open bull championship with Earl of Shadeland 22d and the open female championship with Lady Wilton 26th. The fruits of the importation of 1881 were being gathered.

The Big Show Moves West.—The impressive display made at Toledo and Columbus was practically duplicated at Indianapolis. Fowler's ill luck remained with him, the first prize in the older bulls going again to Prince Edward 2d. In two-year-olds Earl of Shadeland 22d, easily the sensation of the year, was quickly slated for first, and in yearlings Earl of Shadeland 30th duplicated his Columbus winning.

Among the cows the Columbus awards were shaken up, first going to Peeress by Anxiety 5th, second to Edwina, third to Miss Mize, and fourth to Flossie. A different order was also arranged in the two-year-old heifers; Polly Pink moved up from second to first place, Countess of Rossland stood next, and Peerless 3d, that had been first at Columbus, was turned down to third. Other arrangements were also made among the yearlings, Lady Wilton 27th, fourth at Columbus, going to the front, Peerless 2d following and Mr. Earl's Erica 16th coming third. May Fowler duplicated her Ohio victory in heifer calves. In the bull championship open to all breeds Earl of Shadeland 22d carried the honor by direction of Mr. Imboden. In the open herd competition partisans of each of the three leading breeds served on the committee. By agreement of the Shorthorn and Aberdeen-Angus judges the first and second prizes went to the Shorthorns and the "doddies," Dr. Bush, the Hereford representative, voting for the VanNatta Herefords in each case. In young herds first went to the Shorthorns and second to the Shadeland Herefords.

The four leading herds on the eastern circuit put in an appearance at the Illinois State Fair at Olney, supplemented by a good lot sent by Tom Ponting. The entire Ponting exhibit, including Defiance 21849, by Lord Wilton, were direct descendants of old Gertrude 2d by Ponting's Anxiety 3d, and the old matron herself was there on view.

Fowler was at last preferred to Prince Edward 2d upon this occasion, Defiance tying the latter for second place. Earl of Shadeland 22d won by general consent among the two-year-olds, with Cherry Boy second as before. In the yearlings the Earls of Shadeland 30th and 26th won in the order named. The Garfield bulls were now making an even greater "hit" than the Bartle Frere heifers.

In the cow class Edwina at last forged to the front, and Flossie stood next, leaving Miss Mize, the Columbus winner, and Peeress unplaced. Peerless 3d was made best two-year-old with Countess of Rossland second, Polly Pink being unplaced. Lady Wilton 27th continued to find high favor and was first in the wonderful yearling class, with her stable mate, Lady Wilton 26th, second. Mr. Henry won the herd prize, and Mr. Earl scored in the class for five animals the get of one sire, with his Bartle Frere group after a hard battle with Clark's Anxieties. It will be remembered that Sir Bartle Frere and Anxiety 3d were half-brothers, both being out of old Tiny 4467. On the following week most of these cattle were seen at St. Louis, meeting at that point entries from the herd of F. W. Smith. Fowler had first prize as best senior bull, and the Earls of Shadeland 22d and 30th completed an unbroken string of victories. It is worth noting in this connection that in the yearling class an Anxiety 4th bull, Don Carlos, of which we shall hear more later on, shown by Mr. Smith, stood next to the great son of "Bartle."

Western Shows of 1888.—The trans-Mississippi fairs of 1888 did not develop the strength exhibited elsewhere. The show at Des Moines was made chiefly by George Fowler, Maple Hill, Kans., the Kansas Hereford Cattle Co. of Lawrence, A. A. Crane, Osco, Ill., and Fielding W. Smith. Mr. Smith received first in bulls on Dictator 2d; The Grove 3d bull Plutarch 14410, imported by George Leigh and shown by the Kansas company, had second; Mr. Crane's five-year-old Sir Cherry 7295, bred by Mr. T. L. Miller's son, T. E. Miller, and sired by Ivington Wilton, stood third. Murdock 28545, by Beau Monde, was the only two-year-old bull. Smith's future-great Don Carlos at twenty months had to make way for Fowler's Beau Real 3d, by Beau Real and out of Bertha by Rudolph. Another Beau Real baby was first in bull calves.

"TOM" PONTING.

One of the best cows ever seen in America, Lady Wilton 19498, of the Fowler herd, an imported daughter of Lord Wilton, was easily first among the matrons. The Kansas company's Nutbrowne 5th 17243, by Anxiety 4th and an own sister to Hawes' famous Nutbrowne 4th, was second. Mr. Fowler succeeded in beating Miss Beau Real in two-year-olds with Curly 17th, daughter of the Merry Monarch cow Curly 16th. Dictator 2d was champion bull, and Lady Wilton the best female any age. The Fowler entries won the young herd championship, open to all breeds.

Lady Wilton vs. Princess Alice.—At Lincoln the judge, John Gosling, gave Plutarch first over a Garfield bull shown by Howard Bros., Edgar, Neb. Murdock and Beau Real 3d were first in the two-year-old and yearling rings. The cows were placed as at Des Moines, as were also Curly 17th and Miss Beau Real. Murdock was made champion bull, Lady Wilton preserving her Des Moines honor among the females.

In the annual breed contest at Lincoln the Shorthorn herd of Luther Adams of Storm Lake, Ia., was given championship honors, with the famous Cupbearer at its head. Included in this herd was the beautiful heifer Princess Alice, later to acquire celebrity in the Linwood herd of Senator Harris. In the open class for females of all ages she was preferred to Lady Wilton. "Newt" Winn, a Missouri Shorthorn breeder, Alexander Legge, and John Gosling constituted the awarding committee;

the decision went to the Shorthorns over Mr. Gosling's protest. George Morgan, who was present, became particularly wrathy over the verdict, claiming that Lady Wilton's equal had never appeared in a western showyard. "Uncle Willie" Watson, then with Mr. Harvey at Turlington, said to that rare old "brither Scot," William Miller, manager for Mr. Adams, the owner of Princess Alice, "Weel noo, ye've just beaten the best coo on the grounds."

Beau Real came forward at Topeka. An accident had cost him loss of flesh and bloom, but Beau Real out of form was better than the good Plutarch, and the ribbons were placed accordingly. Murdock won again by default. Beau Real 3d beat his Beau Real brother, Shockey's Nimrod, in yearlings. Lady Wilton was first in cows by everybody's consent. Makin Bros. of Florence, Kans., most capable men, who had by this time become established in Herefords, contributed good cattle to this show. They did not win, but there came a day later on when they had to be reckoned with. Miss Beau Real was ordered by the judge, Capt. Huber, ahead of Curly 17th in the two-year-olds.

The open bull-and-get championship at Topeka was awarded by Col. J. F. True and the late John McDiarmid to the Beau Real contingent, over the Shorthorn Scotland's Heros and Col. Harris' Baron Victors. The Shorthorn Cupbearer was made champion bull over all, whereupon the Hereford breeders rebelled and sought their tents, Princess Alice becoming female champion and the Shorthorn group of which she was a member receiving first honors in a one-sided herd contest.

New Alignment in 1889.—Mr. Earl's whirlwind campaign of 1888 had certainly sufficed to advertise sufficiently the claims of Shadeland as a nursery of prize-winning Herefords. Fowler & VanNatta, too, had won such stores of ribbons that it seemed as if they could well afford to remain under cover for a time, so far as showing was concerned. Neither of these herds was on the circuit of 1889. Some new blood came forward, however, more especially west of the river. The recession of the tide following the "hurrah days" of the big importations had forced a number of concerns to "shorten sail," and others to go out of business entirely. The Iowa Hereford Cattle Co. had over-extended itself, and the show herd was sold to go to California where it was successfully exhibited. The Early Dawn people had transferred their interests to others. New men were taking up the cudgels; conspicuous among them at this date in the west was C. H. Elmendorf of Kearney, Neb. The Makin Bros., Florence, Kans., were also now coming to the front. E. E. Day of Weeping Water, Neb., also moved up into the limelight.

At Des Moines Elmendorf, Day, the Makins, the Cosgrove Co. of Minnesota, and Alex. Moffitt & Son of Mechanicsville, Ia., made the Hereford presentation before William Stocking of Rochelle, Ill., as judge. In the aged bulls it was easy to send Makin's

Vincent 16691 to the fore. He was a four-year-old son of Sir Evelyn out of imp. Princess of Wales 12073—a wide, compactly built, evenly turned bull, with good quarters and a nice touch. Day's Province, another son of Sir Evelyn, was second.

Earl of Shadeland 30th.—Elmendorf had bought Earl of Shadeland 30th from Mr. Earl and in his two-year-old form he was a show bull of the first rank, albeit without competition at this particular show. Speaking of this bull at that time "The Breeder's Gazette" said:

"Hereford breeders will not need to have this bull recalled to mind, as he is well known as the yearling which was counted a 'coming youngster' in Adams Earl's herd through the eastern circuit last year. Well, he has 'come.' He had started to 'come' when he tied his companion, the phenomenal Earl of Shadeland 22d, for sweepstakes bull at St. Louis at the close of the 1888 campaign, and since that time he has moved along evenly until today he must be pronounced one of the most charming bulls ever seen in an American showyard. There have been bulls stronger in this or the other point, but all in all he is beyond question one of the most uniformly good bulls of the breed. He is absolutely smooth, with shoulder beautifully laid, neck-vein nicely filled, ribs well arched and deep, quarters long, level and well filled, top and bottom lines perfect, while back and loin are packed deeply and smoothly with mellow flesh."

Besides buying this bull Mr. Elmendorf had secured from Tom Clark Flossie, from Mr. Henry Edwina, and from Fowler & VanNatta Polly Pink, all familiar showyard favorites. All three were thrown into the cow class at Des Moines. However, they had already passed their zenith; Flossie was heavy in calf and Polly was growing somewhat rough in her flesh. On the other hand the Makin entry, Mayflower 4th by Fortune, was shown in fine form, and as she was broad, deep, full through the girth and even in her lines, she drew premier place. Day got second on another daughter of Hawes' Fortune named Cressie, richly furnished, broad-topped and low-legged. Day supplied the winner among the two-year-old heifers in Bright Lass 3d, by Anxiety 4th, the smoothness and quality of the Gudgell & Simpson bull's get being in evidence. Elmendorf's Elena 10th from Shadeland was second. The Day herd drew the blue in yearlings with Mable by Sir Evelyn 2d; Elmendorf's May Fowler (from the VanNatta string) was in this class, but Mr. Stocking did not seem to appreciate her fully. Earl of Shadeland 30th was champion bull and Bright Lass 3d was sweepstakes female.

Gosling Upsets Stocking's Work.—At Lincoln there was a notable upsetting of the Des Moines ratings. John Gosling as judge had by this time become a fixed habit with the Nebraska State Fair management and exhibitors, and in overturning a number of Mr. Stocking's decisions of the week before he undoubtedly reflected the best judgment of unprejudiced men.*

---

*Mr. John Gosling was born in Staffordshire, England, in 1844. He came to the United States in June, 1870, after having had practical experience in the buying, feeding and slaughtering of meat animals. His first work in this country was in John Taylor's packing house in Trenton, N. J. This employment he

JOHN GOSLING.

The exhibits of Messrs. Day, Elmendorf and Makin were supplemented by entries from the local herds of Howard Bros., Milliken Bros., Leighton, Harrington, Moon and the Havens Farm, owned by Mr. Arthur Havens, a son-in-law of Mr. C. M. Culbertson. It was the largest turn-out of Herefords seen in the state up to this date.

No less than eight aged bulls awaited judgment, but Vincent proved invincible. Nevertheless, he was pressed for the place by Mr. Havens' Star Wilton 4th, by imp. Lord Wilton out of a Grove 3d

was obliged to discontinue on account of illness, and in the summer of 1871 he made a trip back to England, returning with a few Shropshire sheep, paying $25 per head as ocean freight upon them. These he exhibited at the New Jersey State Fair, and then took them to the St. Louis show, but at that time there was no class for such animals. A few years later he brought over 68 Shropshires to Lexington, Ky., and exhibited a part of them at various state fairs for their owner, Mr. George Allen, now deceased. From 1872 to 1880 he was engaged in the butchering business at Rockford, Ill., although during that time he had some connection with both T. L. Miller and George Morgan in the way of assisting in the introduction of Hereford cattle throughout the west. When Mr. Culbertson began importing he made use of Mr. Gosling's services in various ways during Morgan's absences in quest of cattle in England. It was then that Mr. Alexander H. Swan hired Mr. Gosling to take charge of a herd of Herefords which he had established at Indianola, Ia. It was from this establishment that the famous Fat Stock Show heifer Grace was developed, becoming the champion Hereford heifer at the Iowa State Fair and tying Mr. Culbertson's crossbred bullock Dysart for champion honors at the Chicago Fat Stock Show.

From October, 1881, to September, 1884, Mr. Gosling was transferred to the Omaha distillery barns, and while there fed 6,000 head of cattle, for some of which record prices were obtained and remarkable dressings reached, as high as 64 to 65 percent being obtained on grade range-bred "white faces." From Omaha Mr. Gosling was sent back by Mr. Swan to Indianola, where he developed the bull calf Storm King, afterwards sold for $1,000 to go to Wyoming. From this source also came the famous champion carcass winner Plush, referred to elsewhere in this volume in connection with the early Fat Stock Shows.

About 1887 the failure of the Swan Bros. caused Mr. Gosling to return to Rockford, where he got together a few Hereford cattle, and fed that splendid steer Sensation for the carcass competition at the Fat Stock Show. Although failing of recognition on the block at the hands of the judges, Mr. Gosling's friends among the Hereford breeders evinced their appreciation of his skill in producing such a carcass, by raising a purse of $237.50 which was turned over to the exhibitor by way of consolation. This carcass was the subject of much discussion at the time, and was purchased by A. C. Terry who for many

dam. Earl of Shadeland 30th had things all his own was among the two-year-olds.

There were sixteen aged cows forward. Polly Pink drew the blue. She might have been smoother in her flesh, but it was exceptionally thick in the right spots, and she also had plenty of scale with a marked show of substance. Another cow that had been unnoticed at Des Moines, Day's Aurora, was drawn for second. She was of a good Hereford type, level and near to the ground. In two-year-olds there was another shake-up. Bright Lass 3d, the Iowa champion, was passed over for Elena 10th

---

years maintained one of Chicago's best retail meat markets on the corner of what is now Jackson Blvd. and Dearborn St. The chef of the Richelieu Hotel had the handling of the beef for the table, and pronounced it the best he had ever served up to that date. This steer was fed oats and barley meal, and did not consume five bushels of corn in all his life.

In 1892 Mr. Gosling associated himself with the Chicago, Burlington & Quincy R. R. at St. Joseph, Mo., in which capacity he served until 1896, when he removed to Kansas City, Mo., and engaged in the buying of bulls for the western ranges. He did much of the buying at one time for the LS range at Tascosa, Tex., the calf product of which herd was afterwards so successful at the Kansas City Royal shows. He also bought bulls for Mrs. Adair's JJ herd at Paloduro, from which stock Mr. Dan W. Black of Ohio acquired calves and fed them to championship honors at the International Exposition at Chicago. Mr. Gosling also bought bulls for the Fowler & Tod outfit from 1897 to 1909 when the herd was dispersed. He also acted as buying agent for the SMS, Spur, Bell and other prominent range companies. His services were also utilized by Mr. Murdo Mackenzie for the buying of the northern-bred contingent of bulls shipped a few years since to Brazil.

John Gosling's great hobby during all these years has been the beef carcass, and he has made himself one of the recognized authorities of his day and generation upon this subject. During the past ten years the instructors at many of the leading American and Canadian agricultural colleges have taken advantage of his intimate knowledge and felicity of expression in this regard, and his lectures to students and various gatherings of farmers and stockmen, discussing the relation of breed and feed to flesh and fat, have been regarded as among the most valuable practical contributions of recent years to the available fund of information upon that question.

Some years ago at Fargo, N. D., he was giving a meat demonstration, and was called upon to answer some questions, among others: "What kind of a beef bull should one select?" His answer was, "One with a Napoleonic expression. Cloudy! 'Bully'! A Duke of Wellington physiognomy indicating character! To this join a King Solomon disposition, and you have the bull you are looking for." This sally was followed by applause and

by Elton 1st 11245. Mr. Earl had given the Elton name at Shadeland to a line of bulls owning Sir Richard 2d as their sire, and Elena 10th had surely inherited from "old Dick" some of his deep natural flesh. Her ribs were beautifully sprung, and her loin wide. The Anxiety heifer that had beaten her before was second. Again, in the yearling heifers the previous judgment was disapproved. May Fowler, although set below Mabel at Des Moines,

laughter, and in recalling it Mr. Gosling remarked to the author of this volume, "Such bulls were Don Carlos and Beau Brummel." Mr. Gosling always insisted that these and other Anxiety bulls strengthened the Hereford type in respect to the deep backthigh, "a formation," according to his view, "which insures the legs getting into the beef quick." This merit, Mr. Gosling insists, was lacking in many of the Herefords prior to the extensive use of the Anxiety blood throughout the west. He urges at all time that what is needed is an increased supply of inner muscle or flesh, as contrasted with carcasses carrying too much outside fat.

One of the most notable tributes ever paid to the subject of this note was that written some years ago by Mr. Cecil Palmer, at one time a leading live-stock artist, making a special study of Hereford form and character. Mr. Palmer said:

"Mr. Gosling's relation to the Hereford cause has been that of an expounder and defender of 'the faith.' Born in Staffordshire, England, the son of an artist, he brought to his mission the eye of an artist, the inclination and capacity of mind for thorough investigation and complete knowledge of his subject in all its relations, the boundless and untiring enthusiasm of a crusader of old and the gift of speech. He has been an advance agent of the Hereford man's present prosperity. Like another John of old, he has been a voice crying in the wilderness, 'prepare ye the way'. He has not only been priest and prophet of the Hereford religion, but he has been an educator as well, and has helped to improve the Hereford by helping to educate the breeder.

"He could see the faults of the Hereford, if he had any, and he always admitted them. He could see the faults of the Shorthorn or Polled Angus with certainty, and he sometimes mentioned them too. Mr. Gosling is a judge of the beef animal, whether on the farm or in the showring. He knows when to feed and how to feed and what to feed and how much—and he has always been an advocate of the liberal and discriminating use of the knife. He knows how to breed, feed and butcher; and also how to cook a beef steak or roast, and when it is on the table he is an epicure.

"Years ago he advocated the idea that two years was long enough for a steer to live. Years ago he advised in a letter addressed to the Illinois State Board of Agriculture the abandonment in the Fat Stock Show of all classes over two years, and the following year, in proof of his theory, fed and exhibited at that show the two-year-old crossbred steer Plush that won not only the two-year-old prize, but sweepstakes on the block."

had been made champion female in the Hereford association specials at the Iowa show under the judgment of "Willie" Watson; Gosling now decorated her with the blue badge of superiority in an exceptionally fine class of heifers. Mabel was second. There were onlookers, however, who would have preferred her for first, among these being Mr. Culbertson, who insisted that she was the best Hereford female of any age on exhibition.

Col. W. A. Harris, whose great herd of Scotch-bred Shorthorns at Linwood, Kans., had by this time come to rank as the best of that breed in America, was recognized on all hands as one of the soundest judges of beef cattle in the west, and it was under his examination at Lincoln that Earl of Shadeland 30th was made champion Hereford bull, and Polly Pink champion over all females. Messrs. Harris and Gosling tied the ribbons in the open championships, the first herd prize going to Williams & Householder's Shorthorns and the second to the Elmendorf Herefords. Vincent was made champion bull and the famous Shorthorn show cow Fall Creek Rose was preferred to Polly Pink for the female championship of the yard.

At the Topeka fair Elmendorf, Makin and Day had it out again. Vincent and Earl of Shadeland 30th won their ribbons as usual. Mayflower 4th was restored to her Des Moines position as head of the cow class. Elena 10th won again in her division, as did May Fowler among the yearlings.

East of the Mississippi River Clark, Henry,

Clough and George O. Holcomb & Son of Troy, Pa., shipped show cattle to a special event at Buffalo, N. Y., where they were joined by J. S. Northrup of Westfield, N. Y. The latter beat Cedric in the aged bull class with Valiant 25071, bred by Clough and sired by his imported The Grove 3d bull Alexander 9821. Holcomb was second in two-year-olds with the Garfield bull Earl of Shadeland 24th 30721, defeating Clark's Anxiety Wilton 30272. Peerless 3d headed the cows and her stable companion, Horatia 3d, was best two-year-old heifer. Clark won the herd prize, and that for get of bull.

There was also a good show at Detroit this same year, where Mr. Clough and R. G. Hart of LaPeer, Mich., submitted Hereford entries to Mr. Van-Natta's judgment. Clough's herd was in best form, and with Sylvester at the head gained most of the honors. On account of these new shows the Indiana State Fair of 1889 had neither Hereford nor Aberdeen-Angus entries.

Clark's "Clean-Up" at Peoria.—The Illinois State Fair of 1889 was held at Peoria. The Hereford show was not large, consisting of but 36 head, entered by Clark, Elmendorf, Henry, Ponting, W. J. Lewis and Frank Crane. Ponting's Defiance won in the first class shown, Lewis receiving second on Quantrille 10774, son of Clark's Anxiety 3d. Earl of Shadeland 30th beat Anxiety Wilton (son of Peerless) and Crane's Emerson, the latter by that very excellent stock bull Grimley. In two-year-olds Henry was first with Caractacus Wilton, son of his

old show bull Caractacus that had meantime been sold for export to South America.

In aged cows Elmendorf's Etiquette, by Anxiety 6th out of Flirt by Rodney, wore the blue. In three-year-olds Peerless 3d beat Polly Pink, and in two-year-olds Clark won again, this time with Horatia 3d by Anxiety 3d. There was no denying the blue among the yearlings to the same exhibitor, the prize falling to Cora Belle, a heifer that had been bred by Mr. Clark's neighbor, McEldowney, from Peerless Wilton and Crystal Belle by Cedric. Just by way of "rubbing in" his skill at the game Clark walked off with the second prize on Lottie by Anxiety 3d. May Fowler was unplaced. When on top of all this Clark's Horatia 4th headed the heifer calves, his cup was full and running over. The Hereford association specials for best bull and best female were sent by the judgment of Mr. Gosling to Earl of Shadeland 30th and Peerless 3d.

Cherry Boy Champion.—The persistency with which Fowler & VanNatta, Adams Earl and Thomas Clark followed the great shows of the period under review was one of the most interesting phases of this era in Hereford progress in the United States. Fowler & VanNatta banked specially on the practical every-day character of the descendants of Tregrehan. With ample bone, massive girth, heavy quarters and general show of constitution, the VanNatta cattle appealed always to those who had in mind the exacting requirements of the farm, feedlot and the open ranges of the west. The Shadeland stock undoubtedly displayed superior refinement. Neat heads and horns, fore-and-aft finish, quality and Wilton character distinguished most of Mr. Earl's well kept cattle. Mr. Clark, with less capital to back his work, applied his practical knowledge with extraordinary success to his Anxiety-Peerless combination, and year in and year out he held his own with marked success against all competition with thick smooth-fleshed cattle of his own breeding and feeding, coming back with unfailing regularity to challenge all America.

The first tilts of the interesting campaign of 1890 developed the fact that the Clark and VanNatta herds, contrary to their usual custom, had journeyed westward to try conclusions with the trans-Mississippi country. These were days of intense interest to all who were following the fortunes of the "white faces"; the excitement attaching to the annual competitions attained unusual heights in the autumn of 1890 because the giants of the eastern circuit went out of their way to cross swords with their brethren of the west. Nothing could better illustrate this than the painstaking character of the reports made by the press at the time. It all seems like a dream at this distance, but as the author recalls the subjoined notes on the Hereford class at Des Moines, written from personal observation at the time, it is but yesterday. John G. Imboden was the judge.

"The ring for aged bulls brought out Mr. VanNatta's Cherry Boy 26495, Mr. Yeoman's Beau Real

3d 30769, Makin Bros.' Vincent 16691, and Elmendorf's Earl of Shadeland 30th 30725. The Indiana bull possesses much of the same character that has carried his famous sire Fowler 12899 to victory in many a hard-fought field—great show of constitution, a wide deep chest, broad chine, well sprung rib, and an abundance of firm flesh, standing squarely on his pins, moving with remarkable freedom of action, and showing a head and horn of most attractive character. He scarcely has the scale of his sire, but has size sufficient and is of that commanding presence which never fails to impress. He was selected by the judge, after an examination of unusual thoroughness, to head the class. He has for dam Cherry Pie 2d 17849 by Horatius 7163. A second was found in the first-prize winner of this same ring a year ago—Makin Bros.' Vincent, by Sir Evelyn 9650, as wide, thick and low as ever, his quarters of exceptional weight, but lacking of course the exceeding freshness of Mr. VanNatta's active, yet heavy, bull. Mr. Yeomans had hoped to get much further forward with his great son of Beau Real than third, and with a breeder instead of a butcher upon the bench his chances for preferment would certainly have been exceedingly bright; but the bull was finally relegated to a lower rank on account of a lack of firmness in his flesh—an unpardonable fault of course under the hand of the experienced buyer for the block. As to his depth and even disposition of meat throughout the carcass there can be no exceptions taken, and he shows a loin and covering over the hips that reveal clearly the impress of his wonderful sire. He comes of the same family of cows as did Sergeant Major and his dam was a daughter of 'the mighty Rudolph.' Three years ago upon the same spot in a close contest

Beau Real himself had met and vanquished Fowler, and the revenge wreaked last week by the son of the defeated bull over the progeny of the victor upon the former occasion will be cited as affording fresh proof of the proposition that 'all things come to those who wait.'

"In two-year-olds Day scored with Cameo, a son of Beau Real, remarkably mellow and deep in his covering. The VanNatta herd supplied the yearling winner, Rare Boy by Cherry Boy out of Rarity by Assurance, a sturdily built, well grown and level-fleshed young bull and, like all the Tregrehans, active as a cat.

"The aged cow class was a strong one, and was deservedly headed by Harry Yeomans' Flora, by Godolphin, a grand big yellow quality cow, with a top of the rarest sort, and a fine cow calf at foot by Bellman. The purchase of this great-backed cow at Mr. Fowler's Nebraska sale, in thin condition at an absurdly low price, and the honors reaped by her at this show, afford striking illustration of the opportunities often presented by these dull times in cattle for good judges to make capital investments. Bought for a song because not sold in show fix, she has easily developed, under the skillful hand of Mr. Yeomans, into the best Hereford cow seen at this exhibition. The same owner had forward the Ponting-bred Moweaqua Lass, by Defiance, a gem of a cow with a world of flesh carried near to the ground. Mr. Elmendorf was represented by the well known Anxiety cows Flossie and Horatia 3d, but the former should have been left in pasture. It is unfortunate that a cow that had been almost invincible in her bloom should be forced into the fighting after all her prospects of winning have long since 'gone glimmering.' Flossie had served her time.

"In two-year-old heifers Elmendorf was placed first with Lily by Autocrat, VanNatta second with Jessie by Fowler, and Day third with Mabel by Sir Evelyn 2d. No decision of the week created more discussion than this, and it is not likely that any large number of people could agree as to how the heifers should be ranked. Many thought Mabel entitled to head the class on account of her great substance, evenness and quantity of flesh, and superior quarters, but Mr. Imboden faulted her forward as being too open at the top of the shoulder-blades. Lily was certainly more thoroughly feminine in her forequarters, her blades being nicely laid. Though not so strong behind as Mabel, she was rather neater in her bone, and while perhaps a bit hollow in her loin, showed neither the dimple of Jessie in her back nor the slight, very slight, disposition to droop in the middle seen in Mabel. After selecting Lily, for the reasons named, for first, Mr. Imboden hesitated long between Jessie and Mabel for second. He recognized the strength of Mr. Day's fine heifer and evidently disliked to set her so low in the list, but he regarded the open crops as sufficient in the case of a breeding female to warrant him in scaling her down to third. The VanNatta heifer that stepped into second has the same broad chine and good rib seen in all the members of the Indiana herd.

"In the yearling ring six of the twelve head shown were superlatively good, viz.: Elmendorf's Dazzle and Tottie, VanNatta's Gipsy Girl, Makins' Geneve, Yeomans' Melody 19th, and Cosgrove's Wilton Le Sueuress 43d. Mr. Yeomans' Washington twins, Fair Maid and Maiden Fair, were also heifers of fine promise, but the company was 'fast.' Mr. Imboden worked faithfully upon the shapely six and finally drew the Elmendorf entries for first and second and Makin Bros.' Geneve for third. Dazzle, the blue ribbon heifer, is a granddaughter of Mr. Parmelee's Sir Garnet, and is well grown, good backed, and level-quartered. Tottie by Saracen, the second-prize winner, is a remarkably thick-loined low-legged heifer with a fine coat. She might be a little nicer in her touch, however, and is a bit rough at the tailhead. Geneve (the unbeaten calf of 1889), by Duke of Hesse, a son of Caractacus, has made a decidedly meaty yearling, but has a trace of unevenness in her flesh. Mr. VanNatta's Gipsy Girl is by Blondin, a son of Sidney (he by Sir Bartle Frere out of the great cow Lark and now at the head of Hon. James M. Turner's herd), and is an elegant heifer, though lacking the age of some of her rivals.

"One of the prettiest shows of the week was the line of twelve snappy white-faced heifer calves. Early maturity was written all over them, from their broad heads and protruding neck-veins to their well covered rumps, but VanNatta's big level Nancy, by Lord Fowler, was scarcely to be denied premier place. She is a good bodied calf with fine back and full lines all around. The Iowa Hereford Cattle Company's Maiden Fair 2d by Washington, a smoothly covered seven-months calf, made a good second, and Cosgrove's Wiltona 15th, by Wild Eyes, a tidy level lassie, claimed the third. Makin Bros.' Minerva 2d, by Don Carlos from an Anxiety 4th dam, a good fleshed, yellow red, was 'blanked.'"

The male championship of the class developed an interesting duel between Cherry Boy, the first-prize aged bull, on the one side and Mr. Day's blue ribbon two-year-old Cameo on the other. It certainly is a high tribute to old Beau Real to record that with but two of his get on exhibition at this fair each in turn was found richly enough furnished to give the showy son of Fowler a hard wrestle for his honors. In females there was delay in adjusting the relative claims of the Yeomans cow and the Day and Elmendorf heifers, but the matured animal showed too much spread of top for her younger adversaries and received the ballot for best Hereford female.

At Lincoln, with Mr. Gosling on the bench, Vincent was the winner over Earl of Shadeland 30th, Beau Real 3d, and Star Wilton 4th, second place being assigned to the son of Beau Real. Mr. Gosling had never altogether shared in the popular estimate of the Elmendorf entry. Cameo was the only two-year-old on exhibition. Makin's Good Luck was first in the yearlings, and Mr. Havens drew the blue in the bull calf class with a son of Star Wilton 4th out of a The Grove 3d mother, shown at nine months and after this victory sold to Mr. Sears.

Flora was again first in the aged cow class, and Elmendorf was second with Horatia 3d. In two-year-olds Mabel was chosen, being preferred upon this ocasion to Lily. Mr. Havens won a prize at this show for get of sire on four richly furnished calves by Star Wilton 4th.

At Topeka on the following week Vincent was again preferred to Earl of Shadeland 30th. Elmendorf had both prizes in the cow class, and Lily was given the premier place among the two-year-olds, although later, in the competition for a Hereford association special, Horatia 3d was placed above her.

Messrs. Clark and Fowler & VanNatta had shipped their herds from Des Moines to Hamline, where they met local Minnesota competition, furnished mainly by the Cosgrove Co. Cherry Boy defeated Peerless Wilton, no two-year-olds were shown, and in yearlings Mr. VanNatta had first on Rare Boy. In the cow class Peerless 3d received first, with VanNatta's Celandine second. Peerless 3d was shown this year at a weight of 1,750 pounds. Celandine was the dam of Chicago, a promising bull calf even then in training for the forthcoming World's Columbian Exposition. In two-year-old heifers Mr. VanNatta's Jessie, by Fowler, was the victor, and among the yearlings Clark was first with Horatia 4th, own sister to Horatia 3d.

The Eastern Circuit of 1890.—The shows east of the Mississippi commenced this year at Detroit, where Merrill & Fifield, Sotham & Stickneys, James M. Turner and R. G. Hart put up an interesting fight. In aged bulls Mr. Turner's Sidney 16574, by Sir Bartle Frere out of Mr. VanNatta's show cow Lark by Rodney, was sent to the front, the second ribbon being placed on Sotham & Stickneys' Perfection 30079, a deep-fleshed and attractive son of Sir Wilfred out of Lemon 2d by The Grove 3d. His owners exhibited at this same show the more highly fitted bull Harold 21141, that had been shown so successfully since a yearling as a member of the Curry herd, and expected to win with him. This was in fact his first defeat.

In the cow class Merrill & Fifield had a popular

first in their beautiful Lovely 2d 21977, as yet unbeaten. In a class of nine two-year-old heifers Sotham & Stickneys were first with Miss Archibald A 2d, daughter of the young bull Archibald A, previously alluded to in these notes, that had in the meantime been exported to South America. In yearling heifers Sotham & Stickneys had first on Purity, second on Mystic and third on Gay Lady. The herd prize and the bull-with-two-of-his-get ribbon both went to Merrill & Fifield. These same herds came together again at the Michigan State Fair on the following week with somewhat varying results, the herd prize falling to Sotham & Stickneys.

At Columbus there was a very light show in 1890, Sotham & Stickneys and Elijah Field, Camden, O., being the only exhibitors. While the Sotham & Stickneys cattle were not seen further in this season's circuit, this year marked the beginning of a long series of exhibits on both eastern and western circuits by Mr. Sotham, who set out to devote his energies largely to the refinement of the Hereford type, more particularly in respect to head and horn. We shall meet him again.

At the Wisconsin State Fair in Milwaukee Thomas Clark and the Cosgrove herd came down from Hamline, and were met by the herd of J. J. Williams, the Clark cattle receiving most of the first and championship awards.

There was little doing at the Indiana State Fair of 1890 in the Hereford class, exhibits being made only by two local firms with no special pretensions so far as showyard fitting was concerned. These were the herds of Parkhurst & Son and G. W. Harness & Son. Messrs. Harness had first in the aged bull class with Oregon, sired by Careful out of a Prince Edward 2d cow. This bull had formerly been in service in the herd of Seabury & Sample. The Parkhursts showed the Garfield bull, Earl of Shadeland 20th, in breeding condition, receiving second prize. The Messrs. Harness had first in the two-year-olds on Earl of Shadeland 41st. Parkhursts had first in aged cows with Elzina, and Messrs. Harness second with Perfection, a daughter of old Romeo.

At the Illinois State Fair Cherry Boy had a walkover again, defeating Earl of Shadeland 30th and Peerless Wilton. Mr. Earl received first in two-year-olds on Earl of Shadeland 47th, by Garfield out of a Sir Bartle Frere dam, broad at the chine, with a good head and well sprung in the rib. VanNatta had second on a low-legged bull of good scale called Armour 36916, by Blondin out of Fancy Arrow 2d. Rare Boy was first in yearlings, and Earl Wilton 36th was second. Mr. Earl's Captain Grove by Earl Grove 4th out of Cordelia by Colorado, a rich, low-bodied, strong-backed, wide-headed calf, was first among bull calves; he was then sold to Capt. Scarlett, who had some time before succeeded Mr. Yeomans in charge of the herd of the Iowa Hereford Cattle Co. and was now managing a new Iowa enterprise.

Mr. J. B. Camp, of Harristown, was the judge

upon this occasion, and in the Hereford cow class created some consternation by sending the blue to Elmendorf's Etiquette 11247, an extra good cow by Anxiety 6th, but now wanting in bloom. Peerless 3d, that stood second, was the almost unanimous choice of the spectators for the head of the line. She was at this time a show cow of the best type, with nobly arched ribs thrown well down, thick fore-roasts, and full loins. Elmendorf's Lily headed the two-year-olds, Mr. Clark's Bess standing second.

Nothing more attractive than a line of well fitted Hereford heifers is ever seen in our western showyards, and the 15 yearlings at Peoria this year excited universal admiration. Mr. Earl's Shadeland Cora, by Earl of Shadeland 22d out of a Colorado dam, drew the blue, with Elmendorf's Tottie slated for the red. Next came nine beautiful calves, the honor falling again to Shadeland, this time on Shadeland Fairy, also by Earl of Shadeland 22d—a calf with exceptional rib, full of flesh and hair, decidedly superior along back and loin. Her weak point was her quarters. Clark was second on Plum, one of the best calves of any breed out this season, a daughter of Peerless Wilton out of Peerless 3d. Senator Harris was called in to award the group and championship prizes. He ranked VanNatta's herd first and made Cherry Boy champion bull. The class decision on cows was reversed, Peerless 3d being adjudged best Hereford female in the showyard. The prize for best four animals under four years the get of one sire drew out a strong display, representing the progeny of Peerless Wilton, Earl of Shadeland 22d, Defiance and Grimley, the Peerless Wiltons winning out, with the Ponting cattle second.

The Shows of 1891.—On the western circuit a new Richmond now entered the field—Thomas Higgins of Council Grove, Kans., who had collected a valuable lot of both breeding and show cattle, purchasing among other celebrities Cherry Boy.*

---

*John Steward prepared for the author about this date the appended sketch of Cherry Boy which is deemed worthy of permanent preservation in these pages:

"Cherry Boy was calved April 5, 1886, so is now in his fourteenth year. His dam was imported Cherry Pie 2d 17849, bred by Mr. Stephen Robinson, sired by Horatius, one of old Horace's best sons, second dam Cherry Pie, by Docklow, by Above All, bred by Mr. J. Hewer and tracing back through this Cherry Pie family to Sir Thomas and Sir Benjamin. Cherry Pie 2d was a medium-sized cow with an extra fine head, a splendid coat of hair, and altogether showed much breed character. She was a good breeder, having produced to the service of Fowler not only Cherry Boy, but Cherry Lad, many years used by Mr. Ohl, Iowa City, Ia., and Cherry Girl, Mr. J. M. Curtice's fine breeding cow. I mention these last two animals to show Cherry Boy was no freak, but the result of mating two good individuals backed up by a long line of well bred prize-winning ancestors on both sides. While here I will call attention to the fact that Cherry Boy and Fowler were both spring and early summer calves; this is worth noticing, for while most breeders mate their animals to have them produce calves in late fall and early winter, and so have long ages for the show calves, I could mention many instances where the best calf of a season's crop was dropped in the spring or summer.

"Cherry Boy did not have special care or handling until over a year old, which proved to be a mistake. As he was a high-strung nervy fellow, it took considerable time and patience to get him used to handling for the showring. He was from the start a great feeder and had a good milking mother, which very essential trait has been transmitted to the majority of his daughters. Any one who has seen him after studying the make-up of a beef animal, noticing the masculine head, strong jaw and extra wide muzzle, could tell he was a feeder. Add to this his graceful drooping horn and large full eye, his round, well balanced body, on straight short legs, wide deep chest, extra fullness through crops and heartgirth, an oval padded loin, smoothly laid-in hips, full thigh, bulging twist, deep rich dark-red-colored coat of curly hair, and lordly carriage, and the majority of breeders would esteem him as an impressive sire. He was a bull all over, proud as a peacock, active as a kitten; he needed neither whip nor prod to shape himself for inspection. I saw him in the paddocks a few weeks since, and while four years had passed since I cared for and fed him, he recognized my voice and was on dress parade immediately. Although of course only the shadow of his former self, there was still the same proud bearing, or what we used to call Cherry Boy 'get-up,' which he always had and which marked attractive showy appearance he has transmitted to all his offspring. In a recent conversation with his

With this son of Fowler at the head of his line Mr. Higgins made his initial show at Des Moines. Cherry Boy had beaten Earl of Shadeland 30th twelve months previous, but now the tables were turned. The son of Garfield in the capable hands of George Mason had put on a lot of flesh which was still smoothly carried, whereas Cherry Boy came back after a heavy season's service with some loss of bloom. The mellow-fleshed Cameo came along third, and as usual delighted the hands and eyes of good feeders. Cosgrove pulled the blue in two-year-olds with Wildy 15th, by Wild Eyes out of the big cow Bonnyface by Rudolph. He was shown at a weight of 1,665 pounds. John Gosling exhibited a few good cattle this season, from his place at Rockford, Ill., and contested with "Ned" Scar-

---

owner I was told it is the same now as it used to be when I had charge of the herd—the visitor or prospective buyer, nine times out of ten, selects a Cherry Boy whether it be in stall or pasture.

"Of his many showyard victories I shall only mention the championship at the Iowa State Fair in 1890, and at the Illinois State Fair at Peoria, where he was champion of his breed and tied the Shorthorn champion Young Abbotsburn for sweepstakes. While fitted for the showring several years in succession, he was always a very sure breeder, and I believe has as many calves recorded to his credit in the Hereford record as any bull of the breed. Nearly, if not quite, 300, with still some more to register, makes his record as a breeder remarkable. Not all of these calves were bred at Hickory Grove, as in the spring of '91 Mr. Thomas Higgins, of Council Grove, Kans., fell in love with Cherry Boy (then in his very best form) and after considerable parleying, purchasing a carload of females and paying $1,200 for him, took him down to Kansas, where for two years he did very heavy service.

"In the spring of 1893, while preparing an exhibit for the World's Fair, the bull we intended heading the show herd met with an accident which knocked him out, and as Mr. Higgins, on account of ill health, had by this time disposed of his entire herd to Mr. Anthony, Mr. VanNatta started me on a trip to Council Grove, with the result that when I returned to Hickory Grove Cherry Boy came back with me. At that time he was seven years old and owing to long continued heavy service and climatic changes we failed to have him in his old-time form and place at the Columbian; he stood at the head of the second-prize herd there, his yearling daughter Cherry Lass being one of the group.

"Of a few of the great number of his prize-winning sons and daughters I shall make brief mention: Cherry Boy 2d, sweep-

lett for the red ribbon in this class. Mr. Stocking sent that trophy, however, to Scarlett's Mountain Dew by Washington. The same exhibitor had an easy first in the yearling ring in Captain Grove, first-prize calf at Peoria a year before, now developing into a well grown, level bull. Elmendorf's Crusader, by Earl of Shadeland 30th, was second. Cosgrove was strong in bull calves, winning with Wildy 37th over Elmendorf's St. Louis, son of the famous Lily.

In cows it was Lily vs. Mabel again. As three-year-olds they were about as evenly balanced as they were in their two-year-old form. It will be remembered that Mr. Imboden preferred Lily and John Gosling went to Mabel at the Des Moines and Lincoln shows of 1890. The result of Mr. Stocking's examination this year confirmed the Imboden rat-

---

stakes in calf and yearling form at Illinois and Iowa state fairs; Cherry Lass (full sister of last named,) and Wallflower with Cherry Duchess, winners at Des Moines in 1892, as all were the following year at the Columbian. Cherry Lass afterward being purchased and exhibited by Mr. Sotham; Wallflower and Cherry Duchess, purchased by Mr. George Redhead, the last named being a winner in cow class for him several successive years; Erma, exhibited by Mr. Funkhouser; Rare Boy, Elvira 3d and Minnie's Cherry, of Sunny Slope fame; Columbus, used several years by Mr. Tom Ponting; then, as recently as the late Kansas City Hereford show, Lincoln 2d, Clodia and Miss Betsy 2d were all winners. To go on with this list would make too long a story, or to mention the many valuable breeding cows of his get scattered through most of the prominent herds of the country, there being very few herds which have not some of the descendants of this famed sire and are proud of the ownership.

"I cannot close this article without some brief mention of two of Cherry Boy's greatest sons, the steers Kodack and Cherry Brandy. The former was first in his class, in the first prize Hereford herd, and also a grand sweepstakes herd at the Chicago Fat Stock Show in '91. Cherry Brandy was sweepstakes of the breed at the World's Fair Fat Stock Show and also at Chicago the following year, and was conceded by all good judges to have been one of the most perfect steers ever exhibited. When the history of the noted Herefords of this decade is written, who will deny one of the most prominent pages for this grand old bull? Well may any breeder be proud to breed and bring out two such animals as old Fowler and his greatest son."

ing, and in the subsequent showing for the Hereford specials Col. Harris of Linwood passed a like judgment. However, it was practically a toss-up between the two. They were a pair of grand good cows in any company. Lily showed rather more finish and feminine character forward than did Mabel, and upon this one point the decisions rested. Lily still showed that trace of hollowness in her loins, but her competitor was a bit uneven in her top, and did not stand so well behind. Mabel was a thick massive cow, carrying a great wealth of flesh. A good third was found in Mr. Higgins' Maid of Orleton, not large, but nicely covered and neat. The same exhibitor's Ashton Beauty, a cow of marked substance, and Gosling's yellow-red Fantasma, not large, but meaty and full of quality, were good beasts unplaced.

Cosgrove again came to the front in a ring of six two-year-olds, capturing first with Wiltona 15th and second with Wiltona 22d. Wiltona 15th was an exceptionally well furnished heifer, wide, low and smooth, although she might have been more refined about the head. She was round and well covered, carrying her meat well down on the rib, and was an entirely satisfactory first. Wiltona 22d was a larger heifer but not so evenly filled, her size and weight bringing her the red. Elmendorf's Tottie, with her nice head, good neck-vein, and twist well filled, drew third position. In selecting Mr. Day's growthy daughter of Sir Evelyn 2d, Mayflower, for premier honors among the yearlings a choice which many failed to approve was made. Mr. Higgins' Miss Wilton, a rare thick Beau Real heifer out of the celebrated Lady Wilton, named for second place, would have made a more popular winnner. Elmendorf's Hypatia, by Peerless Wilton, and a March calf, was third. Critics thought that the ignoring of Scarlett's good Washington heifer, Fair Maiden 2d, ripe, tidy, and smooth, but at some disadvantage as to age, was a palpable error. She let down a trifle in her back, but was all quality. In heifer calves the Cosgrove people duplicated their remarkable record in two-year-olds, drawing the blue on Wiltona 31st and the red on Wiltona 33d, Elmendorf following with a daughter of Earl of Shadeland 30th.

For the best bull in the class Earl of Shadeland 30th and the yearling Captain Grove were the chief competitors, and the big bull won. Lily was champion cow.

At the Lincoln state fair of 1891 James A. Funkhouser, Plattsburg, Mo., who was soon to become prominent as a breeder and exhibitor and who had already been elected President of the Hereford association, was called as judge. The herds of Higgins, Day and Elmendorf were before him, reinforced by an exhibit made by John S. Carlyle of Chicago.

John S. Carlyle.—It is not often that Scotchmen become enamored of the "white faces," but Carlyle was for years one of their greatest admirers. He was a grocer by trade, but made the acquaintance of the early exhibitors at the Chicago shows. During the fairs and fat stock competition we fear his customers often wondered what had become of "the boss". Mr. Carlyle was a keen witted, close student of Hereford character, and was never happier than when arguing with owners or herdsmen as to the relative merits of the cattle he so enthusiastically supported. At length his ambition to become an owner, breeder and exhibitor was gratified. While still retaining his Chicago business he bought a farm near Vesta, Neb., and made selections of foundation stock, largely from Shadeland and from the herds near Beecher. While his venture probably did not prosper financially, it is doubtful if any man ever lived who found greater pleasure than John S. Carlyle in the companionship of good cattle. This much is said concerning him because he was really one of the characters developed by the era of which we now speak. Carlyle made a brave fight at Lincoln, and later at the Illinois State Fair, against veteran showmen, and carried home prizes that served him as themes of conversation for months afterwards.*

---

*Mr. Carlyle's Earl of Shadeland 12th was a son of Garfield and Tiny, a low-down, deep bodied, thick flesh-carrier, with good head, good back and loin, well filled at heart and girth, and of altogether very taking character. He was presented in everyday breeding condition only, and consequently was "not in it" with his better-fitted rivals. His cow Crystal Belle was of the same stamp, low to the ground, deep and wide, thick in her flesh and sweet in her general appearance. She was a seven-year-old daughter of Cedric by The Grove 3d, and was the dam of Clara Belle, the blue-ribbon winner at Peoria in 1889. For two-year-olds Mr. Carlyle showed Eletta 2d, by Peerless Wilton out of an Anxiety cow, and Princess Louise 5th, by Cedric out of a daughter of Lord Wilton. This latter heifer was nicely conditioned, with a pretty head and neck, extra back and well filled loin, with ribs richly and evenly covered, and an extra good hindquarter. The plums of this string of cattle, however, were the yearlings

Cherry Boy was preferred by Mr. Funkhouser to Earl of Shadeland 30th, and Cameo came third. Crusader was first in yearlings and Carlyle had first on his bull calf Bobbie Burns over St. Louis. Lily was moved up above Mabel in cows. Carlyle had first and fourth in two-year-olds on Princess Louise 5th by Cedric and Eletta 2d by Peerless Wilton. Tottie was second. In yearlings Carlyle had the great satisfaction of winning with Favorite by Anxiety 3d, one of the most charming heifers of the old bull's get. In calves Elmendorf had first on Blue Belle and Carlyle second on Annie Laurie. Earl of Shadeland 30th had the bull-with-get prize. Carlyle's Crystal Belle won in the cow-with-produce class. Day took the herd prize. Lily was champion female and Cherry Boy champion bull.

Claude Makin judged the Higgins, Day and Elmendorf herds at Topeka in the fall of 1891, reversing Funkhouser on aged bulls and yearlings. Lily beat Mabel again, for the fourth time that season.

Favorite and Bonnie Sadie. In general appearance Favorite was one of the most attractive heifers we have ever seen, and if early maturity was sought it could be found in this almost perfectly formed daughter of Anxiety 3d. Her head and neck, smoothly covered shoulder perfectly laid, her broad chest and beautiful brisket, combined to make her particularly charming as she met you, and if faulted somewhat back of her hips, one could still apply to the balance of her well developed form a description of a perfectly formed little cow and in no way overestimate her. She was refined in character, and no term so fitly describes her as "sweet." Bonny Sadie was a granddaughter of Lady Godiva. She was not so fully matured as Favorite, but had a beautiful coat, was very straight in her lines and as smooth as an egg. Four calves completed the lot—two heifers and two young bulls. Of the former Annie Laurie was the better fleshed, but Mr. Carlyle thought much of Heather Belle. The bull calves Bobbie Burns and What Care I were sons of Fanny and Crystal Belle respectively. Bobbie was a thick-bodied calf with good back and loin and full in the crops. His companion was not so meaty, but was exceptionally neat and clean-cut in his make-up, and could easily be put in extra form.

Eastern Circuit of 1891.—H. H. Clough, Eugene Fifield, and the Sotham Co., all of Michigan, and Elijah Field of Camden, O., met at Detroit. Mr. Fifield beat Harold and Peerless of Rockland with Alger. Sotham took the two-year-old bull ribbon with Harold 2d, and had first in cows with Miss Archibald A, besides first in two-year-old heifers on Mystic. At the Michigan State Fair Alger and Harold 2d again won. Fifield's Accacia was first in senior cows and Sotham's Miss Archibald A was second in three-year-olds. Mystic repeated her Detroit victory.

At the Ohio State Fair of 1891 Clough, Sotham, Elijah Field, Harness of Indiana, and John Savage of Elyria, O., went before Mr. R. Baker of Elyria as judge.

As both the Clough and Sotham herds were now about to come into national prominence we reproduce below "The Breeder's Gazette" comment on this important show. It will be observed that we here meet for the first time, in the bull calf class, Mr. Sotham's afterwards famous sire Corrector:

"Mr. Clough's Peerless of Rockland, by Peerless Wilton out of Jessie by Anxiety 3d, led the ring for aged bulls. He is a bull of considerable scale, with a head full of Hereford character, a grand loin, very heavy quarters and a good hide. Mr. John Savage, Elyria, had second ticket with his Peerless Wilton bull, somewhat smaller, but truly fashioned and full of quality. Messrs. Harness had brought forward Earl of Shadeland 41st, and Mr. Field was represented by Hero 2d. In the two-year-old Harold 2d Mr. Sotham has something which approximates his ideal of a 'white face' bull. He has a beautiful head and horn, refined and yet masculine, he is much neater at the throat than the generality of bulls of the breed. He has a well ribbed back which is deeply fleshed, and his quarters are of the improved sort. He was the only entry of his age. Yearlings were headed by the same exhibitor's Harilton, by Harold out of Gaybird, a daughter of the famous old Gaylass. He is a little shorter in the rib than Mr. Clough's Kodax, in fact there is a little less of him, but he has the advantage of the Rockland bull in quality and especially in the character of the his head and horn. Kodax is a bull of depth and substance, with fine loin and level quarters. He had second honors in this ring.

"Corrector, the latest calf from old Coral, dropped by the service of Harold, carried the winning ribbon among the calves for Mr. Sotham, and was followed by Mr. Clough's Florida, by Peerless of Rockland. Corrector is a very neat, smooth, evenly fleshed calf, well finished about his head and neck and holding a well sprung rib and good loin.

"Miss Archibald A is a good cow not to show against. Probably this negative putting of the case will obviate the necessity of a detailed description. It is enough to say that she is one of the great young cows of the breed. She is a 'big little one' of the stamp which every feeder knows is the most profitable to handle and is full of flesh of prime quality disposed with rare smoothness. She easily stepped to the front in this competition and Mr. Clough's Millie of Rockland, by Romeo, a cow with beautiful head, grand loin and table-back, stood second. There were in this ring Mr. Sotham's Lemon 3d, Mr. Clough's Jessie, Mr. Field's Ida Wilton, and Messrs. Harness' Perfection and Jessie, the latter big useful cows but not conditioned for this competition. Of Mr. Sotham's two-year-old heifers—Mystic by Royal Grove Jr., and Purity by Archibald—we rather prefer the first-named, and such was the rating they received at Detroit, but Mr. Baker reversed positions on this occasion, which more nearly squares with their owner's estimate of their merits. It is questionable whether the Michigan exhibitor was entitled to both ribbons in this ring, for Mr. Clough's Jewel 3d, by Sylvester, is a heifer of much substance, thicker and wider than either of the others but lacking a little of their quality and scarcely so refined in head and horn. Messrs. Harness had in Pet a heifer with a very handsome head and horn got by Earl of Shadeland 41st, and Mr. Field showed Duchess 2d, a growthy heifer of Wilton blood.

"From Mr. Clough's trio of yearlings—Cocoanut by Emperor of Rockland, New Years of Rockland by Sylvester, and Lady Frances by Washington—the first named was selected to wear the red, which here indicates first prize, and this was another overturning of a Detroit decision. Mr. Sotham's Beaubois Beauty, a handsome little yellow-red, had second place. From the six calves Mr. Clough drew both ribbons with a pretty pair—Actress and Jewel 5th.

"Mr. Sotham's herd, headed by Harold 2d and including Miss Archibald A, Purity, Mystic and Beaubois Purity, was so strong in each member and so uniform in nearly all respects that it proved a little too heavy metal for the excellent collection from Rockland which was headed by Peerless of Rockland. Messrs. Harness had the ticket for three cows each with her own calf, with Perfection, Jessie and Pet, the calves being by Earl of Shadeland 41st. The award for the get of a bull fell to Mr. Field on the progeny of Hero 2d, which is a son of Constable's Hero, the Regulus bull once owned at Beecher. It was rather questionable as between this exhibit and that made by the Earl of Shadeland 41st. Mr. Clough had the ribbon for cow with two of her calves on Jessie 5th, with her massive son Peerless of Rockland by her side, together with her young heifer by Sylvester.''

At Indianapolis on the following week the Clough and Sotham herds met again, this time with Col. T. S. Moberly of Richmond, Ky., a leading Shorthorn exhibitor of the day, as the arbiter of Hereford fashion. Rightly or wrongly, as one pleases to take it, he reversed the Columbus awards in every class in which there was competition, save only that for aged cows.

In a notable breed contest here Sotham's Harold 2d gained the two-year-old ribbon—the only one saved to the Herefords out of a nerve-racking contest all along the line with a great lot of Shorthorns and Aberdeen-Angus.

At the Illinois show of 1891 John Imboden had one of the hard days of his long career in the jury box—especially when the cows and heifers came in view. Clark, Elmendorf, Sotham, Carlyle, John Steward, ''Ned'' Scarlett and Fowler & Bassett, Long Point, Ill., supplied the trouble. Fowler & Bassett were showing Armour 36968, exhibited twelve months before by VanNatta. He was a bull showing more quality than Earl of Shadeland 30th, but the old hero of so many contests was still impressive enough to win. Probably no greater contrast could readily be imagined than that presented by these two antagonists in the matter of color. Garfield's son was a very dark red, too dark in fact, while Armour was a real golden yellow. In two-year-olds Sotham's Harold 2d forged to the front over Clark's Phil Armour by Anxiety 3d. Scarlett's square-ended mellow-handling low-flanked Captain Grove was the winner among the yearlings. In bull calves John Steward (Fowler & VanNatta's former herd manager of whom we shall hear more as our story progresses) took first with a youngster, his own personal property, by Cherry Boy out of a Star Grove dam, that proved too much for Sotham's Corrector to handle.

The cow class, a memorable one, was headed by Lily, her chief rivals being Peerless 2d and her six-year-old daughter Peerless 3d and Sotham's "big little one," Miss Archibald A. Peerless 3d was placed second. The heifer classes were exceptionally fine, and as indicating the type and quality being produced at that date by home breeders we quote again from "The Breeder's Gazette":

"In two-year-olds Imboden rather crossed the judgment of a majority of the Hereford-breeding contingent present by sending Fowler & Bassett's big Victoria Belle to the head of the list. She is a grand-topped heifer of tremendous scale, but plain at both ends. John S. Carlyle's Princess Louise was slated for second money, leaving Clark's Horatia 4th, Carlyle's Eletta, and Elmendorf's Tottie unrecognized. The first-prize heifer, while lacking in refinement, is richly furnished all along her back from fore-roasts to loin and is besides big and deep through her heart. Princess Louise and Horatia 4th are not cast in so big a mold. They are of a neater, better-finished type and full of flesh as well. Horatia, with her pretty incurving downturned horns and sweet countenance, meets you more effectively than either of her rivals, is shown with a great coat, has nicely arched ribs and exceptionally neat bone, but in her quarters is not so good as the Princess. The latter is carrying plenty of flesh, but shows that inclination to roll on the rib that is so apt to be developed by quick-feeding cattle. She is of a meaty type, near to the ground, with good width, and deserved all she won, if not more. Those who thought that a richly meated table-back should not carry a roughish head and unsatisfactory rumps to the post of honor would have set Horatia second.

"It was a charming array of yearlings (eleven in number) that were moved into position, and when it is said that Carlyle had an outstanding winner in his beautifully brought out Favorite little more need be added. This splendid daughter of Anxiety 3d was shown in the very height of showyard form. To the bloom that always attaches to a well fed and finely modeled yearling is added the catchy embonpoint of the heifer six months gone with her first calf. Nothing is lacking to complete the picture save a little lightness of thigh and a trace of unevenness about the loin. Generally speaking, however, she is a wonderfully furnished, smooth, thick-fleshed heifer of much breadth and depth, and remarkably developed for age. Indeed, as a specimen of early maturity she is one of the sensations of the year—practically a finished cow in her yearling form. After she had been set aside to head the class there was a spirited contest for second place between Clark's Plum (dam Peerless 2d) and Scarlett's Fair Maiden 2d, by Washington, the position finally being held by the latter. The former wore one of those shaggy coats which Clark succeeds so well in growing. Moreover, she begins well and ends well, her pretty face and wide forehead attracting one from the front and her good thighs satisfying the eye as she goes away.

"The decision which made Carlyle's Annie Laurie first in heifer calves, and Elmendorf's Bluebelle second, with Clark's Annie (by Anxiety) and Fowler & Bassett's thick Bonnie (by Orphan Boy) unplaced was not altogether satisfactory. Bluebelle is a furry-haired, mellow-handling, well grown calf with typical head and should have been first, with Clark or Carlyle second. Carlyle and Clark both have good calves and richly bred (the former by Earl of Shadeland 12th from the famous Felicia), but we cannot grant the license of either of them to win in the presence of the Elmendorf entry."

Elmendorf received the herd prize at the capable hands of David Fyffe, Sotham receiving second. R. C. Auld made Earl of Shadeland 30th champion male and Miss Archibald A champion female.

Death of C. M. Culbertson.—While the "white faces" were carrying all before them at the Fat Stock Show held at the Chicago Stock Yards during the first week in December, 1891, Mr. C. M. Culbertson, the man who had done so much towards the successful introduction of the breed, passed away at Arkansas Hot Springs, whither he had journeyed in the hope of securing relief from a complication of disorders incident to advanced age. He had passed the four-score milestone.

Hereford Constitutions in Evidence.—The fact that Vincent and Earl of Shadeland 30th were able to come back again at the shows of 1892 afforded fresh proof of the staying qualities, the constitution, and the vigor of the Hereford. Ability to stand up under the pressure of long-continued high feeding for show demonstrates the reverse side of the claim made for the "white faces" as the hardiest of all the improved breeds of cattle of the beef type. Second only in point of practical interest for cattle-growers to the demonstration of constitution being made during this same period by the Hereford bulls on the open ranges of the far west was the record of such bulls as Fowler, Vincent, and Earl of Shadeland 30th at the shows of the cornbelt states. Animals lacking in real virility would deteriorate quite as rapidly under the adverse influences of over-feeding as under the effects of hardships suffered under the conditions prevailing in the arid storm-swept areas of the treeless plains and intermountain grazing grounds that supported the cattle industry beyond the 100th meridian.

At Des Moines, in September, 1892, the sturdy son of Garfield, the Earl of Shadeland 30th, so often mentioned hitherto in these notes, was again awarded pride of place as best aged Hereford bull, but the perennially popular Vincent, now in his seventh year, pushed him for the honor. One of Iowa's most practical stockmen, Peter Mouw of Orange City, who had espoused the cause of the Herefords, won third in this competition with Castello, by Dromio, at six years. As good a bull as Cameo had to fail of recognition. The spectacle of these bulls presenting such form at such ages, after extended periods of high feeding and active service, recalls those early days in Herefordshire when their prototypes were hurling their weight into the yoke and were in the very prime of their usefulness at from six to ten years of age.

Meantime, new recruits were constantly joining the Hereford forces. Redhead Bros., of Des Moines, had been convinced of the merit of the breed and had established a good herd with the prize-winning Shadeland-bred Captain Grove, purchased from Mr. Scarlett, at its head. He was sent into this competition somewhat out of form, but still good enough to win. Cosgrove captured the red with Wildy 29th, and VanNatta was third with Chicago. In yearlings Cherry Boy 2d came to the front, a well developed young bull with rather prominent hips, and Makin Bros. were second on Anxiety Boy. In bull calves the Makins were first with Vincent 9th, a level short-legged son of their rare old bull of that name.

Speaking of the power of the senior bulls to hold their own, we have now to record that Lily once again led the aged cows, Cosgrove following with the dark-colored massive Wilton Le Sueuress 43d, and our old friend Mabel coming third. Makin Bros. had bought Mr. Henry's Countess of Rossland, by Prince Edward, and they sent her into this class along with Julia Wilton and Stately 10th, all of good Hereford type and quality, but here unplaced. Day got first in the two-year-old heifers on Hypatia by Peerless Wilton, Redhead Bros. were second on Lulu, by Ponting's Anxiety 3d from a Blenheim dam, very thick but rather over-done. The Makins' Stately 13th by Washington, unplaced, would have been preferred by many. VanNatta had both blue and red on a remarkable pair of yearling heifers—Cherry Duchess by Cherry Boy and Annabel by Star Grove 1st, the latter destined to rare honors in 1893. John Letham was at this date working with his good friend Steward on the VanNatta show cattle, and handled here the blue ribbon heifer calf Cherry Lass, own sister to the first-prize yearling bull Cherry Boy 2d and a heifer with a lot of hair and flesh, extra in her spring of rib.

Earl of Shadeland 30th and Lily had the class championships, and we shall see the old bull yet another year in the biggest battle of his career.

At Lincoln in 1892 Mr. Funkhouser again officiated as judge, the show being made up of the entries of Messrs. Day, Elmendorf, Carlyle, and Makin Bros. Earl of Shadeland bested Vincent; no two-year-olds were shown. Carlyle won in yearlings on Bobbie Burns, and Vincent 9th was calf winner. In cows it was Lily once again, with the Makins' Stately second, and Mabel third. Carlyle had both first and second in two-year-old heifers on Bonnie Sadie and Favorite; the Scotchman was first also in yearlings on Fanny Mack, while Elmendorf drew the blue in calves on Lady Daylight. The championships were as at Des Moines.

The Kansas show of 1892 was not up to the usual mark. Makin Bros. and Elmendorf divided the ribbons, with Earl of Shadeland 30th and Lily again in the stellar positions.

Tom Clark and Harry Fluck, the latter a high-class English herdsman now starting into Hereford breeding on his own account, exhibited at the Minnesota State Fair in competition with Cosgrove. It was rather a light show, reflecting the slackness of the trade which had now been in evidence for some time. Clark's Sanhedrim 46180, an in-bred The Grove 3d two-year-old of Culbertson's breeding, got by Star Grove 10th out of Grove Maid 18th, was champion male and Cosgrove's wonderfully deep-fleshed cow Wilton Le Sueuress was adjudged best female. These same herds were seen at Milwaukee the following week, Sanhedrim again heading the bulls, and Clark's two-year-old heifer Plum was set above the Cosgrove cow.

Death of Anxiety 3d.—During the first week in September, 1892, Anxiety 3d, whose daughters had been so phenomenally successful at the great shows for a number of years, was humanely killed to end his suffering from rheumatic afflictions that had for some months rendered the old veteran's life a burden to himself. He was in his twelfth year.

The Eastern Circuit of 1892.—As usual in those days the ball started rolling east of the river with the annual battle between the Michigan and Ohio herds at Detroit. Clough, Fifield, Sotham and Turner were still in the running. Alger beat Harold 2d, while Turner's Peerless Wilton 13th outranked Clough's Kodax. At the New York State Fair George N. Bissell of Milford, N. Y., and George O. Holcomb of Troy, Pa., presented excellent herds, the ribbons being tied by John Vanderbilt, manager for Erastus Corning.

In Ohio it was Clough vs. Sotham, with John Hooker of New London as "runner-up." L. P. Sisson, a West Virginia breeder of Devons, allotted the prizes. Harold 2d was sent to the front as senior bull over Hooker's Grover Morton. Kodax had no competition in two-year-olds and Corrector scored among the yearlings. Miss Archibald A 2d headed the cows; Clough's Cocoanut, a Wilton-Anxiety combination, with handsome front and splendid loin, was best two-year-old, and the same owner's Actress was the blue ribbon yearling.

The Clough and Sotham herds, supplemented by entries of West & Duncan, Windsor, Ill., made up the Hereford show the following week at Indianapolis. Harold 2d, Kodax and Miss Archibald A 2d were again honored.

Funkhouser Enters the Lists.—A Missouri breeder who now came rapidly to the fore in Hereford cattle breeding circles put in his first appearance as an exhibitor on the big circuits at the Illinois State Fair of 1892 at Peoria. This was James A. Funkhouser of Plattsburg, whose cattle will come in for frequent mention before our narrative ends. It was a large and good class that Harry Yeomans was here called upon to judge.

First of all there was the inevitable Earl of Shadeland 30th, accompanied by his old traveling companion Vincent; also Harold 2d—all familiar "white faces." Their right of way was challenged, however, and, as is turned out, successfully blocked by new antagonists. Mr. VanNatta had always put his money on the Tregrehan blood. He knew its prepotency and the stamina that went with it. But he was wise enough also to know that in the get of Anxiety 5th he possessed another valuable blood element. His imported cow White Spark 2d, of Stephen Robinson's breeding, got by Horatius 7163, he by old Horace, had produced to the cover of Anxiety 5th in 1885 the bull Saracen 23188, that was used in the herd quite freely. He sired among other good calves the bull Hengler 37003, dropped in 1888 by the imported cow Lady Hartington, by Hartington 4010, son of The Grove 3d. This calf had, therefore, a double cross of Horace, besides carrying the old Anxiety blood, and he developed into a bull good enough to win the blue ribbon in this Peoria competition—not a sensational show bull, but with good Hereford character, broad ribs, and well fleshed, although not just as even along his back as might be wished. Mr. Funkhouser had come into possession by purchase from Tom Clark, of one of the

Hereford treasures of his time—Hesoid 2d 40679, bred by George W. Henry from imp. Hesoid (he of the guy ropes referred to in our notes on one of the early Chicago shows) and of Curry's Anita by Harold. We have already spoken of his sire Hesoid as one of the richest of The Grove 3d bulls. Harold we recognize as Mr. Curry's good show bull, and sire of Sotham's Corrector. Harold's sire was Highland Laird, son of Horatius, so we observe that Hesiod 2d, as well as Hengler, had two lines to Horace, the sire of The Grove 3d. Like most bulls of this blood Hesoid 2d seemed to lack stretch and scale, but at three years old he here tipped the beam at 1,950 pounds, being compactly fashioned. He was drawn for second place, and in after years made the reputation of the Funkhouser herd as a sire of good Herefords.* The old Earl of Shadeland 30th fell back to third.

This was something of a Grove 3d day, for in two-year-olds Clark's Sanhedrim, with two lines to the old bull, went to the top of the two-year-olds. Clough's Kodax was second. Captain Grove and Chicago were passed over. In yearlings Cherry Boy 2d came first, but the second prize winner, Mr. Tod Benjamin's Wilton Grove, by Sir Wilfred out of Lemon 2d by The Grove 3d, was greatly admired and might have been first instead of second.

---

*Speaking of Hesiod 2d, Thomas Clark tells this interesting story:

"I bought his mother Anita at one of G. W. Henry's sales. She was carrying the calf which I called Hesiod 2d. He was dropped an immatured calf at seven months, not larger than a jackrabbit. We had to hold him up to suck for three weeks. I sold him with two other bulls to Funkhouser at eight months old—$1,000 for the three."

Vincent 9th was best bull calf, with Clough's Actor second. Clark's Lars, a youngster with a future, was one of the "also rans" in this company.

Few better cow classes had been seen in the west than that which was finally led by old Lily, with Peerless 3d in second place. As flesh carriers they were certainly great Herefords; but for quality one could not but go in raptures over Funkhouser's Petunia 3d or Clough's Jewel 3d. Clark's great two-year-old Plum was preferred in the next class to Clough's Cocoanut. In yearlings VanNatta's beauties, Cherry Duchess and Annabel, were again first and second. In calves the same herd won with Wallflower. Mr. Funkhouser gained the herd prize.

At the Illinois shows of those days there was a "sweepstakes by ages" open to all beef breeds. At this exhibition of '92 Moberley's celebrated Shorthorn Young Abbotsburn won in the aged bulls, Clark's Sanhedrim in the two-year-olds, VanNatta's Cherry Boy 2d in yearlings, and the Makins' Vincent 9th the bull calf championship, all at the hands of that sterling Shorthorn breeder Charles B. Dustin. In an open class for best cow with calf at foot Funkhouser's Petunia 3d, by Anxiety 4th, was chosen.

By this time all interest was beginning to center upon the World's Columbian Exposition at Chicago, to be held in 1893. With that event the curtain fell upon a decade that had witnessed a wonderful development of interest in the white-faced breed.

World's Columbian Exposition.—This exhibition still unmatched in the field of international events of like character was held at Jackson Park, Chicago, in 1893. As a great live stock department had been provided by Chief Buchanan of the agricultural division practically no attempt was made to hold the usual fairs that autumn in the middle western states.

The Hereford association and the leading exhibitors of that period planned a presentation of the "white faces" that should fittingly commemorate the success the breed had now achieved. The display was particularly notable for the fact that the champion bull was, for the first time in some years at western shows, of English breeding. This was Ancient Briton, from the herd of the late William Tudge of Leinthall. Likewise noteworthy was the fact that Messrs. Gudgell & Simpson for the first time participated in the big battles of the ring. We deem this competition of sufficient importance as an incident in American Hereford history to warrant presenting substantially in full an account written at the time for "The Breeder's Gazette." We quote:

Ancient Briton.—" 'It is the general opinion that Ancient Briton (15034) is the best Hereford bull that has gone out of the country for a good many years.' Such was the report which came to America last spring along with H. H. Clough's importation of 'white faces' selected and shipped from Herefordshire by W. E. Britten, at the head of which stood the bull just named. American breeders are fully prepared to concede that Mr. Clough has in Ancient

Briton the best bull of the breed now on this side the pond, for they have run afoul of him in the Columbian showyard and struck their colors on sight. The decision which placed this admirably fleshed and richly bred three-year-old first in the ring for bulls three years old or over met with unanimous approval. Most of his competitors were bulls of maturer years, stock sires of repute and ex-champions, all lacking the freshness and bloom of the imported bull. He was bought of Tudge of Leinthall, combines the blood of the two celebrated Adforton Royal winners Lord Wilton and Regulator (4898), being by a son of one and out of a half-sister to the other, and is a bull of fine scale and quality, with his flesh smoothly carried. Next to him was ranked Gudgell & Simpson's well known Anxiety 4th six-year-old breeding bull Don Carlos 33734, a trifle deficient perhaps behind his shoulders, but with the real Anxiety rib and loin and of better quality and character than the third-prize winner, Elmendorf's massive Earl of Shadeland 30th. The latter has 'come again' in surprising form, heavier than ever before, and with substance unsurpassed. Fourth honors fell to Makin Bros.' good three-year-old Vincent 2d 42942, by their famous old Vincent out of Berrington 2d 28255, with Cosgrove's young Wildy 29th, by old Wild Eyes out of Bonny Face, fifth, and Fleming's Commodore sixth. Mr. VanNatta sent two valuable bulls into this ring—one the famous Cherry Boy 26495, by old Fowler, and Hengler, by Saracen. The former was once a rival of Young Abbotsburn for championship honors at Peoria, but while in service in Kansas was necessarily let down considerably, and after passing out of Mr. Higgins' possession little effort was made to keep him up. Mr. VanNatta bought him back some months ago, but the time was too short to restore his wonted condition. His sturdy old sire, the veteran Fowler himself, might better have been sent into this 'hornet's nest' than his honored son with such a handicap. Hengler is a bull of strong parts but Columbian winners had to be cast in even a more heroic mold.''

Sitting Bull.—''Two-year-old bulls were a small class of six, at the head of which the judge placed Mr. Fluck's heavy-fleshed Sitting Bull, not of extra quality but an exceedingly deep level bull with good ribs, extra flanks, and heavy quarters. He is not a bull of as nice character as Mr. Todd Benjamin's second-prize winner, Wilton Grove, seen at Peoria last fall as a yearling. While not so heavily fleshed as Sitting Bull, Wilton Grove is much nicer in his head and more satisfactory to the touch. Third prize fell to Makin Bros.' Anxiety Boy 47708, a son of Juryman 30279 from Ellen Wilton 12710, a low fleshy bull of good scale, exceptionally full in his twist. Fourth honors went to C. H. Elmendorf's Eureka and fifth to Gudgell & Simpson's Beau Brummel by Don Carlos.''

Lamplighter.—''The yearling bulls were headed by Gudgell & Simpson's Lamplighter 51834, by Don Carlos out of Lady Bird 3d. He is of a very low meaty type, with good head, well rounded chine, nicely fleshed loins and ribs, covering nicely over the hips on to good quarters. Tom Clark was second with Lars 50734, whose dam was the famous show cow Peerless 3d. Lars is only a February calf, and consequently lacked the scale of some of his competitors, but he is of a nice tidy type with level well filled quarters. Makin Bros. were third with Vincent 9th 52705, got by old Vincent out of Barbelle. He is wide and low, with a very deep body, but sags a bit in his top. H. H. Clough's Actor was fourth and VanNatta's Chicago Lad fifth.''

Anxiety-Peerless Again.—"In bull calves Mr. Fluck again scored with Monitor F., the last fruit of the loins of old Anxiety 3d, and as his dam was a Peerless Wilton cow this youngster represents a doubling up of the blood of Clark's celebrated sire of prize stock. He is a good fleshy deep-ribbed calf, with wide head, well covered shoulders, and plenty of substance. Mr. Cosgrove was a strong second, however, with the Wild Eyes calf Minnesota 2d, a beautiful little bull with almost perfect top and bottom lines, strong in his flanks and quarters and excelling the first-prize calf in the twist. Mr. Clough was third with De Forrest, a very sweet December calf by Kodax of Rockland, showing perhaps the most perfect head in the ring, but too young to go further forward in the winning. The same herd also supplied the fourth-prize winner, Col. Davis. Fleming was fifth with Barman and Elmendorf next with St. Tristram."

Miss Beau Real 3d.—"It was a great lot of eighteen cows that were subject to inspection and in some unaccountable manner the judge made his first leet without including one of the best Hereford females in the yard, namely, Lady Tushingham 3d, property of H. D. Smith of Compton, Quebec, Canada. Before making his final ratings, however, he discovered that he had omitted that great three-year-old and finally sent her into the prizelist, although many would have ordered her further forward than fourth place, the position ultimately assigned to her. Mr. VanNatta's wide-ribbed, compact, short-legged Miss Beau Real 3d had the blue ticket. In her foreribs and loins she is certainly an altogether remarkable cow. She is full of substance and quality, although soft in her handling and not standing well on her hind legs. A fair idea of her conformation can be gained from Mr. Morris' sketch appearing in the frontispiece illustration to this week's Gazette. Mr. Funkhouser's next choice was Tom Clark's famous Plum, by Peerless Wilton out of the great Peerless 2d. Her fine middlepiece sufficed to carry her thus far forward in competition with cows that are rather better in their shoulders. Third place was assigned to Mr. Clough's rich but rather roughly fleshed Cocoanut 40726, and fourth as above stated to the big Canadian three-year-old. Some would have placed Lady Tushingham 3d at the head of the class, but when one considers the fact that Miss Beau Real 3d is six years old and is now well along in calf it must be conceded that she has strong claims to the position assigned her upon this occasion. A cow that some would have liked to have seen recognized was Elmendorf's Miss Wilton that was ranked fifth in the line. She is a daughter of the great Beau Real out of the magnificent and very famous Lady Wilton and is a cow of most beautiful character and quality. Gudgell & Simpson furnished the sixth-best cow in Myrtis 16180, now in her eighth year. Next below her came Cosgrove's short-legged, thick Wilton Le Sueuress 43rd. The fact is, this was about as hard a ring to judge as was the Shorthorn cow class, and, as was the case in that competition, there were unplaced cows in the lot which in the opinion of some good judges were the equals of the more successful animals. They were a grand good class and an animal had to be a very 'tip-topper' to secure any position in the leet.''

Annabel.—''The richness of the Hereford exhibit, so strikingly foreshadowed in the preceding ring, developed itself fully in the class for two-year-old heifers, where two of the best beasts in the entire beef cattle section contended for pride of place. We refer to Clough's imp. New Year's Gift and Van-Natta's Annabel. The former, a winner as a yearling at last year's Warwick Royal, was brought out in better bloom than the writer hereof had regarded as possible. She seemed ripe as a peach in England a year ago, and she is of such a refined type that it is surprising that she came out last week so fresh and good in her flesh. Forward she is as sweet as heart could wish. Her lovely countenance, full bosom, perfectly laid shoulders, smooth well rounded chine, deep ribs, and short neat legs combine to stamp her a heifer of altogether extraordinary quality. Annabel is not quite so 'ladylike' in her general make-up, but as a specimen of the sturdy buxom white-faced lassie, she is, to use the current phrase, simply 'out of sight.' We have seen heifers wider-spread than she, but when one considers her uniform depth and thickness of flesh, her substance and almost perfect balancing of parts, she has to be written down as near a model of her kind as American showyards have ever seen. She is furnished at every point, and, while a heifer of stouter build than New Year's Gift, has yet no suspicion of grossness in her marvelous make-up. She shares with the Angus Abbess of Turlington, the roan two-year-old Shorthorn heifer from Canada, and a yearling Hereford heifer soon to be named, the honor of being one of the three or four real sensational animals of the show. Star Grove 1st has to be credited with having sired this predestined champion of her class. With Annabel first and the imported heifer second, the Cosgrove's excellent Wiltonie 33d fit snugly into third place, and Clark's Jingle was fourth. Fifth and sixth positions in the line were held respectively by Gudgell & Simpson's Donna Anna 7th and Makin Bros.' Lady Maud Vincent.''

Lady Daylight.—''Second only in outstanding merit to the great Annabel of the preceding class came Elmendorf's superb Lady Daylight, an easy winner among the yearlings. This exceptionally grand heifer, a daughter of Earl of Shadeland 30th, is from a cow by Beau Real, and she from a daughter of Beau Monde, the dam thus being an inbred Anxiety 4th. She is laid out on a little lengthier scale than Annabel, so that she will never impress one as being quite so blocky as VanNatta's heifer, but she is modeled on such low level lines and rounds out so beautifully in her barrel and flanks, fleshes down so wonderfully fore and aft, that she is simply a 'flash' heifer in any company. John J. Steward brought up from Hickory Grove, along with VanNatta's string, a sweet little bundle of Hereford femininity known as Fowler Queen 2d, got by old Fowler out of Wilton Queen. She is a charming little witch with her famous 'daddy's' thighs, and landed herself in second place among these Columbian yearlings. Elmendorf had another string to his bow this time—Lady Laurel, by Earl of Shadeland out of his champion show cow Lily, and she was no 'tail-ender' either. She was easily third and Mr. Clough's growthy imported Merlin heifer Dorcas had to stand scaling down to fourth. Mr. VanNatta's Cherry Lass was pegged at the fifth notch, same owner's Fairy Browny sixth, Gudgell & Simpson's Normette seventh, Clough's Autumn Leaf eighth, and Redhead's Wallflower ninth—four out of the nine being of Mr. VanNatta's breeding."

Bright Duchess 15th.—"The calves were as pretty a show as seen on the tanbark during the week. They were about fifteen in number and so evenly good that some of those that were left out of the prizelist might safely be substituted for the quartette of winners without falling below a World's Fair standard. Messrs. Gudgell & Simpson were first with Bright Duchess 15th, by Earl of Shadeland 47th—good on her back and carrying her flesh well down. Makin Bros. were second with the ripe Roberta, by Beau Real out of Bertha; Clough was third with the pretty Primrose, a December calf by Kodax of Rockland, and Fleming came fourth with Lady Fenn 2d. Mr. VanNatta's Grove Lassie set down to fifth, was of winning shapes, and Makins' Prairie Flower, listed sixth, was remarkably full of flesh and near to the ground. They were a sweet lot throughout and the rear guard was better than the winners at some shows heretofore seen in the west."

The Championships.—"The male championship of the class was conceded to Mr. Clough's imp. Ancient Briton and the female championship went to Mr. VanNatta's great two-year-old Annabel with as little discussion. It may be interesting to note in this latter connection that the next best females of the class were rated in the following order: Miss Beau Real 3d, Lady Daylight and Lady Tushingham 3d.

"The herd prize fell to H. H. Clough on the following lot: Ancient Briton, Cocoanut, New Year's Gift, Dorcas and Princess; second prize to W. S. VanNatta on Cherry Boy, Miss Beau Real 3d, Annabel, Cherry Lass and Grove Lassie; third prize to Gudgell & Simpson on Don Carlos, Myrtis, Normette, Bright Duchess 15th and Donna Anna 9th; fourth prize to C. H. Elmendorf on Earl of Shadeland 30th, Lily, Belle Mode, Lady Daylight and Fair Nell. Relative rank beyond this point was assigned to the various herds in the following order: Cosgrove, Clark, Makin Bros., Fleming, Redhead and Day.

"The young-herd prize fell to Messrs. Gudgell & Simpson with the following animals: the bull Lamplighter and heifers Normette, Bonnnie Lulu 13th, Welcome 10th and Gertrude 5th. Second went to W.

H. H. CLOUGH'S COLUMBIAN EXPOSITION CHAMPION HERD, HEADED BY IMP. ANCIENT BRITON—From the drawing by Cecil Palmer.

S. VanNatta on Venture, Cherry Lass, Fairy Brownie, Grove Lassie and Alberta; third to H. H. Clough on Actor, Chestnut Leaf, Autumn Leaf, Jewel 6th and Nutty; fourth to Makin Bros. on Vincent 9th, Lady Wiltona Vincent, Lady Geneve Vincent, Stately 14th and Stately 10th; and fifth and sixh rank were assigned to Cosgrove and Elmendorf respectively.

"In the class for four animals of either sex under four years the get of one sire, Elmendorf was first with the progeny of his well known Garfield bull Earl of Shadeland 30th, the winning animals consisting of Eureka, Lady Daylight, Lady Lavender and Lady Laurel. Second went to the progeny of Don Carlos, shown by Gudgell & Simpson, including the two bulls Beau Brummel and Lamplighter and the heifers Donna Anna 7th and Normette; third to the Cosgrove Company on the get of Wild Eyes, including the bull Wildy 29th and three Wilton heifers; fourth to Clough on the progeny of Nutcracker.

"The first prize for best two animals of either sex, the get of one cow, went to Makin Bros. on bull Vincent 2d and heifer Lady Maud Vincent; second to Cosgrove on bull Bert C. and cow Wilton Le Sueuress 43d; third to F. A. Fleming of Canada; fourth to Elmendorf."

Dark Days.—The year 1893 will not soon be forgotten by those who were in debt or lacked working capital. A financial panic swept the United States from end to end. Money went in hiding. Banks failed. Credits were destroyed. Cash could not be had at one time even with Government bonds as security. The cattle business suffered its full share as a result of this catastrophe. Failures were numerous in all lines of business. Confidence was temporarily destroyed, and the recovery from the shock was a long and tedious process. Owners of pedigree stock did not escape the general gloom. Values fell rapidly. High-class registered animals sold in many cases at their mere value for slaughter at the yards—a state of affairs which put some people out of business, but which at the same time put others in at a bargain-counter basis. Those who had a little money and plenty of nerve took advantage of such a situation to stock up. What was one man's misfortune was another's opportunity.

It came to pass, therefore, that the great Chicago show marked the zenith of achievement in Hereford cattle breeding circles during the period of their first great enjoyment of popularity in the west. With this description of that event we enter the shadows of an era of profound depression in all branches of pedigree cattle breeding in the United States—a period which brought many enforced changes in the personnel of those engaged in the industry, but an era during which the foundation for a more enduring prosperity was laboriously but successfully laid.

A Desperate Depression.—Ten years had now elapsed since the great importing movement had been at its flood. The reaction from the boom had set in around 1885. Although at first a slow or creeping decline, it had been expensive to some of those who had allowed their enthusiasm or their cupidity to run away with good judgment. Prices had fallen steadily, beginning with that date except in the case of top cattle. This, however, was not without its redeeming feature. It gave the western ranchmen just the opportunity needed to extend rapidly the sphere of Hereford influence on the open range, and this was taken advantage of to the fullest possible extent. Many speculators and "butterfly breeders," as someone happily termed those who are active only when prosperity's sun is shining brightly, already had disappeared from the ranks when the great panic of 1893 fell upon the country practically without warning.

CHAPTER XIV.

DEFENDERS OF THE FAITH.

It is easy to swim with a tide that is flowing free. Working up-stream is quite another story. Yet this was the task now before those who fought to maintain the herds they had developed at such cost during the golden days that had preceded. We are unfortunately prone, in this western country, to run to extremes. The atmosphere of the prairies, the mountains and the plains breeds optimism. Else we would not have done and dared those deeds of might that have characterized our wondrous growth. We had a little too much steam on in our western cattle breeding. The crash of '93 brought us up to an era of liquidation in breeding stock which had to be got through with sooner or later, and while it left wrecks in its pathway it was the real starting point of the great constructive era upon which we now enter.

Men of faith, men of strength, men of dogged persistence were still behind the Hereford. The names of the more prominent ones weathering the financial gale of 1893 will still figure in our narrative, and we wish in passing to pay tribute to that patient, but for the most part inconspicuous, body of farmer-breeders who from Maine to California held fast to that which they knew to be good through all these evil days, keeping alive the fires of Hereford patriotism through this time of storm and stress. It is obviously impractical, however, for us to go up and down all the by-ways that led to the firesides of these steadfast but modest keepers of the faith. Their names are in the records of their national association, and the work they did during the dark days of which we write still lives.

The main thread of our story is still best followed along the trail of the leading shows.

At the Fairs of '94.—Ancient Briton had been bought at Mr. Clough's dispersion sale of 1894 at $1,025 by Redhead Bros., of Des Moines, and was the first-prize and champion bull of the year. Mr. Sotham had by this time established himself at Weavergrace Farm, Chillicothe, Mo., where he had collected a valuable lot of richly bred cattle from various sources. He had bought the bull Alger, that had been a frequent prize-winner in Michigan, and at Des Moines and Lincoln he was ranked next to Ancient Briton. He was sired by the Grove 3d-Spartan bull Clarence out of the Tudge-bred cow Greenhorn 5th. Funkhouser exhibited this year as a yearling a very remarkable young bull called Free Lance, sired by Beau Real out of the famous Lady Wilton, that only lacked stronger condition to take highest rank. He was second at the Iowa show to Sotham's Cordial by Harold 2d, out of a daughter of Coral. Sotham had pinned his faith to Corrector as a great sire, and was rewarded at Des Moines by receiving first in bull calves on one of his sons, Chillicothe out of Cherry 24th by Cedric. The Columbian champion female, Mr. VanNatta's Annabel, had now become Sotham's property, and headed the cows at Des Moines, defeating that other noted VanNatta product Cherry Duchess, now owned by the Messrs. Redhead. Elmendorf's Lady Daylight, commonly called "Baby," was easily the best two-year-old and beat Annabel for the female championship. The yearling heifers by Hesiod 2d shown by Mr. Funkhouser were up to the best standards ever set in western shows.

At the Illinois State Fair, held at Springfield, where it had now been permanently located, Clark, Sotham, Elmendorf and Funkhouser fought for place under Imboden's judgment, Ancient Briton leading the senior bulls. Clark's Lars, son of Peerless 3d, as a two-year-old weighed near 2,000 pounds and was generally allowed to be the best bull Mr. Clark had ever bred. He had an Anxiety loin, a Peerless head and a Grove 3d shortness of leg. Cordial again beat Free Lance, and Chillicothe was first among bull calves. Annabel led the cows, and Miss Wilton, own-sister to Free Lance, was drawn for second. Lady Daylight continued her victorious career in the two-year-old ring, and her stablemate Lady Laurel, daughter of old Lily, was second. Redhead's Bright Duchess won the blue in the yearlings, and Clark was first in a strong class of calves with Jessamine, by Peerless Wilton.

Some Notable Transactions.—In the spring of

1895 William S. VanNatta bought the entire herd of S. W. Anderson of Asbury, W. Va., consisting of 40 females and 10 bulls, with Earl Wilton 31st at the head. The herd of John S. Carlyle, deceased, was closed out at Vesta, Neb. Thomas Clark sold Sanhedrim 46180 to W. S. Ikard of Henrietta, Tex., and T. F. B. Sotham acquired the herd of Samuel Weaver of Forsythe, Ill.

Charles S. Cross Begins Showing.—In 1895 at Des Moines, the scene of so many notable contests, a new name appeared in the entry list, that of Charles S. Cross of Sunny Slope Farm, Emporia, Kans. Taking advantage of the ruinous prices that had been prevailing he began accumulating breeding cattle of a superior stamp, including the great Anxiety bull Beau Real and several of his daughters obtained from Mr. Fowler of Maple Hill, who had purchased most of the Shockey & Gibb cattle. Beau Real died soon after, however, and one of his sons, Wild Tom, was placed at the head of the herd, which a few years later came into national prominence. At two years old the badly named Wild Tom was sent to Des Moines accompanied by a string of heifer calves. The Messrs. Redhead had divided their show material, sending Ancient Briton and females to the Wisconsin State Fair. This left the Iowa show lamentably weak. Harry Yeld was feeding the Redhead cattle at this time, and he certainly made the most of his opportunities in their behalf. Wild Tom was a wide-ribbed, short-legged bull of good substance and was made champion, and Red-

head's Venus 6th, by Sir Wilfred, was female champion.

Lars and Free Lance.—Illinois undoubtedly had the best show of Herefords in 1895, well fitted entries being contributed by Clark, Sotham, Funkhouser, Redhead, Todd Benjamin, Fluck and Elmendorf.

Clark's Lars, now three-year-old, came into the ring weighing 2,400 pounds, deep, round and level, after having beaten Ancient Briton at Minneapolis, and was ranked above Wilton Grove. Alger, now beginning to age, had third. Ancient Briton was not shown. Next to Lars the best Hereford bull on the road in 1895 was clearly Free Lance. Mr. Funkhouser's manager, Will Willis, had handled this bull with consummate skill from a feeder's standpoint, and sent him into the ring a fit pattern of a high-class show bull. He had style, flesh and finish, and gave Lars a good fight for the championship of the class. Sotham's yearling Protection, by Corrector out of Coral, a bull of exceptional promise, made a satisfactory leader in the yearlings. Hesiod 20th was best calf.

The great Lady Laurel was the blue ribbon cow, now even a finer type than her illustrious dam had been. Lady Daylight stood second. Both had calves at foot. Funkhouser's Lorena here turned the tables on Bright Duchess. Both were extraordinary two-year-olds. The yearling contest was between Clark's Jessamine and Sotham's Grace, the former gaining the judicial favor. Funkhouser's Dewdrop, by Hesiod 2d, topped the heifer calves. Clark won on herd, with Funkhouser second. Sotham captured the young herd prize, with Protection, Benita, Grace, Lady Chloe and Lady Plushcoat, all Correctors but one, and a finely finished group they were. Clark's Peerless Wiltons had the get-of-sire ribbon, the Correctors coming second and the Hesiods third. Lars was senior, and Free Lance junior champion. bull. Lady Laurel and Jessamine were the female champions.

Trade Slowly Revives in 1896.—Prices were still unsatisfactory. Sotham sold in April at Weavergrace 19 bulls at auction at an average of $200, the highest price being $500 for Exemplar to Mr. Tuggle. Twenty-nine females sold at an average of $145. In October Gudgell & Simpson and Mr. Funkhouser sold 73 head at Kansas City for an average of $168.75, 24 bulls averaging $196 and the tops being $665 for Hesiod 30th to N. W. Leonard and $425 for Hesiod 29th to Scott & March.

On the western fair circuit Sotham, Redhead, Elmendorf and Funkhouser were the leading exhibitors. In Minnesota Sotham's Protection and Grace were champions. In Nebraska Elmendorf led the aged bulls with St. Louis and Funkhouser the youngsters with Hesiod 29th. Lady Laurel was champion cow and Funkhouser's Dewdrop best heifer. These same cattle were in stellar roles at Des Moines.

Clark's Lars was champion bull at Springfield, and was thus described:

"Lars is one of the outstanding animals. His

shoulders are too prominent, he needs a little filling about the tail and his hair is perhaps inclined to be harsh, but when that is said the bag of the stone-throwing critic is empty of missiles. Such massiveness on such short 'pegs' has rarely if ever been seen in an American showyard. The bull is shaped like a barrel, 'rotund' is the word; barring his shoulders and his bit of a dip at the tail he is round and smooth as an apple. As a flesh-carrier he presents one of the most striking illustrations of the deep-fleshing qualities of this great breed. It need hardly be recalled that he is a son of Capt. Kidd (of Grove 3d blood) and Clark's great show cow Peerless 3d.''

VanNatta's Actor, with his Anxiety blood clearly revealed in his great loin, was second. Sotham had no competition on Protection in two-year-olds, and in yearlings Clark scored with Littleton, son of Lars. Cherry Duchess by Cherry Boy headed the cows, and Clark's Jessamine ranked the two-year-olds, with Sotham's Grace second. Clark won both herd prizes, and Funkhouser had the get-of-sire prize on his Hesiods.

Ancient Briton Goes To Texas.—In the spring of 1897 values and public interest in Herefords began to expand throughout the entire west. There had been four lean years sure enough. Those who had held on and those who had accumulated good breeding stock at the low prices prevailing now began to reap the benefit.

Col. C. C. Slaughter of Dallas, Tex., owner of one of the leading southwestern herds, bought a big lot of good bulls, including Ancient Briton at $2,500. After this sale $1,000 each was refused for three of Ancient Briton's sons—Christmas Gift, Country Gentleman and Little Briton.

Kirk B. Armour, whose herd at Excelsior Springs, Mo., included a lot of fine old Culbertson cows bought the bull St. Louis at $800, Lady Laurel at $1,000 and Dimple, a daughter of Lady Daylight, at $700.

F. A. Nave of Attica, Ind., shortly to become very prominent in the trade, bought the bull Dale for $1,000 at Harness & Graves' Chicago sale, where 24 head sold at an average of $226.

Sotham sold 56 head at an average of $214, including Sir Comewell to Mr. Hornaday of Ft. Scott, Kans., for $840, Col. Slaughter securing Protection, then four years old, at the comparatively low price of $450. Scott & March sold 72 calves, from eleven to fourteen months old, in the spring of 1897 for $11,400, for range use.

The Shows of 1897.—The Hereford classes at the leading fairs of 1897 were well filled. New exhibitors entered the lists and the average quality of the entries was exceptional. In the west Mr. Cross contributed largely to the success of the Hereford presentation. Mr. Funkhouser's entries were of outstanding excellence and the newly organized firm of Steward & Hutcheon came forward for the first time with well fitted cattle of an admirable type. John Steward had been for many years Mr. VanNatta's trusty manager. Will Hutcheon had been with Hon. M. H. Cochrane at Hillhurst, and latterly had assisted Steward at VanNatta's. They had now formed a co-partnership, and engaged in the breeding of Herefords on their own account at Greenwood, Mo. What they may have lacked in capital they made up in sound judgment and practical knowledge of all the "ins and outs" of the fitter's art. Another real artist in the business of selecting and fitting showyard material, Mr. Ed. Taylor, was in charge of Mr. Sotham's cattle, and his entries were always presented in the best of bloom.

The trouble began at the Minnesota show when Mr. Cross, Mr. Sotham, and Steward & Hutcheon first crossed swords. Sunny Slope's sturdy son of Beau Real, Wild Tom, ran away with senior bull honors, and Sotham was second in two-year-olds and yearlings with the Corrector bulls, Sir Bredwell and Thickset. The former weighed 1,900 pounds, was from a Grove 3d-Spartan dam, and joined fine breed character to rare scale. Thickset was a grand type, rich in his flesh, evenly fashioned throughout, with faultless head and horn, gay carriage and shown at a weight of 1,600 pounds at eighteen months. He was out of Grove Lassie by Star Grove 1st, and his grandam was Lassie by Mr. VanNatta's Fowler. Sotham scored again in bull calves with the double Corrector Excellent, by Exemplar out of the famous Grace.

A Memorable Minnesota Contest.—The females at this show were of extraordinary merit. In fact, the female classes were strong throughout. Public interest in the judging was at fever heat with Prof. C. F. Curtiss on the bench. The writer hereof witnessed most of the contests of this period, and

THOS. F. B. SOTHAM.

the subjoined description written at the time will not only serve to reflect the character of the animals shown, but will indicate the efforts made by "The Breeder's Gazette," then as now, to keep the public fully advised as to what was transpiring at the great shows of the period. We quote:

"Six superb cows started an argument which was still going on at last accounts, and made the strongest show of mature females of the beef breeds seen in the yard during the entire week. Messrs. Steward & Hutcheon drew forward a pair of Anxieties that would ornament any pasture in England or America—the six-year-old Maud of Mr. Wm. S. Van-Natta's breeding, by Anxiety 5th from a cow of C. K. Parmelee's, and Pretty Lady by Don Juan, from the great Gudgell & Simpson herd. Mr. Sotham rested his case upon the beautiful three-year-old Benita, by Corrector from an Archibald dam, and Mr. Cross complicated matters by offering Annette, by Eureka, Robertha, by Beau Real of World's Fair fame and Makin Bros.' breeding, and the massive Mary Benjamina, by Richard Grove. Director Curtiss said: 'First to Benita, Annette second, and Maude commended.' Equally good authority revised this to read: 'Maude first, Benita second, Annette commended,' but the ribbons of course followed the fiat of the awarding judge. Robertha's peerless head and beautiful forward finish is marred by lack of levelness behind the hips. Mary's wonderful scale and great quarters could scarcely prevail against the superior smoothness and refinement of the three favorites named. Benita is a cow of splendid quality, with fine head, neck and shoulders, and a table-back. She is a bit upstanding as compared with such as Annette, Robertha and Maud, but is in nice bloom, handles well, and has an elegant heifer calf at foot by Protection. Annette is scarcely as breedy a type as Benita or Maude, but on the beef proposition she is a hard nut to crack. Just a little inclined to roll, she is yet compactly fashioned and so full of flesh that she cannot be denied position. Maude is a Hereford cow such as breeders often dream about and now and then produce. She is rather soft in her handling and somewhat gaudy about the tail-root, but is marvelous in her smoothness everywhere else, extraordinary in her shoulders and heart, strong in her back, has fine width, great depth, good length, low short neat legs, and a good head carried on a thin breeding-cow's neck. She has also been a prolific breeder.

"Beau Real's Maid did what it has been thought she was capable of accomplishing all summer. In fact she did more. She was not only the blue ribbon two-year-old, but later on was crowned queen of the white-faced females two years old or over in competition with the grand cows just described. Such recognition is sufficient to give Sunny Slope place at once high in the list of nurseries of top-notch Herefords. A sweeter or more symmetrical white-faced maiden has not been thrown by America's breeding herds in recent years. Evenly good from horns to hoofs, criticism becomes virtually disarmed as her rare character develops under close examination. Neat in her head and horn and beautifully filled in her bosom, she shows a pair of elegantly modeled shoulders, a well rounded chine, wide thick-meated ribs, captivating wealth of loin, and good quarters. In making up her showyard raiment the feeder did not forget his Shakespeare. At any rate the advice of Polonius to Laertes had been heeded. 'Rich not gaudy,' in her covering, she brings to the Hereford camp this fall as handsome a body of beef as connoisseur could covet. Mr. Sotham's good Lady

Plushcoat, by Corrector out of a daughter of Dr. Grove, made a strong second. She carries a world of flesh along her rib and across her loin, although not quite true behind. The same stable supplied the third-prize heifer Lady Chloe by old Alger, that went to Texas two years ago to show the Southrons how big and massive a Hereford may be made. Chloe's dam is a daughter of the celebrated Coral and she is worthy of her high-class ancestry. She has a refined front and is deep in her flesh, but a bit uneven in her back.

"A month ago Sunny Slope's buxom Wild Tom heifer Miranda would probably have topped the yearlings seen at this fair, but she was not herself last week and failed to put up her customary show. A good substitute was found, however, in her half-sister (by same sire) Pretty Maid, chosen by Prof. Curtiss to wear the blue. Not so blocky as Miranda, she is yet a strong, well grown, firmly fleshed heifer. Sotham's Lady Coral (own sister to Lady Chloe), that has inherited a grand loin from her sire Alger, was drawn for second and Steward & Hutcheon were relegated to third with Salina (bred by Mr. VanNatta, and shown by him as a calf last fall), a daughter of the Sotham-bred Eureka. She is of a very wide-ribbed, tidy, low-down type and was slated by some of 'the boys' to top the class. She is indeed a beefy one, but somewhat uneven in her back and at setting on of tail. Seven heifer calves were quite as hard to judge as were the cows. Three of these were genuine Klondyke nuggets—rich, yellow, and good as gold in the present state of Hereford trade. There was Sotham's Georgina (own sister to Grace and Sir Comewell), same owner's Benison (by Corrector from the first-prize cow Benita), and Sunny Slope's Diana, by Archibald 5th. The Gazette passes up the task of determining the relative merit of these three ripe, sappy, low-legged beauties. It is probable that the judge had a majority of the onlookers with him, however, when he drew Georgina for first. At seven months she is probably the best-developed calf seen at leading fairs in many years. Mr. Sotham and herdsman Taylor are indeed entitled to warmest congratulations upon the production of such a grand specimen of early maturity. Benison, by the famous Protection, has a truly wonderful back and is also a great triumph for Weavergrace principles. Such a pair are rarely produced in any one herd in one season. Diana is a fully developed cow in miniature, a little wonder in her way. This trio were of a type and had the ribbons gone to them it would have made little difference, so far as the equities were concerned, as to relative ratings. Prof. Curtiss realized this, but found such a lot of good flesh on the growthier Wild Tom calf Dorcas that he braved criticism long enough to pull her in between Georgina and Benison for second, leaving Diana hunting for honors elsewhere, which, by the way, she found in good shape a little later in the contest for champion calf of any beef breed, as appears below.

"Wild Tom was declared champion bull two years old or over and Thickset was made junior champion. Beau Real's Maid and Georgina were given the senior and junior female championships respectively. Sotham won the young herd prize with Thickset, Lady Coral, Lady Brenda, Georgina and Benison, and also the get-of-bull contest with a company of Correctors."

A Typical Breed Battle.—These were still the foolish old days of breed competitions, now happily a thing of the past. It may be interesting, therefore, as illustrating what happened under the system then in vogue to reproduce our account of the "grand sweepstakes—open to all beef breeds" at this Minnesota State Fair of 1897:

"Profs. Curtiss and Shaw and Charles Kerr presided at the drawing in this important distribution of cash, and after the revolutions of the wheel had ceased it appeared that on the whole a fairly even divide was secured.

"Sweepstakes by ages came first and Mr. H. F. Brown's Cruickshank Shorthorn Victor of Browndale pulled down the first plum—that for best bull of any beef breed three years old or over. Goodwin & Judy's Blackcap King was declared best two-year-old, and Mr. Sotham's Thickset claimed the yearling bull championship of the yard. Honors were therefore easy up to this point, but Sotham closed up the bull classes with a calf victory on Grace's sappy son Excellent. The first round therefore ended rather to the advantage of the 'white faces.'

"In the cow class the problem was about like this: 'Here is a peach, a pear and a plum, all luscious specimens; which is the best fruit?' A nice query for a state fair association to propound, isn't it? Reminds us of our school-boy debates upon such weighty questions as, 'Which is the most destructive agent, fire or water?' or, 'Which is the most dangerous calling, that of a soldier or sailor?' The jury said they preferred plums. That is they awarded the palm to that model of 'doddie' neatness and compactness—Goodwin & Judy's round ripe Zara 5th. The best two-year-old heifer in the yard was found in the comely Shorthorn Browndale's Ella Kennedy. The best yearling turned up in McHenry's Pride 7th and the crack calf was declared to be Mr. Cross' Diana, which as mentioned in our review of the Hereford class above, failed of recognition the day previous. It thus appears that the Herefords and Angus had the best of the fight up to this point, winning six (three each) out of the eight rounds. Mr. Brown had two falls to the credit of the Shorthorns, however, and bided his time.

"In the class for best herd under two years, to consist of one bull and four females (latter to be bred by exhibitor), the 'bonnie blacks' repeated their remarkable performance of last year, drawing both first and second; Mr. McHenry had the honor of holding the right of the line with his Blackbird bull and blooming bevy of rich-backed heifers, Goodwin & Judy receiving second and Mr. Sotham third for the Herefords—more 'soup' for the Shorthorns.

"When Goodwin & Judy plucked the prize for best four (or more) cattle of any age or either sex the get of one bull with Blackcap King, Zaras 5th and 9th and Blackcap 13th (own sister to the King), all by Black Monk, it looked still blacker for the rival breeds, and to add to the gloom that seemed settling down over the Shorthorn camp Sotham found second on his Correctors (Sir Bredwell, Thickset, Benita, Lady Plushcoat and Georgina). The Browndale Golden Rules (Spicey 4th, Ella Kennedy, Golden Princess and a Waterloo heifer) were third.

"When the grand finale was reached, however, the $650 capital prize, the sun rose bright and clear over the Shorthorn host and equilibrium was restored by a decision which sent the grand prize of all to Mr. Brown's Victor of Browndale, Spicey of Browndale 4th, Ella Kennedy, Waterloo of Browndale 7th, and Golden Minnie. The blacks were close in at the death, Goodwin & Judy claiming the red with Blackcap King, Zaras 5th and 9th, Rosebud Rho, and Blackbird heifer calf. Third honors rested upon the Sunny Slope Herefords, consisting of Wild Tom, Annette, Beau Real's Maid, Pretty Maid and Diana, with Sotham's Thickset, Benita, Lady Plushcoat, Lady Coral and Georgina fourth.''

"The Gory Hill of Hamline."—It was at this same show that a famous fight over a "breeder's stake" occurred, calling out the following comment, made at the time by the author:

"This association has in its prize-list another big bone of contention known as a breeder's stake: 'For the best beef herd of cattle, six in number, any age, of any breed or sex, owned and bred by the exhibitor. Conditions: One hundred dollars entrance fee and $100 added by the society. The whole amount of the stake to be divided as follows: To the best lot, 50 per cent of the stake; to the second best lot, 25 per cent of the stake; to the third best lot, 15 per cent of the stake; to the fourth best lot, 10 per cent of the stake.' Four exhibitors concluded to go out after this Friday morning, making the value of the stake $500. Prof. Shaw and Mr. Kerr were called and sent the $250 to Mr. Sotham's Herefords—Sir Bredwell, Thickset, Excellent, Benita, Lady Chloe and Benison—three bulls and three females, placing the Browndale Shorthorns second, Goodwin & Judy's Angus third, and Mr. Westrope's Shorthorns fourth, so that each participant had a place. The judges acknowledged the great merit of Mr. Brown's Shorthorns as individuals, but awarded the first place to Sotham because, as they expressed it, 'of their uniformity in the ideal type of a beef animal'; adding that as 'representing the skill of the breeder in molding refinement, type and finish the winning herd proves Mr. Sotham unequaled in results.' The even division of the sexes was another point in favor of the Sotham entry, showing good work in breeding both bulls and heifers. There was but one bull in each of the other herds. The jury further reported: 'In a breeder's exhibit uniformity of ideal type should and did outweigh a collection of prime animals of different types.' In this respect the Judy Angus entry was backed for second place.

"And so the battle of the breeds was ended. Each interest had received 'distinguished consideration,' and, while all were not entirely happy, white-winged peace brooded that night over the beef cattle barns on the erstwhile gory hill of Hamline."

At Des Moines Funkhouser appeared with a great string of show cattle headed by Free Lance, and Scott & March of Belton, Mo., added to the fame of the Plattsburg establishment by exhibiting in capital form the splendid yearling bull Hesiod 29th. Free Lance was the product of the union of two exceptional animals, Beau Real and Lady Wilton. The latter had been bought by Mr. Funkhouser at the Fowler dispersion sale at Kansas City in February, 1893, at a low price. It was not certain that she was still a useful breeding proposition, but as these were dolorous days, dollar-wise in the cattle trade, one could afford to take a chance upon almost anything at the prices current. Steward & Hutcheon, George Redhead, Z. T. Kinsell and others rounded out a strong white-faced entry. Funkhouser had the male and female championships with Free Lance and Cherry, by Cherry Boy, both herd prizes, get-of-sire and produce-of-cow. Will Willis' cup was truly overflowing.

First Appearance of Dale.—In the east a new

sensation was sprung by Harness & Graves of Indiana. At New York State Fair they had the bull championship on Columbus, by Earl of Shadeland 41st, and at Indianapolis Mr. I. M. Forbes, the well known Shorthorn breeder of Henry, Ill., acting as judge under the Governor of the state, Hon. Claude Matthews, as superintendent, had placed this massive bull ahead of Wild Tom. In the yearling ring these same exhibitors presented a son of Columbus named Dale that not only won first in his class, but the male championship as well. He was the phenomenal youngster of the year, and only at the commencement of a career equalled by few American-bred cattle of his day and generation.

Tom Clark's Jessamine easily led the cows at this Hoosier show, but in two-year-olds his grand heifer Juno gave way by Mr. Forbes' direction to Mr. Cross' big, smooth, broadtopped Beau Real's Maid. Jessamine won the female championship and Mr. Clark's brave array of the get of Peerless Wilton drew the much coveted get-of-sire award.

John Lewis and His Troubles at Springfield.—At the Illinois State Fair of 1897 "Uncle John" Lewis, Shadeland's "grand old man" tied the ribbons on one of the best Hereford shows of the period of which we write. Clark, Cross, Funkhouser, and Sotham furnished competition that supplied the "thrills." Here is our comment on Free Lance and Wild Tom, the aged bull antagonists, as written at the time:

"In the senior bull class it was Free Lance 51626

JOHN LEWIS.

vs. Wild Tom 51592, a little family affair as it were, both animals being sons of the celebrated Anxiety bull Beau Real, and both having been bred on the same farm—the Fowler ranch at Maple Hill, Kans. The Funkhouser bull is possessed of such an overpowering bulk that Wild Tom with all his weight looked a veritable David alongside the Missouri Goliath of bulls; but the giant in this case won. They are animals of such a materially different type that it is difficult to rate them. Tom is short-legged and thoroughly masculine. Free Lance is projected on a bolder scale and his head and horns have even more refinement than usually characterizes the Wilton family, to which his famous dam belongs. Tom is six months older than his half-brother and has done heavier work as a stock bull. He rests his claims for recognition rather upon his business capacity than showyard finish. Free Lance had one of the best mothers ever seen in a Hereford herd—imp. Lady Wilton. He had a back like an English billiard table and a heart girth such as is not seen more than once in a decade. Tom is also possessed of all necessary substance, shows breadth of rib and loin proportioned to his inches and had for dam a daughter of Bredwardine by old Horace. The same breeder who would feel compelled to give Mr. Funkhouser's remarkable bull a prize over Tom might prefer the latter for breeding purposes, but as to Free Lance's showyard strength there can be no dispute.''

Sotham's Sir Bredwell was easily first in two-year-olds, but in yearlings there was battle royal. Let us quote again from ''our favorite author'':

''In yearlings two compact thick-fleshed bulls of outstanding merit had to be reckoned with—Hesiod 29th 66304 and Thickset—the former of Mr. Funkhouser's breeding and now the property of Messrs Scott & March of Belton, Mo., and the latter bred and owned by Mr. Sotham of Weavergrace. Thickset is the stronger-backed bull, but Hesiod has the greater depth of body. The Corrector has the usual good head and horn of the Sotham stock, but the Hesiod is also faultless in the same particulars. Thickset has a grand chine and rib, but is fairly matched by Hesiod's well covered shoulders and strong heartgirth. They are indeed a royal pair, and no showyard decision can add to or detract from the fair fame of either. Hesiod 29th was given first and Mr. Sotham's bull second.''

Beau Real's Maid, Juno, and Dewdrop.—In the cow class Clark's Peerless Wilton-Anxiety 3d marvel Jessamine, with her furry coat, wonderful substance, flesh and finish, was unapproached, but in two-year-olds there was approximate perfection in several quarters. This ring witnessed the meeting of Beau Real's Maid, Juno and Dewdrop—three of the best white-faced females bred in this country during the period under review. Sotham's Lady Chloe was in the fight also, but was scarcely thought equal to the job of turning down either of the three first-named. The glorious uncertainty of the showyard, however, here found fresh exemplification. Sunny Slope's daughter of Beau Real had met and defeated Lady Chloe at Hamline. Juno had been seen at Indianapolis and Milwaukee and in each case judgment was rendered for Beau Real's Maid. Dewdrop had met neither of these heifers at Des Moines. Our comment at the ringside follows:

"Mr. Lewis began by throwing Mr. Funkhouser's

broad deep daughter of Hesiod 2d entirely out of the running. She has not held just level in her marvelous back. That must be admitted; but to cast Dewdrop altogether for that fault was a piece of judicial severity such as is rarely seen in a great showyard. This being decided upon, Lady Chloe was listed for third and the question was narrowed down to Beau Real's Maid and Juno for the blue and red. If some little lack of bloom cost Dewdrop all her chances then by that same token the freshness and finish of Beau Real's Maid should have landed her where she had been placed twice before this season, in advance of Juno; but the air seemed full of cobwebs about this time all the way down the long line of two-year-old Hereford and Shorthorn heifers being judged simultaneously in front of the grand stand, for while Brother Boyden was mixing up the 'red, white and roans' in a way that startled the assembled company Mr. Lewis upset things among the 'white faces' by sending first to Juno, second to the Maid, and third to Chloe! Juno is a heifer such as any man might well be proud to have produced, and there is of course ample room for honest difference of opinion as between her and Dewdrop and Beau Real's Maid. The fine scale and beautiful finish and refinement of the Cross heifer have seldom had a counterpart in western showrings, and much as we appreciate Juno and Dewdrop we can but defend the right of Beau Real's Maid to head these 'crack' two-year-olds of 1897. They were a great lot and we congratulate Mr. Clark upon his good fortune here in beating probably the handsomest heifer he has ever shown against. When his list of winnings for the past twenty years upon cattle of his own breeding comes to be made up what a story of showyard success will be unfolded!"

Sir Bredwell, Benita, Lady Chloe, Lady Brenda and Georgina, drew the blue for Sotham as best graded herd, uniformity of type being the rock upon which Lewis took his stand. Free Lance was champion bull and Jessamine best female.

Death of Adams Earl.—The founder of the Shadeland herd died in January, 1898. The part he had played in the introduction and successful dissemination of the Hereford blood has been outlined in preceding chapters, but the influence of his work with the "white faces" was so far-reaching that he is by common consent accorded a permanent place in the American Hereford gallery of fame.*

---

*Mr. Earl was born in Fairfield Co., O., in 1819, and came of New England stock. His parents removed to Indiana in 1836 and settled upon the fertile Wea Plains, upon the borders of which the famous farm of Shadeland is located. Arriving at his majority he undertook about 1844 the marketing of farm products at New Orleans by means of flat-boats floated upon the Wabash, Ohio and Mississippi rivers. He subsequently engaged in merchandizing upon quite an extensive scale at Lafayette and ultimately became associated with the late Moses Fowler in various important enterprises, such as wholesaling groceries, banking, etc. In 1860 he engaged in pork and beef packing and a few years later became a partner in the Chicago house of Culbertson, Blair & Co. About 1870 he became the moving spirit in the building of a railway from Lafayette to Kankakee, which is now a part of the Big Four System, Mr. Earl being the president, general manager, and builder. Meantime, in connection with Mr. Fowler and A. D. Raub, he had purchased 36,000 acres of land in Benton county and spent large sums of money in tiling, fencing, building, etc., and so important were the operations of this syndicate that on their tender of $40,000 to build a courthouse at the new town of Fowler the county voted to move the seat of local government to that point.

CHAPTER XV.

CLEARING SKIES.

Early in the year 1896 it became apparent that values were rapidly recovering from the low levels established after the financial panic of 1893. Mr. C. S. Cross of Sunny Slope Farm, Emporia, Kans., sensing the advent of better days in the cattle trade, had sent John Steward to England in the autumn of 1897 to select a high-class lot of cattle for importation and sale. Needless to add, the purchases were made with strict regard for quality; Steward was commonly recognized as one of the best judges of Herefords of that period. Not only that, but he had a reputation for integrity that insured a faithful execution of his trust. Moreover, Harry Yeld, who had in the meantime gone back to his native land and who was in close touch with the best breeding establishments in Herefordshire, had been advised in advance of Steward's mission and requested to co-operate in locating and securing options upon some of the best young cattle on the market at that time.

The Cross Importation.—The importation, consisting of 26 bulls and 15 heifers, was brought out in Mr. Yeld's charge and passing through quarantine at Garfield, N. J., arrived in time to be put on the market, along with a lot of well fitted home-bred stock, at Sunny Slope in March, 1898. The event aroused intense interest in American Hereford cattle breeding circles. It had been many years since any importations of consequence had been made. Prices had not only been so low as to discourage enterprise in that line, but the $100 fee for the registration of imported cattle was still in force, and the owners of large herds descended from the earlier importations were not slow to deny the necessity for any further recourse to the old-country stock. It was strenuously insisted that there was little if any occasion for any such extensive patronage of the English herds. It was claimed that better cattle were being bred and shown in the States than were being produced on the other side of the Atlantic. The extraordinary excellence of the "white faces" being produced in the herds of such pioneer breeders as Gudgell & Simpson, Clark and VanNatta and by the owners of valuable cattle bred from the Culbertson, Earl & Stuart and later importations, as evidenced by the leading shows of that time, certainly gave color to the contention that America had really passed the motherland in the matter of level-quartered, finely finished Herefords.

Notwithstanding this natural opposition to the importation and sale of cattle brought out with speculative intent, there was now such a widespread wave of enthusiasm in behalf of good Herefords, and so insistent was the demand of the western range for white-faced bulls, that on the 2nd and 3rd of March,

1898, when Mr. Cross exposed his 150 head of imported and home-bred cattle for sale at auction, a crowd estimated as high as 3,000 people faced the auctioneers, Col. James W. Judy, Col. Fred M. Woods and Col. J. W. Sparks, when the selling began. Mr. Cross reserved for his own use the imported bull Keep On.

$3,000 for Salisbury.—The highest price paid was $3,000 for the imported two-year-old bull Salisbury, bred by John Price. He was taken by Mr. Murray Boocock of Keswick, Va., who was at that time engaged in the formation of a Hereford herd, after a sharp contest with George W. Henry of Chicago and C. N. Whitman, the latter representing the owners of the Lucien Scott herd. W. S. VanNatta & Son secured the imported yearling March On, bred by Ed Yeld and sired by Lead On, a famous English stock bull that was unfortunately lost by accident just as he seemed to be entering upon a great career as a sire in the old country. This proved a fortunate purchase, and more will be heard of the bull and his get later on. One of the promising young bulls of the importation was the Turner-bred Saxon, that was sent into the ring with a reserve bid of $1,000. Others would have offered more money, but it was generally known that Mr. Cross really desired to retain the bull for his own use, and with the consent of the company he was therefore withdrawn.

Good Buying by George H. Adams.—Bidding on the best females was active at strong prices, the best price being $1,500, paid by George H. Adams of Crestone, Colo., owner of a 100,000-acre range in the San Luis Valley. This top figure was given for the two-year-old imported heifer Luminous, sired by Post Obit (11542). Mr. Adams was a persistent and liberal bidder throughout the entire sale, among his other selections being the three-year-old imported cow Leominster Daisy 2d by Lead On, taken out at $1,205. He also bought the good cow Miranda, by Wild Tom, and of Mr. Cross' own breeding at $905, his total purchases at the sale including 20 head at an average of over $500 each. Mr. Adams was an enthusiastic advocate of the Herefords for use on western ranges, and maintained a fine herd of purebred cattle in addition to some 5,000 head of high-class grade "white faces." He had bought some 50 head of good breeding cattle when the large and superior herd of Thomas J. Higgins had been dispersed in Kansas.

Over $400 Average for 144 Cattle.—This sale injected new life into the American Hereford cattle business. The 144 head sold for $58,585, an average of $407, in many respects one of the most extraordinary results ever attained on either side the water. Higher averages had been made, but not upon such a large number of animals. The 23 imported bulls brought an average of $616, the 14 imported females an average of $563, and the 107 home-bred lots fetched an average of $341. After the sale a number of the lots changed hands at advanced prices. Mr. Whitman, who had bought the imported bull

Randolph, of John Tudge's breeding, on the first day for $600, refused an offer of $1,000 for him on the following day.

Mr. Cross was a prominent figure in the trade at this time. He was President of the First National Bank of Emporia, and a man of great enterprise, deeply interested in good cattle. He had first engaged in the business of breeding pedigree Herefords at a time when he secured valuable foundation stock at beef prices. His relations with leading breeders of the cornbelt, as well as with the owners of the largest outfits on the western range, were intimate, and he did a large business at private treaty as well as at public auction. Shortly before his phenomenal sale of 1898 he had sold one lot of $3,500 worth of bulls, headed by the show bull Climax, to go to Texas. Unfortunately, as was afterwards developed, Mr. Cross had inherited certain burdens and responsibilities in connection with the business of his bank which ultimately involved him in such loss and humiliation that in a moment of desperation in November, 1898, he took his own life at Sunny Slope Farm, his death being deeply mourned by the entire Hereford cattle breeding fraternity. Fortunately Mrs. Cross had participated in an active personal way in nearly all of his Hereford cattle transactions, having a herd drawn mainly from Sunny Slope sources. Mr. C. A. Stannard succeeded to the ownership of Sunny Slope Farm and became for many years a prominent figure in the trade.

Other Sales in the Spring of '98.—Business was now brisk all along the line. K. B. Armour sold a good lot of bulls to go in service in the herd of the Matador Co. in west Texas. Sotham sold 50 head at auction on April 13 at an average of $342, upon which occasion Wayne Ponting paid $1,575 for Excellent, a two-year-old bull by Corrector. On April 15 Scott & March of Belton, Mo., sold 93 head at an average of $215. In May William Humphrey of Ashland, Neb., bought $9,000 worth of cattle of O. H. Nelson, including 34 cows, at $200 each. H. M. Hill sold 34 head at Kansas City for an average of $393, and Gudgell & Simpson made an average on 60 head of $479.

Beau Donald Shown.—The event of the year 1898 in showyard circles was the Trans-Mississippi Exposition at Omaha. Exhibits at the earlier state fairs were light, owners preferring to hold back for the more important event. Still there were some interesting developments elsewhere. Sotham was without competition in Minnesota, but east of the river few ribbons were won by default. Frank Nave and John Hooker appeared at the York State Fair. Dale had gone on famously and backed up his New York championship by beating down all opposition later at Indianapolis, where the herds of Tom Clark, Clem Graves, Hooker and W. H. Curtice of Kentucky were entered. Curtice was showing the massive, heavy-quartered, five-year-old in-bred Anxiety bull, Beau Donald 58996, by Beau Brummel 51817, and in the senior bull class won over Graves' Cherry

Ben. Dale had a walk-over in two-year-olds, and was subsequently made champion bull. Clark was strong this year, as always, in heifers of his own production, and gained the female championship of the Hoosier state with the great yearling Everest, daughter of Lars. At the Ohio State Fair Murray Boocock of Virginia came forward with a herd headed by his $3,000 purchase at the Cross sale, imp. Salisbury, and won most of the prizes.

Dale vs. Sir Bredwell.—At the Illinois State Fair of '98 Mr. Nave's deep-fleshed Dale had graduated into the three-year-old class, and met Sotham's Sir Bredwell, with Imboden on the bench. These bulls were of totally different types. Dale was broad, short-necked, thick and deep—as compact a block of beef as any breed ever throws—quite lacking in style and gayety of carriage. With a butcher-feeder as arbitrator it was no surprise that he here found favor. Sir Bredwell had scale, stretch, imposing presence and quality. As a breeding proposition most critics would have preferred Sir Bredwell at the time, but Dale certainly lived to vindicate his own prepotency and to confound all critics.

Two In-bred Toppers: Everest and Benison.—The feature of the female classes was the struggle between Everest and Benison. This is the story as it was written at the time:

"Tom Clark never bred a better one than Everest, and those who have followed our western shows for the past twenty years will understand what such a statement means. But Sotham never produced a more perfect heifer than Benison; so here was a repetition of that memorable day when Grace and Jessamine met as yearlings in 1895—with the tables turned. In that great trial of strength Mr. Clark won; in this instance the tide of battle turned in Sotham's favor. Little things sometimes decide such contests. The Clark heifer was bulling the day of this showing. Benison might be bigger but not better. She is a heifer of exquisite finish shown in great bloom. She has a back and loin of marvelous perfection, capital quarters, a twist filled to a finish, model shoulders, and short neat legs. Everest is bigger and thicker, with handsome head, nobly arched ribs deeply covered, and carries her burly body on well set 'pegs.' She is wonderful in her wealth of flesh, and barring a little inclination to bunch at the tail-root, is smooth and true in all her lines.''

The fact that these top heifers were products of blood concentration is of interest. Everest had double lines to both Anxiety 3d and to the great cow Peerless. Her sire, Lars, was the result of the coupling of those animals, and her dam, Eletta 2d, was by Peerless Wilton, a son of old Peerless, out of a daughter of Anxiety 3d. Benita was a double Corrector, her sire, Protection, and her dam, Benita, both being by old "Dad"—Sotham's pet name for the bull that made Weavergrace famous.

The Omaha Exposition.—The Herefords were the outstanding feature of the live stock department of the great exposition held at Omaha in 1898, and we feel warranted in again quoting from our own notes on certain phases of this big show:

"The American Hereford Cattle Breeders' Association added $3,000 to the exposition company's rather meager prizes. This bonus, together with the prevailing activity in the west in white-faced cattle, drew out an incomparable display in this section—the largest and best of its kind ever seen in the United States and eclipsing the average exhibit of the breed seen at the annual meetings of the Royal Agricultural Society of England. The size and quality of the classes throughout—excepting only that for aged bulls—aroused the enthusiasm of visitors to the highest pitch. The great amphitheater was packed while the Hereford judging was in progress, the spectators evincing keen interest in the work. The difficult proposition of passing upon this record-breaking exhibit was assumed by Mr. Claude Makin of Florence, Kans., and it is a pleasure to be able to state that this trying task was discharged with singular accuracy and impartiality. A more satisfactory piece of work of this character has rarely been seen in American showyards. Awards were given by wire in our last, except the group and championship prizes, which were not assigned as last week's Gazette went to press. We now supply details as to the showing throughout.

"The list of exhibitors included C. S. Cross, Emporia, Kans.; T. F. B. Sotham, Chillicothe, Mo.; George H. Adams, Crestone, Colo.; F. A. Nave, Attica, Ind.; William S. VanNatta & Son, Fowler Ind.; Gudgell & Simpson, Independence, Mo.; James A. Funkhouser, Plattsburg, Mo.; Scott & March, Belton, Mo.; Cornish & Patton, Osborn, Mo.; C. G. Comstock, Albany, Mo.; Peter Mouw of Sioux Co., Ia.; Z. T. Kinsell, Mount Ayr, Ia.; C. H. Elmendorf of Nebraska; Stanton Farm Co., of Nebraska; Steward & Hutcheon of Greenwood, Mo., and E. E. Day of Cass Co., Neb."

Sir Bredwell Beats Free Lance.—"There was but one light show in the entire section—that seen in the senior bull class. Dale went back here among the two-year-olds under the rules for computing ages at the exposition, so that the tourney was opened by a tilt between Funkhouser's Free Lance and Sotham's Sir Bredwell. The pitcher that goes to the well each day is sooner or later broken. The big son of Beau Real and Lady Wilton has dared defeat on many a hotly contested field the past four years, and has borne back to Plattsburg in triumph spoils of showyard war that will furnish a theme for many a fireside tale in the years to come as his many battles royal are recalled. But here he fell before the superior freshness, bloom and character of his younger antagonist. The doughty old warrior's weight could not prevail against Sir Bredwell's superior front and smoothly carried flesh. Such is the way of the world. Show bulls meet the common fate. Repeated fittings and passing years render it difficult for even the kingliest of them all to hold their own indefinitely against the rude assaults of active aspirants for leadership among the younger element. At five and one-half years of age Free Lance relinquished showyard sovereignty at Omaha last week to the three-year-old son of Corrector and Beatrice. Bovine monarch never possessed a more regal presence than the newly chosen champion. It may seem a somewhat heartless proposition, that nerve-jarring vivat of the French, but in it is condensed the whole philosophy of the inevitable: 'The king is dead, long live the king!'"

Dale Wins Again.—"We now approach the most sensational string of young bulls of any beef breed seen in this country since that memorable day when Fowler, Bowdoin, Sergeant Major, Broadbreast, Cedric and the rest of that comely company locked horns at Chicago in 1885. The two-year-old class at Omaha will indeed be long remembered by all who were so fortunate as to be present when the lines were formed. Mr. Nave's Dale 66481 has already had his portrait painted in these columns this season in the warmest colors at our command. He came here with Sir Bredwell's scalp at his belt—presented by Imboden in the three-year-old class at Springfield—prepared to meet Sotham's other well equipped champion Thickset, chief of all bulls of his age of any beef breed at Hamline. He came to meet also that other 'warrior bold,' Scott & March's great Hesiod 29th, brought out by Mr. Godfrey at this show in astonishing form and bloom. He struck also the shield of Sunny Slope's well clad knight Keep On—fetched all the way from Herefordshire to test the mettle of our western Herefords. And if by chance he were able to successfully run the gauntlet of these his most powerful adversaries, there still remained to be dealt with Gudgell & Simpson's Don Carlos bull Douglas, and two sons of the World's Fair champion Ancient Briton. It was a daring undertaking, this single-handed challenge of Dale against the flower of all the great trans-Mississippi herds. Thickset alone of all his foes had been, like himself, doing the grand circuit. The rest had been held in reserve all season for this attack. For weeks and months preparation for this day had been going steadily forward. All the arts known to the feeder's and fitter's craft had been exhausted in an effort to place these favorites in the arena in the pink of perfection; and right here The Gazette desires to pay a passing tribute of respect to the capacity, intelligence and fidelity of the men who had in their immediate charge the 'making up' of these bulls. Such patience, skill and judgment as has been displayed in this work deserves the highest commendation.

"More than any other exhibitor in his class Mr. Nave is indebted to the man behind the bull for success achieved. Dale is a dream—one of those phenomenal feeders that occasionally fall into the hands of careful fitters and by their peculiar capacity for putting on flesh with astonishing rapidity and absolute levelness round out into marvelous perfection of form. As a model carcass he is easily the sensation of the season. Mr. Makin met the general approbation of the great throng that had assembled to witness the contest by assigning the post of honor to this extraordinary bull. Such evenness from end to end, such ripeness, smoothness and rotundity have rarely if ever been seen in the American showyard. As a feeder's and butcher's type he is faultless. With Dale at the head the problem as to what to do with Thickset and Hesiod 29th became a serious one. It was generally conceded that Keep On in his present form, good as he is, could scarcely hope for a better rating than fourth in such a group of high-class bulls."

Thickset and Hesiod 29th.—"After an extended examination the judge drew Thickset in for second, with Hesiod 29th in third place. Makin had clearly the four best bulls to the front. As to that all were agreed; but there were many who would have stood the Hesiod in front of the Corrector. This it was argued would have been the logical arrangement, as the Scott & March bull is nearer the type of Dale than Sotham's. In this connection the weights and ages are of interest. Dale and Hesiod stand each other off at 2,040 pounds. Thickset beats them both, pulling down 2,200 pounds. Dale was dropped in September, 1895, Hesiod 29th on Oct. 6,

FRANK A. NAVE.

1895, and Thickset on Feb. 20, 1896. The latter is big, smooth, mellow, high-styled and strong-quartered, wide between the eyes—indicating the good doer that he is—but perhaps a little thin in horn and muzzle. Hesiod 29th, like Dale, will never be a big one, but his was the one perfect head and horn of the entire class. In point of breeding character, as revealed in head and face, this bull is the peer of Sir Bredwell. In compactness, breadth and depth of carcass he fairly rivals Dale. Not so perfectly padded at every point perhaps—he has been working as well as preparing for show—he has the same general feeding quality coupled with the front of a bull that should make a royal stock-getter, possibly the most valuable of the class for breeding purposes. Although Mr. Sotham had the satisfaction of having Thickset placed one notch above him, Weavergrace was quick to see the superb character of Hesiod 29th and offered $2,000 for him after the show was over, which flattering proposition was declined with thanks by his appreciative owner. Keep On is smooth and mellow in his flesh, round, low and heavy, but he has inherited a wide-spread horn and a muzzle somewhat lacking in breadth. There is ample distance between his eyes, however, and this certain index of a kindly feeder is backed up by a carcass that is both ripe and rich. Nave's Earl of Shadeland 22d bull Gold Dollar 73652 was drawn into fifth place and Steward & Hutcheon's Rose Chief 68945, smooth, low and with plenty of style, was sixth.''

This competition was of such historic interest that we here record also what happened among the yearling bulls and calves. Our descriptions will not only give a clue to the individual character of the competing animals, but these accounts of the leading shows of that period indicate the blood that was producing the tops and the men who were in the van of Hereford progress. We quote:

George Adams' Orpheus.—"A long line of yearling bulls proved very perplexing, really a more difficult class to judge than the two-year-olds. One either had to begin with Mr. George H. Adams' big, strong-backed, broad-loined Wild Tom bull Orpheus 71100 or with one of the low-down, blocky sort, of which there were several fine specimens present. As a result of his preliminary examination Mr. Makin drew out a leet comprising the following in the order named: VanNatta's Lincoln 2d, by Cherry Boy out of Old Lark; Steward & Hutcheon's Bovic 79124, by Benson 64017; Sunny Slope's Climax 4th, by Climax; Sotham's Grandee, by Corrector; Gudgell & Simpson's Dandy Rex 71689, by Lamplighter; Adams' Orpheus, of Sunny Slope breeding. It is worthy of note in this connection that the three bulls at the head were all of VanNatta extraction. Having drawn Dale to the top in the previous class it was not surprising, therefore, that the judge on final examination went to that broad block of 'baby beef,' Bovic, for first choice, sending Lincoln 2d down for the red ticket. Bovic was easily the shortest-legged bull in the bunch, carrying 1,500 pounds in about as small compass and as near to the ground as is ever seen in the showyard. He is exceedingly rich in his flesh, full in his neckveins and remarkable at the twist, a rare feeding type, ripened as nicely as one would expect from two such experienced feeders as John Steward and Will Hutcheon, his owners. Lincoln 2d is a grand-fronted bull, strong in his girth, deep and well spread in his ribs, even and rich at the loin, but with hips a bit prominent. Climax 4th was not disturbed for third place. He has one of the handsomest heads carried by any bull seen at the show, a finely arched back well covered, a great loin, is well let down at the twist, and stands on short neat legs. He fails a bit from hip to tail, but has improved a lot in his handling and is now one of the great yearlings of the day. In retaining Sotham's Grandee for fourth Mr. Makin ran counter to the judgment of most of the outside talent. Gudgell & Simpson's Dandy Rex, with his good back (despite a 'tie' in it), strong quarters and short legs, and Orpheus should probably have gone in next to Climax 4th, but the judge found points of excellence in Grandee and Adams' other entry, Zapola Chief 70034, that led him to list them in the order named, ahead of Rex and Orpheus. We can scarcely approve of this rating; still Grandee looks like coming into an extra two-year-old and Zapola Chief, with his good head, big chest, depth and thickness of carcass, is a bull of strong parts."

Hesiods Again.—"In bull calves Mr. Funkhouser forged to the front, scoring a double victory on his fine pair of Hesiod 2ds, Hesiod 46th and Hesiod 50th. Makin first picked the big stylish 1,120-pound Hesiod 46th for first, but ultimately turned him down to second and moved up his half-brother into first place, a transposition which met with the general approbation of the spectators. Hesiod 50th, the winner, is a brother to Hesiod 30th, now the property of Mr. N. W. Leonard. He is exceptionally wide, low and thick. He carries the splendid head seen in nearly all the get of Mr. Funkhouser's great stock bull, has plenty of hair, stands wide behind and is as neat as he is ripe. Hesiod 46th also carries a great coat, shows beautiful character in his head and face and possesses splendid style. He is not quite so level and true in his lines as the first-prize calf. His breeding is superb, his dam having been Dream by Washington, second dam the great Miss Beau Real by Beau Real, third dam Bertha by Rudolph.

"The third-prize calf, Cornish & Patton's Prince Otto, is one of the very best youngsters seen on the circuit this year, and should probably have had second place. He is evenly good from end to end, having a fine head, well covered shoulders, an evenly spread back, good depth, covers smoothly over the hips, has straight well filled quarters, low flanks and ample scale. He has been sold to Miller & Balch of Missouri. Steward & Hutcheon were fourth on Dixie, a very wide, smooth, low-legged December calf that has been reserved for use in their choice little herd of 'white faces' at Greenwood. Like their first-prize yearling bull Bovic, Dixie is a son of Benson 46017, he by Anxiety 4th. Mr. Nave's Duke of Fairview 4th, good in his flesh, with a furry coat and capital head, was fifth, and Mr. Cross' Elvira's Archibald, a strapping big son of Archibald 5th, with great spread of rib and loin, was sixth. A calf in this ring that attracted considerable attention was Gudgell & Simpson's Beau Dux, especially strong in his quarters and flanks, and sold to Mr. Funkhouser. He was sired by Beau Brummel 51817."

The Great Cows and Heifers of 1898.—We cannot better reflect the character of the Hereford females of this era than by reproducing the following account of the female rings at this exposition as published in "The Breeder's Gazette" the week following the awards:

Dewdrop.—"There were seventeen entries in the cow class, and a hot finish was witnessed between Funkhouser's Dewdrop, by Hesiod 2d, Sunny

Slope's Beau Real's Maid, and Nave's fine Anxiety-Monarch cow Atoka, of Shadeland breeding. The big Maud Muller, that won first for Mr. Nave at Springfield, was properly set back to fourth place. Dewdrop and the Maid gave Mr. Makin a lot of trouble. The superb front of the Cross cow was hard to get over. She has weakened a bit at the rump since calving and might be a bit heavier at the thigh, but her grandly spread and deeply covered ribs and beautiful shoulders have rarely been excelled in western showyards. Dewdrop is of rather a blockier pattern and was shown with plenty of hair. She has the Hesiod beauty of head and horn and is extremely short on the leg, with broad ribs deeply laden, in fact, one of the greatest flesh-carriers of any breed in the cattle department. She wants a little between the hips and tail-root, but conformed so closely to the judge's apparent ideal as respects breadth and depth without height that he at length awarded her premier position. Sentiment about the arena was well divided as between this royal pair. Mr Nave's Atoka, that received third honors, is a cow of beautiful lines, in fact almost a perfect parallelogram, her long, level and well finished carcass being carried close to the ground on neat bone. Like all of Mr. Nave's entries she is shown with a great wealth of hair and is in admirable bloom. She has a fine face, excellent shoulders, good finish at the tail, although wanting a little behind the hips. She shows rather too much 'leather' under the jaws for an ideal show cow, but is so neat, level and symmetrical and is shown in such beautiful condition that she is a prime favorite wherever she goes and had friends here for the blue. She certainly made as strong a third-prize cow as ever held that position in this country. Her stable companion, the massive Maud Muller, was fourth and Sotham's Benita fifth. The latter is a cow of superb breeding character with a table-back, but is criticized sharply in her hind legs."

Dolly 5th.—"Eleven head of two-year-old heifers were next presented. The tops were found in Mr. Adams' Luminous and Miranda, Funkhouser's Delight, Nave's Dolly 5th, Sunny Slope's Pretty Maid and Sotham's Lady Brenda. Close comparison between these fine heifers brought out many differences of opinion. The judge first drew in at the head of the list Delight, but subsequently moved Mr. Nave's Dolly 5th ahead. Dolly is a heifer of fine scale and substance, showing great width of rib and extraordinary depth of body. Although a bit heavy in her horn she is very nice in her shoulders, full in her neckveins, remarkably heavy in her chine, full in the twist, and stands well on good short legs. Dolly was sired by Java 64045, a brother to Mr. VanNatta's champion steer Jack, having been sired by Hengler out of Jewel Fowler 49207. Delight received the red ribbon on account of her scale and great strength of back, her ribs and loin being richly furnished with thick flesh. Her horn is good but she has a trifle too much length of face. She is a daughter of Free Lance. It seemed rather hard to turn down so fine a heifer as Luminous to third place but it must be remembered that she has probably journeyed farther by land and sea during the past year than any other animal at this show. She was imported from England by Mr. Cross last fall and bought by her Colorado owner at the Sunny Slope sale in March. She was shipped to the San Luis Valley and back again for this show to the Missouri River. She is of scarcely as blocky a type as the heifers that had precedence over her in this class, but will certainly grow into a great cow. Fourth honors fell upon Mr. Cross' Pretty Maid, by Wild Tom, a heifer carrying a tremendous lot of flesh upon a very wide back; but she is a little uneven in her top and quarters. Her half-sister, Mr. Adams' Miranda, by Wild Tom, is a great block of beef, as nearly without legs as is possible to breed a beef animal. She is wonderful in her neckveins, but does not carry her back altogether level, is growing a bit gaudy about the rump, and stands badly behind, showing the effect of her long railway shipment. She was one of the very ripest heifers in the lot, and there were many who could not understand why she was turned down to as low a position as fifth. Sotham's Lady Brenda, with her good heartgirth, nicely arched ribs and strong loin, was sixth in the judge's rating."

Diana.—"The yearling heifers were headed by Diana, the remarkable 'chunk' that gained so many victories as a calf for Mr. Cross at the great fairs of 1897. She is almost as extraordinary a carcass as Dale, a feeder's type par excellence, but wanting the finish of head and horn seen in Sotham's brown-eyed beauty Benison, by Protection out of Benita. Makin was sorely tempted to put Benison to the fore. She is not big but is one of the finest models sent into western showyards in many years. Feminine and finished, she is lovely in her neck and shoulders, thick and true in back and loin, evenly filled, well balanced and shapely. The breadth and extraordinary thickness of Diana proved an attraction that could not be resisted however, and the wonderful daughter of that great getter of quick feeders—Sunny Slope's Archibald 5th—was left in undisputed possession of the post of honor. Third place was assigned to Mr. Funkhouser's Olga, calved Jan. 13, 1897, and sired by Hesiod 2d. She has the Hesiod trademark, a beautiful head and face, an elegant back of even width, well filled at all the feeding points, and like all the Hesiods low on the leg. She is shown with a great coat and much wealth of flesh. Gudgell & Simpson were fourth on a Lamplighter heifer known as Mischievous 71758. She is a big, deep-bodied, thick-fleshed one with great chine, finely covered shoulders and yellow skin. Funkhouser's other Hesiod heifer, Level 71470, was fifth."

Carnation.—"There were seventeen heifer calves in line, and Nave had the honor of bearing the blue with the well grown and nicely conditioned Carnation, by Acrobat 68460—of Mr. Earl's breeding—out of Erica 51st 41238 by Garfield, second dam Lady Wilton 26th by Sir Bartle Frere. She shows her fine breeding in her pretty face, and her feeding quality is indicated by her full flanks and neck veins. She has good length and nice quarters, altogether a fine promise for a handsome cow. Mr. Cross got next to the Hoosier heifer with Miss Grove, by Climax, a prime block of baby beef, great in her quarters and twist, nicely spread on the back, and 'pegged' near to the ground. Funkhouser drew third on Rollela, another good-backed Hesiod. Sotham pulled fourth out of this hot fire with Silence, by Corrector, and Steward & Hutcheon fifth with Queenie, by Benson. Mr. Adams' furry-haired January calf Graceful Gift ought to have had rank here somewhere among these sappy white-faced lassies."

Group and Championship Prizes.—The grand finale at this epoch-making show is thus set forth:

"The senior herd prize was awarded to the well-brought-out cattle of Mr. Nave. This enterprising young Indiana breeder should feel very poud of this triumph, achieved as it was in the face of the competition of so many veteran showmen. The Funkhouser herd was second, Cross third, Sotham fourth, and Gudgell & Simpson fifth. There were eight contestants for the young herd prize, the right of the line being held at the finish by Mr. Funkhouser with a lot headed by the handsome young Hesiod 50th. The Sunny Slope entries, led by the showy Climax 4th, were second, Sotham third, Gudgell & Simpson fourth, and G. H. Adams fifth.

"There was a large and interesting show made for the get-of-bull and produce-of-cow prizes. Mr. Sotham succeeded in winning for best four animals the get of one sire on the progeny of Corrector; Funkhouser was second with Hesiods; Cross was third on the get of Archibald 5th; Steward & Hutcheon were fourth on stock by Benson, and Gudgell & Simpson fifth on Lamplighter. The produce-of-cow prize for best two head of either sex was also gained by Mr. Sotham with Sir Bredwell and Benefice. Mr. Nave was second on the progeny of the Shadeland cow Erica 51st, Sotham third on Grandee and Genevieve out of Gaily, Gudgell & Simpson fourth on the progeny of Miss Charmer 4th, and the Stanton Co. fifth on a pair from Hare Bell.

"Sir Bredwell, Mr. Sotham's three-year-old, was made bull champion after a spirited contest with Mr. Nave's Dale. There can be no question whatever as to the marked superiority of Dale at the present time, viewed purely from the standpoint of the feeder and the butcher, but Sir Bredwell's fine breed character sufficed to carry the judge to the older bull. Dale was rated second, Bovic third, Thickset fourth, and Hesiod 50th fifth. Funkhouser's Dewdrop was made champion female, Nave's Dolly 5th second, Cross' Diana third, Sotham's Benison fourth, and Sunny Slope's Beau Real's Maid fifth."

K. B. Armour Active.—Kirk B. Armour of Kansas City had by this time become intensely interested in the Herefords, and had made a large importation from England. He not only brought

K. B. ARMOUR.

ample means and a genuine personal enthusiasm to the work, but had the assistance of such able lieutenants as Frank Hastings, then a member of his "Packing House Cabinet" and subsequently with the Swensons at Stamford, Tex., and of William Cummings, one of the most experienced cattle buyers in the west. Mr. Armour made a number of importations from Herefordshire in succeeding years. Late in October he made the good average of $385 on 113 head of imported and home-bred cattle sold at Kansas City. At this sale Murdo MacKenzie, manager for the Matador Land and Cattle Co., paid $1,000 for the bull Shore Acres. George W. Henry gave $1,000 for the good stock bull Kansas Lad. T. F. B. Sotham took Lady Laurel and Frank Nave got Lalla Rookh at $1,000 each. Scott & Whitman bought imp. True Lass at a bid of $1,025.

Death of George W. Henry.—It was during this two-day sale that Mr. George W. Henry of Chicago died suddenly at the Midland Hotel. Mr. Henry had been one of the most active promoters of Hereford breeding during the "eighties," but after selling Rossland Park had dropped out of the trade for some years. He had subsequently, however, bought the old Reed Farm near Goodenow, Ill., and was engaged in founding a second herd, under the capable management of Mr. John Letham, when suddenly stricken while in attendance at this sale.

Another Gudgell-Funkhouser Sale.—At Kansas City, on Nov. 15 and 16, 1898, Messrs. Gudgell & Simpson and James A. Funkhouser sold 97 head of cattle for $27,000, an average of $278. Buying for western range and Texas account was active, and J. M. Curtice took out the twelve-month-old bull Hesiod 50th at $1,400. William Powell, who was now located in Texas, and Hon. John Sparks of Nevada, and O. Harris of Missouri, a man of whom there is much to be heard later, were liberal buyers.

Death of Charles B. Stuart.—Through the death of Charles B. Stuart at Lafayette, Ind., on the 20th of February, 1899, the Hereford breed, and more particularly the Hereford association, lost an ardent, efficient, intelligent, forceful and resourceful champion. The vital factor in the upbuilding of the Shadeland herd, he had been a member of the executive committee of the herd book society from its first organization, and was serving his seventeenth consecutive year as the "live wire" of that powerful committee at the time of his decease. He had seen the business of the organization grow from next to nothing up to the point where its assets exceeded its liabilities by more than $35,000, and volume 22 of the record published in 1900 contained 10,000 entries.

Following closely upon the decease of Mr. Earl, as already recorded, Mr. Stuart's death came as a distinct shock to the Hereford cattle breeding fraternity on both sides the water. Overwork and incessant application tell the whole story of his breaking down while yet a comparatively young man. Nervous prostration overtook him while in the floodtide of professional and business success, and a career of uncommon brilliancy closed ere it had been fully unfolded.

A son of the late Judge William Z. Stuart of the Indiana Supreme Bench, the deceased took up the practice of law after graduating from Amherst College and the Columbia Law School and quickly attained reputation as one of the keenest-witted attorneys of the Indiana Bar. For many years he was entrusted with the legal business of the Wabash Railway Co., originating in that state, besides being retained in many important cases before the highest judicial tribunals. His wife, who survived him, was a daughter of Mr. Earl, who was one of the leading business men of Lafayette, and Mr. Stuart's fine judgment and acknowledged talent were in constant requisition in connection with the promotion and development of large industrial and financial enterprises. In business and in his professional work Mr. Stuart was equally successful, but he paid a fearful penalty for his assumption of burdens beyond any one man's powers of endurance.

Mr. Stuart had a genius for mastering the details of any subject to which he gave his attention. He became not only an expert judge of Herefords, but as a student of bloodlines and combinations he was confessedly one of the best informed men on either side of the Atlantic. The Shadeland catalogs of his preparation were for years models of their kind and brimming with facts and comments of value to his fellow-breeders. He was partial to the Wilton blood, and the Stocktonbury cattle and this great Wabash

Valley Hereford breeding establishment proved a mine of bovine wealth to the west. In judicious combination with crosses from Colorado, Sir Richard 2d and Horace (through Garfield and The Grove 3d) the daughters of Lord Wilton and of his famous son Sir Bartle Frere gave American state fairs and fat stock shows some of the most remarkable cattle this country has ever seen. The record-breaking bull Earl of Shadeland 22d was the pride of Mr. Stuart's heart. America has known few as good in any beef breed.

Happy indeed were the days the overworked attorney used to snatch away from business and spend among his four-footed pets at Shadeland Farm. Had he devoted more time to the cattle and less to his office he doubtless would have lived a longer life. With a few edibles from the city markets under the seat, and a congenial companion by his side, Mr Stuart liked nothing better than to turn his back to the town, intent upon a day's outing at the farm. "Uncle John" Lewis knew upon such occasions that he had come to take luncheon with Mrs. Lewis, and while the good wife of the kindly old herdsman was preparing a collation fit for a premier of the realm, old "Bartle" or Garfield or some of the boxes filled with sappy white-faced babies would be hastily visited. The newest arrival was always an object of interest and if anything was ailing in any way it was certain to receive an early call. The noon-day meal over and the pug puppies duly discussed, the grand tour was commenced. Mr. Stuart was fond of drawing out his guests when favorite cattle were under examination. Sometimes visitors would hit upon Lewis' favorite and sometimes upon Mr. Stuart's choice. Oftentimes the herdsman and his steadfast friend would already have agreed upon one that was to be put aside as too good to part with. Still there was always ample scope for argument, and Stuart had the lawyer's real relish for debating the fine points. In this respect Mr. Earl was quite different. He was a man of few words but nevertheless enjoyed these Hereford "sessions" quite as thoroughly as any other member of the party.

Memory recalls few fairer scenes than we have witnessed in the Shadeland pastures. The herd was usually kept in strong condition and carefully sorted by ages, sexes and type. The various bands of cows and heifers never failed to make a great impression upon visitors, and in its palmy days Shadeland was easily the great show place of the United States as a Hereford nursery. In the course of all our journeyings to the farm, however, we do not believe that the question as to which was the best cow of the herd in its prime was ever really settled. We once went through the lot with Mr. Earl, Mr. Stuart, John S. Carlyle and John Lewis; we recall readily Mr. Earl's quiet conservatism, Mr. Stuart's keen analysis of form, Mr. Carlyle's brusque opinions (usually dashed with broad Scotch wit), and the modest courteous comment of Lewis. Those sunny summer days will come again. The grass will grow as green. The Wabash in the distance will yet roll its turbid flood through the dreamy woodlands, but the old associations are broken, never to be re-formed amidst earthly scenes.

Spring Sales of 1899.—Sotham opened the ball at Kansas City on March 1 by selling 46 head of cattle for the fine average of $516. Col. C. C. Slaughter of Texas took out the show bull Sir Bredwell at $5,000, Mr. Frank Nave's representative, Mr. Keyt, being the "runner up." Mr. Nave was reported to have offered $7,500 for the famous son of Corrector a few days later, but the bix Texan replied, "Not for $10,000." Nave got the yearling bull Eye Opener, by Protection, at $1,100. This was the second highest average up to date made on Herefords in the United States, Mr. Earl having registered $574.20 on 38 head at Kansas City, Nov. 8, 1883.

On the day following this sale Mr. F. A. Nave sold at Kansas City 49 head at an average of $383, the top price being $1,075, given by Mr. Armour for the heifer Armel, by Columbus, the sire of Dale. Grant Hornaday of Ft. Scott, Kans., followed with an offering of 38 head which averaged $350, Col. Slaughter taking the Corrector bull Sir Comewell at $1,600. Mr. Frank Rockefeller of Cleveland, O., was a free buyer of good lots at each of these sales for his ranch at Belvidere, Kans.

While no sensational figures were reached at the April sales at Kansas City by C. A. Stannard and Scott & March, about 200 cattle were sold at good fair prices. There were 97 head in the Sunny Slope lot that averaged $177.30, the best price being $555 for the heifer Ashton Bloom with a bull calf at foot by the $3,000 imp. Salisbury. The Scott & March offering of 99 head made $192. This was a specially good lot of breeding cattle brought forward in beautiful bloom by the herd manager, Mr. Godfrey.

The Curtain Falls on Stirring Scenes.—As the century drew to its close in the autumn of 1899 the apotheosis of the Hereford in America was reached. All that the fondest admirers of the "white faces" had ever predicted for the breed had now come true. The Hereford had entered into full and almost undisputed possession of the great cattle ranges of the west, thus opening up a field infinitely broader than the Herefordshire fathers had ever dreamed.

By judicious concentrations of the blood of the earlier importations a type of cattle had been evolved that in point of finish, levelness and smoothness clearly surpassed the Herefords of old Herefordshire. The appeal to the magic power of in-and-in breeding by men possessing the experience imperatively demanded for its wise application was now manifesting itself marvelously in every showyard. A realization of this fact added the joy that always accompanies the accomplishment of a sustained purpose to the intense enthusiasm attending the conquest of the grassy empire dominated by the snow-clad summits of the Rockies.

Big men in Texas, big men all through the great

breeding grounds of the southwest, big men in Colorado, big men in Montana and Wyoming, big men in the Dakotas, big men in Kansas and Nebraska, big men in the cornbelt were banded together in the American Hereford Cattle Breeders' Association in proud possession of a captured market. They felt their power and proceeded to use it in effective fashion in promoting the general good.

Inception of the American Royal.—The Hereford association working through efficient committees held a never-to-be-forgotten show at Kansas City, in which 541 highly fitted cattle participated, and nearly 300 head were sold at auction at an average of $317. At this sale John Sparks, afterwards Governor of Nevada, paid $2,500 for the beautiful Armour Rose. Col. Slaughter paid $1,950, after a battle with Mr. Funkhouser, for the young VanNatta-bred bull calf Aaron, and B. C. Rhome of Texas took Beau Donald 2d at $1,200.* A few days later Mr. Armour bought Aaron from Col. Slaughter at $2,000 plus the choice of any bull calf in his own herd.

---

*At this sale an episode unique in the annals of such events occurred when the bull calf Bonnie Prince, the property of Mrs. Kate Wilder Cross, widow of Charles S. Cross, was offered. Mrs. Cross had in so many ways endeared herself to the Hereford cattle breeding fraternity that there was a hearty response to Col. Woods' felicitous appeal in her behalf on the introduction of the calf into the ring. He was quickly run up to $900, at which point the widow of the late Charles N. Whitman announced that she would individually add $200 to the last bid for the calf no matter what it might be. This generous offer was accompanied by a shower of silver dollars tossed onto the tanbark under the leadership of Col. Slaughter, with the compliments of everybody, by way of expressing appreciation of what Mrs. Cross had done for Herefords. Mr. Marshall Field's representative took the calf at $910, and when to this was added the free-will offering of the company it was found that something over $1,200 had been realized. Mrs. Cross subsequently established a herd on her own account which she successfully conducted for some years.

This remarkable event proved the foundation of the "American Royal", that has ever since focused annually the attention of western cattle growers upon the Kansas City exhibition established under that name. This show really marks the beginning of the end of our story of the permanent establishment of Hereford cattle breeding in the United States. What remains to be told relates largely to the Herefordizing of the range, and to the latter-day achievements of the more successful breeders and exhibitors of pedigree "white faces" in the older states. We digress therefore at this point to discuss the introduction and dissemination of the blood on the open range, which after all was the great point towards which all this work with the pedigree "white faces" had really been tending.

CHAPTER XVI.

THE LONG TRAIL.

Western ranching had its genesis in the cattle originally introduced into Mexico by the Spanish conquistadores. The admission of Texas and the Gadsden Purchase of 1854 brought within the boundaries of the United States enormous tracts of arid and semi-arid lands susceptible of a great pastoral development, but insofar as the territory north of the present Mexican border is concerned, cattle-raising as a business (as distinguished from the mere maintenance of the herds as a source of food for their owner and his dependents) was virtually unknown among the rancheros of that period.

After prolonged negotiations and a vigorous political contest, Texas, formerly a portion of Mexico and later an independent republic, was admitted to the Union by joint resolution of Congress, approved by President Tyler on March 1, 1845. As a result of the Mexican War and by the Treaty of Guadalupe Hidalgo, on Feb. 2, 1848, Mexico ceded the territory now covered by California and Nevada, also her claims to territory covered by Texas, Utah, the bulk of Arizona, New Mexico, and portions of Wyoming and Colorado.

The tract of land known as the Gadsden Purchase, comprising territory lying within the present limits of the states of New Mexico and Arizona, was obtained from Mexico in 1854. It embraced 45,535 square miles bounded on the north by the Gila River, on the east by the Rio Grande, and on the west by the Colorado. It had an extreme breadth north to south of 120 miles. The United States gave $10,000,000 for it, and Mexico agreed to cede claims arising from Indian incursions. This land was purchased to settle a dispute and to secure a route for the Southern Pacific Railroad. The treaty was negotiated with Santa Anna by James Gadsden, a South Carolina soldier who was Minister to Mexico, in December, 1853, and was finally ratified on Aug. 5, 1854. The sale caused the banishment of General Santa Anna from Mexico.

Throughout the vast interior regions comprised within the lands acquired from Mexico but few attempts had been made to invade the deserts, plains and mountains that were the hunting grounds of the aborigines. Along the Mexican gulf and the Californian coasts hides had an established value, but even near tidewater there was no market of any consequence for fresh beef.

The Spanish Longhorn.—Cattle of Spanish derivation have never been specially distinguished as flesh-makers. A pair of horns well adapted for purposes of offense or defense, as the case might be, has always been accounted an important characteristic, however, and the Mexican descendants of the animals brought across the Atlantic by the Spaniards neither gained in the one respect nor failed in the other in their new environment. Nevertheless, it is unfair to assume that the blood of the Spanish cattle was base. Good cattle did come out of Spain. Naturally of good size, some of them reached the heroic in stature. There are yet some native Spanish cattle in Chihuahua and other Mexican states that are big, rugged, and of considerable merit as beef animals. Cattle of the longhorned type excel as animals of draft. They have amazing energy and endurance and what may be termed "cow sense." When bands of mixed cattle were common on the plains and deserts of the west it was notable that the longhorns led the herds in their migrations. These cattle felt the "call of the wild," had weather wisdom and knew where to find grass and water. They were admirable mothers and their calves sired by "Durham" or Hereford bulls were excellent. Whatever may have been their faults, judged by the standards of latter-day beef-makers, it must be said that they not only served every purpose required of them at the time, but constituted the best possible material for use by those who first sought to put cattle ranches on the map of our new possessions.

The extension of United States authority over the Lone Star State, and the discovery of gold in California in 1849, resulted in an influx of population and capital that soon exerted a stimulating effect upon the production of cattle throughout southern and north central Texas, as well as beyond the Sierra Nevadas. The herds came to be valued for beef, as well as for their hides, horns and hoofs. And thus the infant industry of cattle-growing in a commercial sense came into existence in the great southwest.

Capt. Richard King.—While the military campaign that carried the American flag to the City of Mexico was in progress, a man who was destined to exercise a far-reaching influence upon the industrial development of our new frontier, was engaged in transporting freights and army stores along the west coast of the Gulf and up the navigable waters of the region that constituted the base of our operations. This was Capt. Richard King. Upon the cessation of hostilities he decided to engage in business ashore, and to this end acquired title to a large tract of wild land lying near the coast between the mouths of the Nueces and Rio Grande rivers. He had conceived the idea that the production of horses and cattle on a large scale in this territory could soon be made a lucrative business, and the idea proved the foundation of not only his own but also of many other fortunes subsequently accumulated as a result of extensive land and grazing operations.

Santa Gertrudis.—When Capt. King first rode across the plain from Brownsville to Corpus Christi it was one vast flowery meadow, lovely beyond compare. There were then no thickets of mezquite or other brush except the occasional bits along the streams. Later occupancy of the land and the keeping out of fires caused the appearance of great thickets of small trees and brush, largely of leguminous nature, such as the mezquite tree. Within recent years the manager, Mr. Kleberg, has cleared again at much expense vast areas of these infringing thickets.

The original tract comprised about twelve setios of 4,428 acres each. This aggregated more than 50,000 acres, for the most part flat, treeless and without streams or springs of fresh water. There were grasses sufficient to support live stock, and the water problem was met in a primitive fashion by means of large tanks or reservoirs built along the few drains, impounding the storm-waters; but as few points could be found where dams would be of any avail these watering places were few and far between. Moreover, the matter of markets had yet to be worked out.

Upon this property in 1854 the headquarters of the now world-famous Santa Gertrudis or King Ranch were established, and here we may fairly say our modern American ranching had its earliest important exemplification. Cattle and horses of the common Mexican types were purchased and roamed at will over the vast arid plain that had the brackish waters of the Nueces for its northern boundary. In the meantime the proprietor made an outlet for his cattle by slaughtering them for their hides and tallow, which products he hauled to Corpus Christi, the nearest port on the coast. The offal was fed to hogs, which in turn were slaughtered and the lard shipped by sea. Of course, there was no market available at that period for fresh meats, except for local consumption, and that was chiefly by the owner's household and his Mexican herders and retainers. The horse breeding soon became profitable, the surplus stock finding ready sale in the developing interior of Texas.

Packing Houses in Embryo.—The growing of cattle for their hides was so obviously a wasteful procedure that the attention of capitalists was drawn to the opportunity for profit afforded by such conditions. It is said that two plants were established near Rockport, Tex., at a place called Fulton, before the outbreak of the Civil War in 1861. One of these was occupied mainly in the canning of fish and green sea turtle, and to this it is stated that there was added a dessicating department for the making of beef extract. The other was called the Coleman Fulton Packing Co., an enterprise carried on by the Coleman Fulton Pasture Co., whose lands are now the property of Charles P. Taft. This company packed beef in salt as pork is packed, their main business being the making or pickling of corned beef. Prominent New York City capitalists, including "Commodore" Vanderbilt, are said to have had an interest in one or both of these concerns.

Before these enterprises were started, however, Capt. King and some of his associates had attempted to preserve the meat of cattle for shipment by the infusion of brine into the veins of the cattle immediately after they were slaughtered. But on account of the lack of transportation facilties and because of this undeveloped method of preserving the beef the effort was abandoned, and only the hides, tallow and offal were saved.

Capt. Kennedy.—Prominent among those who early recognized the possibilities of this new industry in that region was Capt. King's old companion in the river and coast-wise steamboat service, Capt. Mifflin Kennedy, who had also decided to remain upon the border after peace had been proclaimed. Kennedy engaged first in commercial dealings with Old Mexico, but a few years later joined Capt. King in his ranching operations, as will be referred to further on.

First Efforts at Improvement.—While many attempts were made by King and Kennedy to improve the quality of their herds, but little headway was made in that direction for many years. In the first place there were no improved breeds nearer than the distant bluegrass pastures of Kentucky. Transportation was tedious and expensive, and worst of all it was soon discovered that northern cattle taken to those southern plains almost invariably succumbed to a fever, the nature and origin of which was at that time not understood. The longhorn thrived and multiplied untouched by the mysterious plague, but the northern cattle either died or were left mere wrecks of their former selves. We now know that this was the work of the tick that infects the low-lands of the lower latitudes. It may be said in passing that it was upon this same great Santa Gertrudis Ranch in later years that the veterinarians of the Bureau of Animal Industry worked out many of the original proofs as to the real character of the so-called Texas or splenetic cattle fever. To

Mr. Robert J. Kleberg, who succeeded to the management of the great landed estate left by Capt. King at his death in April, 1885, is credited a large share in this important work of discovering and developing the true nature of the disorder that cost American cattle growers so dearly before a correct diagnosis was established.

Capt. King blazed the way for the great cattle business that afterwards brought such wealth to the Texas commonwealth, and which after the Civil War was extended northward and westward until the ancient grazing grounds of the bison, leading up in all directions to the rugged walls of the Rocky Mountains, were at last converted into one enormous open cattle pasture. His business prospered, and he lived to see his landed estate expand to 500,000 acres. At the time of his decease this was enclosed by a good fence, but the huge holding was divided into but two pastures—one the upper or northern in Nueces county, known by the original name of Santa Gertrudis, and the other known as the lower or southern range in Cameron county. This vast property in more recent years was more than doubled in area, so that Mrs. King, who was the sole devisee and legatee of the estate, ultimately became the mistress of a princely domain of more than one million acres, well stocked with highly-bred Herefords and Shorthorns. But that involves the story of Robert Kleberg's stewardship, to be referred to further on.

Breeding Up the Native Stock.—Following the earlier successes of Capt. King and Capt. Kennedy

and their contemporaries in the extreme south of Texas, cattle were introduced into the central and northern portions of the giant state. The foundation herds were longhorns, but in the late '50's and the years just preceding the outbreak of the Civil War in 1861 the owners had made strenuous efforts to improve the breed. Shorthorn bulls, mainly from Kentucky and Missouri, were freely bought, and while the death rate among them constituted a heavy tax upon their enterprising buyers, the persistency with which the policy was pursued at last manifested itself in a gradual betterment of the general cattle stock of that entire region; so much so that when the great expansion in cattle ranching set in after the close of the war the pastures lying to the north of San Antonio contained a leaven of "Durham" blood that ultimately leavened a large proportion of the entire lump, while on the lower ranges the so-called "coast" cattle were still of the distinctly longhorned type.

The Mormon Cattle.—The early Mormon emigration to Utah was a considerable factor in fixing the cattle stocks of that region, for these people took with them good milking cows largely of Shorthorn blood. In the early '80's Utah still had many good descendants of these valuable milch cows, and many a ranch was stocked with cattle bought in the Mormon settlements. These cattle, however, had the habit of milk-giving too strongly pronounced to make them ideal range stock, as the cows frequently lost parts of their udders from having more milk than their calves could take, and they were such persistent milkers that they were apt to go into winter too thin in flesh. They formed, however, ideal mothers for the creation of grade Hereford herds.

Pacific Coast Cattle.—On the western coast the situation was somewhat similar to that in Texas. In the extreme south the Spanish stock still prevailed in its natural state, but a steady stream of "settlers" from the middle west, seeking their El Dorado at the end of the Oregon and Santa Fe trails, had driven many a beast of Shorthorn or Devon extraction across the great divide, where under climatic conditions favorable to northern-grown animals they had planted the seeds of substantial improvement. Thus it came about that in both Oregon and California a start towards a higher standard had been made at a comparatively early period. In all these instances the cross of the Shorthorn on the longhorn had increased the size, leveled the carcass and improved the fleshing capacity of the cattle.

Shorthorn Crosses in Evidence.—The Hereford had no place in the original invasion of the range country. The first great pitched battles with the elements were fought mainly by the Texas longhorns of both the improved and unimproved types. Had they all been of the straight "coast" type, it is possible that the earlier efforts, more especially in the north, might have met with fewer reverses. In those first fierce exposures to unaccustomed rigors the Shorthorns and their grades had to bear an important part, for as already stated the blood had been introduced into north Texas before the first herds hit the northern trails. Moreover, by the time the forward movement got into full swing a considerable stream of Shorthorn blood was pouring into the great drive from herds that had been established in the south of Kansas, in the Indian Territory and the Cherokee strip. Such points as Harper, Medicine Lodge, Caldwell and Wichita were all on the confines of a great cow country that had recourse for bulls to the Missouri and Kentucky Shorthorn herds. Then, too, the westward drift from central Texas into New Mexico, Arizona, the Panhandle and Colorado included some cattle of an "improved" Texan type.

The Great Migration.—As late as the year 1860 the mountains of Colorado still looked down east, west, north and south upon a grassy wilderness that practically knew only the hoofs of the buffalo and the antelope and their pursuers—the hunters and the hunted. Railway iron at length pierced the very heart of this great preserve, however, and the Union Pacific locomotives sounded the end of the old, the beginning of a new regime—the coming of the cattle.

Crossing the Red River the great hegira to the north began in earnest along trails soon to become historic, only to fade away again after the lapse of many years into mere traditions of the past. The herds were headed largely towards El Reno, Camp Supply and Dodge City. From near Muskogee the famous Chisholm Trail followed the valley of the Arkansas as far as Wichita and thence on to Abilene. The pastures of the Territory, the Cherokee strip and southern Kansas, first felt the pressure from the south, but about the same time a drift set in from central Texas up the valley of the Pecos, in which direction trails soon wended their way out into New Mexico and beyond.

Eastern Colorado and central and western Kansas and Nebraska, constituting a vast realm of free grass, were successfully pastured. The tide of immigration was rolling steadily into the Rocky Mountain region across the plains from the mid-west states. The imaginations of the adventurous everywhere were stirred by the stories of fortunes to be made in western cattle. Daring spirits flocked to the scene of the spectacular expansion, and plunged into the game regardless of their inexperience—"the butcher, the baker and candle-stick maker," all anxious to engage in this wonderful new business of cattle ranching. The big pastures and mountain meadows of Montana, Wyoming and Colorado were not long in filling up. Denver, the capital of cowland, was the scene of feverish activities. Big deals capitalizing alluring propositions were easily handled. Goodnight was waking up the Panhandle, and Swan and his contemporaries were enthusing the north. All the way from Helena to San Antonio the pot boiled furiously.

Farther and farther into the interior of this inland empire, the cowmen pushed their way, and the railway and the stagecoach soon sought gateways into the nation's virgin pastoral possessions. Staid Scotch capitalists, scions of the British aristocracy, and "tenderfeet" of nearly every name and nation joined in the chase—the race to put cattle into every nook and corner of the great big Brobdingnagian West, regardless of climatic conditions or possible consequences.

In the midst of it all the new southwest was not forgotten. The advantages of the lower latitudes as a breeding ground were many and obvious. All were ready to listen to new schemes for further development in any direction. Out on the pastures of New Mexico and Arizona soon the cattle found a footing. Far-off Nevada escaped not the hoofs of the on-coming herds, and there was always California. The creatures of a "wild" that was fairly continental in its vast expanse, stupendous in its distances, its heights, its depths and possibilities, gave way in all directions before the grand army of the occupation. The victory was only gained, however, at heavy cost. The gods were at first propitious. Fortune smiled alike, for a time at least, upon the just and the unjust, but the inevitable happened. The bubble of indefinite and unwarranted expansion and improvidence burst. But experience teaches. Better methods gradually supervened, and in the meantime the hardy Hereford had been introduced and cattle ranching took on a more settled character.

CHAPTER XVII.

FIRST HEREFORDS ON THE RANGE.

It is now impossible either to fix definitely the date when the Herefords made their first appearance upon the western range or to locate accurately the place where the earliest experiments in pastures limited only by the horizon were really staged. However, it may be stated with reasonable certainty that the time was somewhere near 1870, and the place Colorado. It can also be safely recorded that the initial buying was cautiously approached by men who had no assurance whatever that the venture would prove successful. But it did. Had it not, this volume might never have been published.

In a letter written by Mr. T. L. Miller in 1877 the statement is made that "it is now ten or twelve years since the Herefords were first taken to the plains," but he gives no names or dates. This would place the period of their introduction at from 1867 to 1869. It is of course easily possible that some of the old Stone, the Ohio or early eastern blood had found its way west at that time. In fact we should think this extremely probable. The state of Colorado, being in the direct line of cornbelt emigration, would naturally be one of the first to receive the blood of improved cattle of eastern origin. The Texans of that day bought almost exclusively from Kentucky, and as that state, at that period, had no

Herefords of which there is trace it is not difficult to realize that Colorado would logically beat Texas to the "white faces," and such was undoubtedly the case. Mr. Miller himself sold three Hereford bulls in 1873 to George Zweck of Longmont, Colo.—a yearling, a two-year-old and a three-year-old afterwards registered as Plato 590. In 1874 he shipped five bulls to Denver, which were sold to Colorado ranchmen. The first purebred Herefords to go to Texas, so far as we can learn, were a bull (Chief) and a heifer by Miller's old Success, sold by William Powell, then of Beecher, Ill., in the spring of 1876 to J. F. Brady of Houston. It is said that about this same date a Mr. Hooker took Herefords from Beecher into southern Arizona.

Making Good.—On being asked, "Why are the Herefords the best cattle for the plains?" Mr. Miller answered: "Because they are the most hardy; they are the best grazers; they mature earlier; they are nearer the ground; they are more compact; they have more hair; they have thicker and softer hides; although shorter on the leg, they are better travelers, and as grazers they become higher-fleshed and riper steers; they carry their flesh to market with less shrinkage; they are heavier-topped steers, and the best animal in the family of Herefords is the steer."

The blood was liked on the Colorado range from the very first, and in 1876 ranchmen who had already tested it there reported as follows:

Judge Downing, of Denver, sold six Hereford

grade steers in June, that were four years old in the spring, weighing 1,800 pounds each, and twelve others and three heifers, weighing a fraction under 1,500 pounds each. None of them had been fed at all, having made their weights on grass alone, except they may have been fed hay at times during storms.

Mr. Church, who lived near Denver, had turned off thirty to forty grade Hereford steers for several years, at three years old, averaging about 1,250 pounds each, that had never been fed anything except what they themselves had taken from the range; and one lot of these steers was sold in Buffalo at 7 cents a pound.

Judge P. P. Wilcox, of Denver, said that his cattle ran with a herd in which there was a grade Hereford bull, and from him he had several white-faced calves, and that these white-faced calves were as good at two years old as his others at three.

Another prominent stockman in southern Colorado testified: "The Hereford cross on my native cattle has been very satisfactory. They stand the winter well, take on flesh rapidly, and are really the best cattle for these ranges that I have ever had anything to do with."

Commenting upon these and similar reports and launching a challenge against Shorthorn breeders, Mr. Miller with prophetic vision said:

"There is now open to the world, and brought into the world, a stock country, the like of which was never before known. It changes or will change the whole system of breeding, and the question must and will be solved as to the breed of cattle best fitted for it."

Speaking of difficulties tending to restrict enterprise in the placing of good bulls on the open range, Mr. Miller added:

"One of the great drawbacks to a more rapid trade has been the difficulty of holding the bulls for use in the owners' herds. The practice being to run their herds on a common range, the cattle of several owners intermingle. Jones, buying thoroughbred bulls, and his herd and Smith's running together, Smith gets the use of Jones' bulls. Very few of the cattlemen have fenced at all. Very few have thought they could herd their cattle, although this is entirely practicable. The introduction of barbed wire has made fencing practicable, and many are finding that herding is practicable.

"The Messrs. Thatcher Bros. & Co. and the Messrs. Swan both intend to select cows upon which to use these bulls, and herd them during the coupling season, and then place their bulls, until the coupling season returns, in pasture prepared for them.

"The late Mr. Iliff had enclosed some ten or twelve pastures, containing from 1,000 to 3,000 acres in each, for use of cows during coupling season, and out of coupling seasons for the bulls. There are many who have adopted this practice, and the number is increasing. The difficulty in introducing fine stock has been, first, the cost, and secondly, the difficulty of getting the use of them."

Whereas Mr. Miller had in the first instance been obliged to "force" the western market by shipping small consignments at large expense, and offering them for sale on their arrival in Colorado, the returns soon began to come in so favorably from all quarters that sales were easily made and at advancing figures. Beecher continued for several years to be the main source of supply. In 1878 Mr.

Miller sold forty bulls to the Swans in Wyoming, and in 1878 Thomas Clark sold twenty young bulls to J. E. Temple, Chico Springs, N. M. The results of the use of the blood wherever tried proved so satisfactory that numerous inquiries came into the market for white-faced bulls. Unfortunately not all of those secured were purebred, and many of the grades had little to recommend them except their white faces. Nevertheless, it was soon made clear that the breed was destined to materially reduce the risks of cattle-raising on the open ranges.

Prominent among those who became identified with the Hereford cause in the new west at an early date, in addition to those already mentioned, were the Culvers of Colorado, Reynolds Bros., John W. Prowers, J. W. Iliff, John H. Hitson, Thatcher Bros., G. F. Lord, Ikard Bros., T. W. Owen, B. C. Rhome, G. H. Curtis, Hall Bros., Geddes & Bryan, R. S. Van Tassel, J. A. Baker, Jones Bros., Joseph Scott, Lee & Reynolds, W. E. Campbell, Towers & Gudgell and Dickey Bros. The earliest owners of pedigree Herefords in Kansas, so far as is shown by the first volume of the American herd book, were C. W. Kimball of Wichita, W. M. Morgan and J. M. Winter of Irving, F. H. Jackson of Maple Hill, T. H. Cavanaugh of Salina and H. Woodward of Blue Rapids.

First Hereford Sale in the West.—On May 23, 1879, Charles Gudgell sold twenty-five young Hereford bulls at auction at the Kansas City Stock Yards. It was the day after one of the big Hamilton

Shorthorn sales. This was the first auction sale of Herefords held at Kansas City, and the first west of the Mississippi River. Nine of these bulls were sold to Towers & Gudgell, a range outfit in which Mr. Gudgell was interested, for use on their herd of the OX brand on the Cimmaron River in what was then known as "No Man's Land," now Beaver Co., Okla. At the same auction sale at Kansas City one bull was bought by Col. Driskill, at that time one of the leading cattle growers of Texas, who was also buying Shorthorns at the Hamilton sale.

About this same date Charles Gudgell sold the bull Picture 1403 to Jones Bros., Las Animas, Colo., for $1,000. This bull had been bought from F. W. Stone, Guelph, Canada. About the same time J. W. Prowers took some Herefords to his ranch near Las Animas.

The Hawes and Campbell Herds.—Major W. E. Campbell of Caldwell, Kans., and J. S. Hawes of Colony, Kans., established large and excellent herds of purebred Herefords, which were drawn upon heavily, not only by those founding new purebred herds in the Missouri River region, but also by ranchmen further west. Mr. Hawes had been breeding Herefords for a number of years at South Vassalboro, Me., and in the fall of 1881 moved his entire herd of about 100 head to his Kansas farm, comprising at that time some 1,200 acres. During the height of the great demand for the "white faces" Mr. Hawes ran his herd up to more than 300 head of well bred pedigree cattle. It was noted for some years as the home of the show bulls Fortune and Sir Evelyn. During the years 1883 and 1884 Mr. Hawes sold $50,000 worth of purebred Herefords.

Major Campbell had considerable interests on the range, and engaged with great enthusiasm in the breeding and handling of pedigree Herefords, buying liberally from the best herds further east and exhibiting at the Kansas fairs. One of his best known bulls was The Equinox 2758.

Hereford Endurance Demonstrated.—The winter of 1880-81 was of exceptional severity and losses on the range were heavy. This was particularly true of the "pilgrims," as the trail herds recently from the south and turned out on the northern ranges were commonly called. The testimony that followed was very largely to the effect that the mortality among the Shorthorns had been greatly in excess of that in the case of the Herefords; and the fact that the "white faces" had passed through this ordeal so successfully now made them hot favorites throughout all parts of the range country.

Writing in June, 1881, Major Campbell said:

"The question is not which is the best beast, the Shorthorn, the Hereford or the Texas bull, but which is the best rustler and most profitable range animal. It does not matter to us what breed of cattle has been most successful in the feedyard or showyard, for we are interested in neither. What we want to know is which breed is best adapted for range purposes and range purposes only, and all this talk about valuable milking qualities amounts to nothing with ranchmen. In fact, they do not want heavy milkers, but cattle that will give enough milk to support their calves and convert the remainder of the feed into first-class beef.

"As you are aware, I have been breeding Shorthorns for years, and I still admire them very much, and have about sixty bulls in use at one of my ranches. At another I am using nothing but purebred Hereford bulls. Experience has proved them to be the hardiest and best range cattle I have ever known; and I do not hesitate to say that hereafter I will never buy another Shorthorn bull for range purposes. I have a small herd of thoroughbred and quite a number of high-grade Hereford cows that were out all winter without feed, and today they are in fine condition, most of them being ready for the butcher's block. I also had quite a number of thoroughbred and high-grade Shorthorn cows that fared the same. Some of them died, and none of them are fat yet. I am now breeding them to Hereford bulls, against the advice of my Shorthorn friends. That I may be fully understood I will say that I intend reserving all my thoroughbred and high-grade Hereford bulls for my own use. My Shorthorns have done me good, and I do not intend to knock them in the head, as Mr. Miller might advise, but I intend putting white heads on them as fast as I can."

This undoubtedly reflected with accuracy the opinion of a large number of those who were at that date financially interested in range operations.

It is manifestly impracticable to detail the operations of all those who in the years following this successful test of Hereford endurance, took part in their introduction into the various parts of the range country. The territory covered was too vast and the operations too general to admit of more than passing references to a few of the firms, individuals and corporations that figured most conspicuously in the movement that placed the "white faces" firmly upon the western map.

On the Northern Range.—As late as the early '80's the "white faces" were not much in evidence in the northwest. Around Cheyenne there was considerable of the blood, but apart from that vicinity probably not 5 per cent of the northern herds were at that date crossed by Hereford bulls. The great bulk of the cattle in Montana and Wyoming had either come direct from the Pacific Coast or from Texas. Numbers of these had been and were still being crossed with Shorthorn bulls.

A. H. Swan was one of the first to introduce the Hereford blood upon the Wyoming range. His firm, Swan Bros., paid Mr. Miller $10,000 for forty head of bulls in the spring of 1878. A second lot of fifty head followed not many months later. They had previously had some of the blood from Culver and Mahony of Colorado and Wyoming. Mr. Swan's was a strong personality, and he had a big following; his example in adopting the Hereford was quickly followed throughout all that vast country stretching away from the Union Pacific Railway to the Canadian border.

A meteoric record, that of Alex. Swan. His quick rise to apparent affluence when fortune smiled upon his ventures on the open range, his promotion of the big Scotch company that still bears his name after years of vicissitudes, his plunging in lands and sheep and cattle, his alliance with George Morgan, the "advance agent" of the English Hereford propaganda, his staggering reverses and final fall—all told would make a tale only too typical of the smiling, frowning, fascinating west.*

---

*Mr. A. H. Swan had gone to Cheyenne from Indianola, Ia., some time around 1876. He had all the instincts of a promotor—and in his time engaged in many different things, and succeeded in inducing others to join with him in his undertakings.

He started in business as a grocer at Indianola shortly after the Civil War, but soon developed a genius for speculation. One of his first schemes in Indianola was in connection with the building of a railroad to Des Moines—now a part of the Rock Island system. Afterwards he became a speculator in Warren county lands. After going to Wyoming to embark in the then new business of cattle ranching he utilized some 2,500 acres as a farm for the purpose of breeding bulls for shipment to the western range. When John Gosling took hold of this farm it was a cattle-feeding plant, but it was soon afterwards changed into a breeding establishment with a cow herd of some 600 head.

In 1881, Mr. David Kauffman took an interest in this business and was made manager of this farm. Mr. Gosling was transferred to South Omaha and placed in charge of cattle-feeding at the distillery sheds, where in the course of three years he handled over 6,000 head.

Kauffman retired in 1884, and the Bosler Bros., of Pennsylvania, who had become interested with Swan in his range operations, took an interest in the Indianola farm. Mr. Gosling thereupon returned to this farm, when the grade herd was reduced in numbers and additional purebred Herefords were bought from Culbertson's and other good herds.

A correspondent of "The Breeder's Gazette," writing of Swan and his Indianola career, says:

"During the years of his prosperity Mr. Swan was connected with numerous enterprises in and about Indianola. He had interests in farms, a coal mine, brick yard, flouring mill, canning factory and a bank or two. He had close business connections with several English capitalists and live stock men, and was fond of bringing them to North Farm to see the cattle. The story was recently told me by an old resident that, when it was known that Swan was coming to town, the word would be passed from one to another, 'Eck's a comin', Eck's a comin' 'smorning', and a spirit of suppressed excitement pervaded the little town as if awaiting a visit from the President of the United States. His partners, various employes and other retainers would repair to the railway station an hour before train-time to discuss what 'Eck' would do on 'this trip' with regard to his numerous local activities. When the train would at last arrive Swan would come off with his following of Englishmen and eastern capitalists and lead the way to the hotel like a lord, passing out greetings and shaking hands on all sides.

"But the end came, when his ambition and self-confidence overran his judgment with the inevitable result. That was early in 1887. North Farm went to the Boslers, who held it for a number of years, selling it in parcels, and finally selling the tract containing the main improvements to Mr. Jacob Piffer, in the hands of whose estate it still lies. Mr. Swan had acquired the farm from an early settler and financier, who had combined its component parts into one body of land. This was D. H. Van Pelt, grandfather of Prof. H. G. Van Pelt, Iowa's dairy expert."

The Swan Land and Cattle Co.—The Swan Land & Cattle Co. was first talked of in the summer of 1882, and was formally organized in Edinburgh, Scotland, in the spring of 1883. Mr. A. H. Swan, accompanied by Mr. John Donnelly, now of the Sioux City Stock Yards, went over and floated the company. Mr. Colin J. Mackenzie, of Portmore, was elected chairman and had under him a strong board of directors. Mr. Finlay Dun was made secretary. The basis of the company was the Swan & Frank Co. holdings. Various other properties were acquired, notably those of H. B. Kelly and E. W. Whitcomb. The headquarters were established at Chugwater, Wyo., and there they have remained ever since. The authorized capital of the company was $4,500,000, but only about $3,250,000 was actually subscribed and put in use. The capital today is $1,250,000, thus showing a loss of $2,000,000. The assets, however, are more valuable than the present capital.

Swan was a "plunger" always, and rising upon what seemed to outsiders a wave of success he embarked in many enterprises, controlling the Two Bar, Double O, Horse Creek, Kingman and other properties, and borrowing money wherever he could. At the inception of the company Swan had as his assistant in the management Zack Thomasson, a very able man, but he left to join the Ogallala Land & Cattle Co. in Nebraska. Mr. Thomasson remained only a short time with that outfit, sold out, and invested his money in real estate.

THE OLD SWAN FARM HOUSE AT INDIANOLA, IOWA.

The seasons of 1883 and 1884 were prosperous ones on the range, and this company, as well as other properties controlled by Swan, showed excellent profits, but a decline of prices came in 1885 and 1886. The summer of 1886 was a dry one. The ranges were overloaded, cattle were thin, and values declined severely. Prices for the Swan native steers in these early years ranged as follows: 1884, $47.06; 1885, $40.24; 1886, $30.15; 1887, $29.43; 1888, $35.24. The great bulk of these were strong in the Hereford blood, many of them being first crosses on Oregon cows.

The Wyoming Hereford Co.—This organization had no connection with the Swan company proper, except that some shareholders held stock in each. The ranch upon which the purebred Herefords were carried comprised some 30,000 acres on Crow Creek, just east of Cheyenne. George Morgan was engaged as the active manager and made a number of importations direct from Herefordshire, as has been referred to in a previous chapter. One of his early operations was the purchase of the entire herd of Mr. J. H. Yeomans of Stretton Court, comprising 200 head which were shipped from Liverpool on April 16, 1883. Another large importation was made in 1884, including 186 bulls bred in England. The herd numbered over 500 head at one time, including more than 300 breeding cows and a sensational array of stock bulls, among which were Rudolph by The Grove 3d, Lord Wilton 2d, Victor by Winter de Cote, and Sir Thomas of G. S. Burleigh's breeding.

From this herd large numbers of purebred bulls went out to spread the fame of the "white faces" throughout the northern range. All the more enterprising breeders of Montana and Wyoming had recourse to it, and while it made no money for its owners it placed within the reach of the cattlemen of the north blood that left its mark for many a year. Under different ownership the herd is still maintained.

First Herefords in the Panhandle.—It seems to be generally allowed that the credit for the revolutionizing of the blood of the Texas Panhandle herds along Hereford lines is largely due to Charles Goodnight, whose career as a scout and pioneer on the old frontier would supply material enough for a stirring volume on the development of the great southwest. He embraced cattle-breeding as a profession in 1856 in Palo Pinto Co., Tex., beginning with 430 head and handling them on shares until the Civil War. He early set about to improve them; the only way open at that time was through selection, but by this primitive means he succeeded in producing what was doubtless one of the best herds in Texas at that time. When the war came on he joined the Texas rangers, and served against the Comanches and Kiowas.

In 1886 he laid off the "Goodnight Trail," by way of the Pecos River through Colorado to Cheyenne, Wyo. He settled in 1870 near Pueblo, Colo. From here he removed to the Panhandle in the fall of 1876, establishing in what is known as the Palo Duro Canyon of the Red River the JA Ranch, with

1,600 graded Colorado cattle and seventy-five head of high-grade Shorthorns as the foundation herd. The latter, known as the JJ herd, were set aside as a breeding plant, and kept entirely distinct from the other herd were bred to purebred Shorthorn bulls. Mr. Goodnight's headquarters were in Armstrong county, but the range covered portions of Donley, Hall, Briscoe, Swisher and Randall counties.

Adair & Goodnight.—In 1877 John Adair, an Irishman of considerable wealth, while traveling in the United States met Charles Goodnight in Denver. The latter was at that date probably as familiar with the southwest as any white man then living, and he persuaded Adair to join him in the Palo Duro Canyon ranch proposition. A partnership was formed by the two men, in which Adair held a two-thirds interest and Goodnight the remaining one-third.

The country at that time was without railroads, settlers or cattle, and teemed with buffalo. The Comanches, who inhabited this country, had been rather thoroughly subdued the year before by the McKenzie expedition and removed to the reservation in Oklahoma, at that time Indian Territory. The partners, accompanied by Mrs. Adair, who was the eldest daughter of Major General James S. Wadsworth of Geneseo, N. Y., made their trips to and from the new ranch for hundreds of miles across country on horseback and with wagons, and on at least one occasion were escorted by a troop

COL. CHARLES GOODNIGHT.

of United States cavalry. Their efforts for the first few years were expended in acquiring the necessary land, herding back the buffalo and bringing in cattle with which to stock the new ranch. The lands were largely acquired from the firm of Gunther & Munson, who had "located" a large territory under the then very liberal land laws of the state of Texas. The greater part of the cattle, as above stated, were originally brought from Colorado and the north and were grade Shorthorns.

From the beginning the active management of the property was in Mr. Goodnight's hands, and the new firm soon began to buy cattle in large numbers. Out of the herds purchased Mr. Goodnight selected the best for breeding purposes, thereby starting what was known as the JJ herd. These he bred first to purebred Shorthorn, or "Durham" bulls, as he still prefers to call them, and as far as possible raised therefrom the bulls for the main range or JA herd.

O. H. Nelson Brings in Herefords.—In the spring of 1883 Mr. O. H. Nelson, representing the firm of Finch, Lord & Nelson, cattle dealers of Burlingame, Kans., bought in Kansas, Iowa and Missouri between 500 and 600 head of as good young breeding cows as could be secured without buying registered animals. He brought them into the Panhandle of Texas, locating them on a part of the Adair & Goodnight range south of Red River on Tule Creek. This was in Swisher county, near where the thrifty town of Tulia is now located. At that date this country was of course still unfenced and practically unoccupied. This herd consisted mostly of Shorthorns, but a few were one-half and three-fourths blood Herefords. The bulls, some twenty in number, were all good registered Herefords that had cost from $300 to $600 per head. Mr. Nelson reached the range with this herd about June 1, 1883, having been on the trail from Dodge City for six weeks. About August 15 of this same year he sold the cows to Mr. Goodnight for $75 per head, counting calves; that is, each cow and calf brought $150, the dry cows and heifers $75, and the bulls were turned over at $250 per head.

This good lot of cattle was turned in with the JJ herd, and the Shorthorn bulls were all taken out and replaced by registered Herefords. These were the first Herefords brought into the Panhandle, excepting a few that Nelson had taken down in the spring of 1882.

The Price Importation.—In the summer of 1883 Finch, Lord & Nelson arranged with J. R. Price & Son of Williamsville, Ill., to make a joint importation of Herefords from England, and sent "Ned" Price over to locate them and see what they would cost. Mr. Nelson was to have joined him later, but owing to press of business did not go, so that Price did the buying and importing. This lot numbered about eighty bulls and twenty cows. Out of this importation there were sold to Adair & Goodnight and delivered at Wichita Falls, Tex., in March, 1884, forty bulls at $400 per head. Finch, Lord &

Nelson took a part of the imported cows as well as some of the bulls to their herd at Burlingame, Kans. Speaking of this purchase Mr. Goodnight in a recent letter to the author says: "Taking them as a whole, they were the best lot of imported cattle I have ever seen."

Mrs. Adair Acquires the Property.—Shortly after this extensive introduction of Hereford blood Mr. Adair died at St. Louis, Mo., in 1885, while on his way out to the ranch from Ireland, and his large interest passed to his wife, Mrs. Cornelia Adair. Two years later the partnership was dissolved, Mr. Goodnight receiving for his interest practically one-third of the land and cattle. The remaining two-thirds has been known ever since as the JA Ranch and is still owned by Mrs. Adair, who though residing in London, England, makes frequent trips to the property in the Panhandle. At the time of Mr. Adair's death the partners owned or controlled for grazing purposes upwards of 1,000,000 acres, and their herd of cattle numbered more than 40,000.

Since this change of blood from Shorthorn to Hereford about 1883 purebred "white faces" have been used continuously on the main or JA herd. Coincident with this change the partners began building up the special JJ herd, resting largely upon the base of the well bred cows bought from Finch, Lord & Nelson. This herd has been crossed exclusively by registered Hereford bulls ever since, and has been the main source of supply for bulls for service on the JA's. To avoid too close breed-

MRS. C. ADAIR.

ing additional bulls are from time to time introduced from good herds in various states.

Mr. Goodnight brought his share of the JJ's on dissolution to his present home in Armstrong county, branding them +JJ. He bred them up to a high standard, selling them in 1896 to C. C. Slaughter.*

Richard Walsh, Manager.—Mr. John Farrington managed the Palo Duro property from 1887 to 1890. Mr. Arthur J. Tisdall was manager for one year, 1891. He was succeeded by Mr. Richard Walsh, who for eighteen years conducted the business of the ranch with the greatest success and became one of the best known and best liked cattlemen in the southwest. He resigned his position in 1910, spent a year in southern Brazil in company with Mr. Murdo Mackenzie, the former Matador manager, and is now managing an immense newly established ranch in Rhodesia, owned and controlled by the British South Africa Chartered Co. The Palo Duro management at present is in the hands of J. W. Wadsworth, Jr., who has held the position for the last four years. The property now comprises 500,000 acres, completely fenced and cross-fenced into convenient pastures. From 1892 to 1910 eminently successful efforts were made by Mr. Walsh to concentrate the property in a solid block. This was accomplished slowly and surely by exchanging lands on the perimeter for those state school lands

---

*Charles Goodnight at this date (1914) is still living and is breeding buffalo and a cross between the bison and the cow which he calls "cattalo". He is successfully farming some 1,200 acres of his ranch, and as always doing all in his power for the upbuilding of the country he knows and loves so well.

THE PALODURO RANCH HOUSE.

Copyright photo by Edward E. Smith

VIEW ON THE SPUR RANCH, AFTERWARDS TAKEN OVER BY THE SWENSONS.

within the range which had been entered upon by settlers in great number during the '90's. This difficult and at times delicate task, extending over many years, was accomplished by Mr. Walsh without incurring ill feeling or serious controversy of any kind, which speaks volumes for his fairness and diplomacy. The solidification is complete, and today there are no "strays" inside the JA fence.

In recent years particular efforts have been put forth in the way of permanent improvements, particularly as to watering facilities. This work is now nearly complete. The JA herd continues to maintain the high standard set in 1901 when its carload lot of steers was awarded the championship at the Chicago Fat Stock Show. And in 1904 when its steers were awarded the grand championship at the St. Louis exposition the Hereford had come into his own on this property. There he thrives, there the management believes that he surpasses all other breeds, and there he will doubtless remain.

Big Demand from Texas.—Finch, Lord & Nelson did a big trade in bulls for the Panhandle herds during the years 1881 to 1888 inclusive, sending into that country during that period no less than 10,000 head. In 1881 the bulls were all Shorthorns, and so they were mostly in 1882, but from that year the proportion of Herefords increased rapidly. Most of these at first were one-half- and three-fourth-bloods, but from 1883 on the firm each year bought registered bulls for their own use and for Adair & Goodnight as well as a few other customers.

In the spring of 1884 Nelson bought about 500 head of the best unregistered cows available in Kansas and Missouri. In this purchase over one-half were grade Herefords, the others being Shorthorns. This herd was put on a ranch in Hall Co., Tex., and established the subsequently well known "Bar Ninety-Six" brand. In a few years this became a very fine high-grade herd of "white faces," and for several years afterwards whenever a "white-faced critter" was seen in that region one did not have to look at the brand to determine ownership, as there were no others in the country.*

Bulls destined for the Texas trade of this period were commonly assembled at Dodge City, Kans., and then driven down the trail. The distribution commenced on the Canadian River, then at Mobeetie, then at Clarendon, and thence as far south as Colorado City—about 600 miles from Dodge City. Finch, Lord & Nelson sold to several large ranches as many as 500 head a year each for several years in succession. These included Adair & Goodnight, the Matador Land & Cattle Co., and the Espuula Land & Cattle Co. They also had many customers taking a smaller number, including W. H. Creswell, the Clarendon Land & Investment Co., which owned the "Quarter Circle Heart" ranch, Nick Eaton of the U—U, Day & Maddox of the YJ, Lee & Reynolds, Lee & Scott, the Hansford Land & Cattle Co., Coleman & Co., Robert Moody, and others.

---

*Mr. Nelson withdrew from the Burlingame firm some years ago, and is now breeding Herefords on his ranch near Romero, in Hartley Co., Tex. He handles many bulls, bred in the cornbelt states, as well as those bred in the Panhandle of Texas.

The Prairie Cattle Co.—This corporation began operations by buying in 1880 and 1881 the herds on three different ranges, with considerable bodies of watered lands in each case. One of the first purchases was that known as the JJ herd from the Jones Bros. This herd ranged in southeastern Colorado, from the Arkansas River down to the neutral strip, now in Oklahoma, and should not be confused with the JJ herd of Adair-Goodnight origin. The herd known as the Crosselle was purchased from Hall Bros., whose cattle ranged from the top of the Dry Cimarron down as far south as the Canadian River. The herd known as the LIT, purchased from Littlefield, ranged in the northwest corner of the Panhandle of Texas, with headquarters at Tascosa on the Canadian River. At that time the country was unfenced, and while these cattle were run in separate divisions during some winters they drifted so far as to occasionally overlap one another. But they were always brought back to their respective ranges in the spring.

When the Joneses and the Halls started their herds they had unlimited range with abundance of grass. The buffalo were about gone and the cattlemen were just beginning to realize what a splendid thing it was to have unlimited free grass and water. The range was lightly stocked, the cattle were not disturbed, and the result was that they did well and their owners prospered. It has been claimed that when the Halls originally went to the Crosselle, a herd of 1,500 head was turned loose in the fall of the year at the head of the Dry Cimarron in Colfax Co., N. M., and in the spring every one of these cattle was found in good condition within 15 miles from the spot where they were turned loose. This seems an almost incredible statement, and yet even if approximately true demonstrates what a splendid cattle country that region was at that date, in respect to feed, shelter and water. Cattle on the open ranges of course drifted great distances in time of storms when there was lack of natural shelter.

These herds were all started with Texas cows driven up from southern and central Texas. The Jones brothers were probably among the first in their country in the early '70's to improve their herds by turning loose pedigree Shorthorn bulls. The Halls a little later did the same. Shortly afterwards the Herefords began to attract attention. But good Herefords were difficult to procure before 1880 and commanded high prices, the result being that thousands of grade Hereford bulls were turned on the range, many of them of inferior quality. There was a keen demand from all parts of the range between 1878 and 1883 for white-faced bulls, and as late as 1884 a good Shorthorn bull without pedigree sold for $50 as a yearling, while a white-faced yearling would bring $75 and often prove a very inferior animal at that.

The Halls had purchased a few Herefords, but very few before they turned over their property. Probably the first large bunch of Herefords bought for this herd was that purchased in 1886 by W. J.

Tod, who was manager for the company from 1885 to 1889. These bulls were turned loose on the LIT range, where the cows were practically all a good variety of Texan. These, although only grade Hereford bulls, were well bred and made a marked impression for the better in the herd. Since then the Prairie Cattle Co. has bought almost exclusively purebred Herefords.

During those years the Prairie company was branding from their three ranges over 20,000 calves a year. In the early '90's and for years before there was a great influx of immigration into southeastern Colorado, and before this time the range was becoming seriously overstocked. The company found that in northern New Mexico and southern Colorado without fences or any control of the range it was unprofitable to run a cow herd. The Prairie people therefore removed all their herds from New Mexico, sold their water rights there, and managed the southern Colorado range entirely as a steer proposition, though still retaining a breeding herd in Texas, where they own the land, and in this way the property is managed today.

The Prairie Cattle Co. owned until very recently 215,000 acres of fenced land in northern Texas, the pastures varying in size from a few sections to 6,000 acres. This ranch carried about 10,000 cattle in the breeding herd, upon which only pedigree Hereford bulls were used, experience having convinced the management that the Shorthorn was unsuitable for the rough conditions the cattle had to undergo. In

Colorado the company owns 32,000 acres of land, scattered over the range, solely with a view of keeping the water open. Up to and before 1886 the company had a small herd of Aberdeen-Angus cattle ranging on parts of the New Mexico range, but it was found that the calf crop was usually disappointing and the herd was closed out. The company started with a capital of $3,000,000, half of which was fully paid up and the remainder debentures. After the dull times and low prices at the end of the '80's and the beginning of the '90's, the capital was reduced to about half of this sum, approximately where it stands today. The company probably owns at this writing about 38,000 to 40,000 head of cattle, principally steers.

Mr. Murdo Mackenzie managed this property for a short time after Mr. Tod left, and in 1889 was succeeded by James C. Johnstone, who held the position until 1906 when he returned to Scotland. Speaking of the use of Herefords on this herd Mr. Johnstone in writing to the author from Edinburg in May, 1914, said:

"During the years I managed the company I purchased for the herd many hundreds of purebred bulls, all Herefords, for I found that they were better than any other breed for range purposes. I bought my Hereford bulls principally in Missouri, Kansas and Illinois, and for two or three years bought all the bulls Mr. Kirk Armour bred on his farm at Excelsior Springs. I remember at one of the big sales of bulls in Kansas City I was passing the auctioneer who was selling a bull which was knocked down to a customer at $500. Kirk Armour happened to sight me, and called out to the auctioneer, 'There is a man who has got as well bred bulls as that running by the hundred on his company's ranch in Texas, and I have seen him buy in the times when he was getting them for from $50 to $60 per head.'

"This method of breeding made a very fine showing, and I left a fine white-faced herd of some 10,000 head on the Romero Ranch in the vicinity of Channing, Tex., when I quit the company's service."

Mr. H. Glazbrook, the present manager, in response to an inquiry as to his experience with the Hereford blood says:

"Since 1903, when I first became connected with this company, I have had considerable experience with Hereford cattle, both on the open range in this state and in the pastures of our Texas ranch. Previous to that, in fact since 1878, I had been engaged in the cattle business, mostly in Texas. During those early years we had little but the old longhorn cattle—now practically extinct in that state—and no fences. Not much effort was made to improve the class of stock there until the advent of the barb wire fence, at least not in the vicinity where I was ranching, and I think this applies generally to the whole of the state. When attention was given to improving the breed it was approached chiefly through Shorthorns, and there can be no doubt that this blood greatly improved the herds, though it might possibly be said that any good blood would have done so. I do not remember when the Hereford first made its appearance in Texas and I cannot remember the first Hereford I saw in that state, but when it did come it came to stay.

"The Hereford is in my opinion best adapted for range purposes, his hardy constitution and 'rustling' qualities being great assets. (By the latter expression I mean his ability to take care of himself.) I never see a Shorthorn on the range without thinking of the Scotchman, who being partial to the Hereford for this business, on being asked if he did not admire some range Shorthorns, remarked, 'Ay mon, they are mighty good cattle for hame', meaning of course for the barn or some place where they could be taken good care of. Our Colorado range is given up entirely to grazing steer cattle, and when purchasing I always endeavor to obtain herds showing strong Hereford breeding. On the Texas range we raise our own bulls from a purebred herd kept for that purpose, though we also buy some. The very best of Hereford bulls are purchased for the purebred herd. We have used nothing but purebred Herefords with our herd for about twenty years, during which time it has not been crossed with other blood. I believe that what has been said about the Hereford deteriorating if bred in line too long is attributable to adverse conditions of the range, and not to the breed. Until recently our cattle received no feed except the natural grasses. Lately, however, we have fed to some extent during the winters."

Conrad Kohrs.—The "grand old man" of Montana, President of the Pioneer Cattle Co., and one of the pillars of northwestern progress and prosperity, Conrad Kohrs, was one of those who availed himself of the opportunity to test out the Hereford blood by purchases from the Swan-Morgan herd. Seven head comprised his original selection at Cheyenne, and while he has always been a staunch supporter of the Shorthorn he has adhered to the Hereford cross ever since it was first used. He has never been prejudiced as between the different

CONRAD KOHRS AT SEVENTY-NINE, AND HIS GRANDSON
CONRAD KOHRS WARREN.

breeds, and has made repeated experiments to determine which would give the best results on the range. He has not only used the Shorthorn at all times, but has tried the Aberdeen-Angus. In his early experience he accumulated on the Sun River Range one of the best herds of non-pedigree Shorthorns in the west. These were descended from good cattle that had been picked up originally in the early days along the old California and Oregon Trail. They were maintained in the Deer Lodge Valley. When the pastures got short in Deer Lodge, he was obliged to move them into the Sun River country. As early as 1879 he branded 4,900 calves on the Sun River Ranch.

The Hereford bulls bought from Swan and Morgan were sent into this herd and the best bull calves produced were kept for breeding purposes. The steers from the first cross gave great satisfaction. In the early days, when cattle were few in Montana and grass abundant, Mr. Kohrs preferred the Shorthorns among these crosses because he found that they would weigh more at four years old than the Herefords. But in those days there were no railroads and the cattle had to be driven a great distance to Laramie City or Cheyenne on the Union Pacific; this put them in bad condition and they never brought a satisfactory price in Chicago, because they were too large for feeders and not fat enough for good beef. When the Northern Pacific was built Mr. Kohrs moved a lot of cattle to Tongue River, about 150 miles south of Miles City, and his first shipment over that line in 1882, consisting of 400 four-year-old steers, was made from that point. As the railroad facilities at that time were not very good the cattle were a long time on the road, but with a heavy shrink they weighed 1,585 pounds in Chicago, and brought the top price at that time for range cattle—$5.85. He shipped 700 three-year-old steers that same year which weighed in Chicago 1,365 pounds and which also sold at $5.85. So it is clear that the herd at that time was a good one.

Herefords Good Travelers.—Mr. Kohrs says:

"I prefer the Herefords on the range because they are great rustlers. They are better on their feet than the Shorthorns and as the grass has grown scarcer and water more inaccessible the cattle have to travel farther than formerly, and we find that the Herefords keep in better condition than the Shorthorns and go through the winter better because they will always hunt for grass when there is any to be had."

A number of years ago Mr. Kohrs bought the purebred Hereford herd of the Childs estate. This was a good lot derived largely from the stock of Adams Earl. The pedigrees were not obtained on account of a dispute between the herdsman and the administrator, so the cattle have been bred as a non-pedigree herd, although registered bulls have been constantly maintained in service.* "Since we

---

*Associated for many years in the management of the Pioneer Cattle Co. has been Mr. Kohrs' son-in-law, Hon. John M. Boardman of Helena, the present general manager of the CK ranch. The author feels certain that western cattlemen in general will be particularly interested in the portrait of Mr. Kohrs appearing elsewhere in this volume. It is a recent one, taken with his grandson, who will probably follow in the footsteps of his fathers. Mr. Kohrs recently celebrated his seventy-ninth birthday at Deer Lodge.

have had the Childs' herd", says Mr. Kohrs, "I have found that the crossbred makes a magnificent steer, even better in the first cross than either the Shorthorn or the Hereford. Our Hereford herd at present numbers about 300 head, while our purebred Shorthorn herd numbers about 700 head. Still, today our demand for Shorthorn bulls is greater than for the Herefords. Many small breeders are coming in. They have pastures and take care of their stock in the winter time, and they prefer the Shorthorn bull. As far as we are concerned, with regard to the cattle we have on the range, we have for the past four years used nothing but the Hereford bull. I have found that those who have used grade bulls instead of purebreds in building up range herds were disappointed. Strong-blooded bulls only should be used."

Asked by the author as to his experience in crossing Shorthorn bulls on Hereford cows, Mr. Kohrs replied:

"I do not believe there is anyone in the state who has to any extent tried that cross, because the Hereford cow has never been plentiful enough in our state and therefore there were not enough to make it worth while to experiment. The only thing I can say so far as breeding the Shorthorn on the Hereford is concerned is that the herds we have received from Texas, where it was claimed that the Shorthorn bull had been used on the Hereford cows, have never been good lots. I presume this is largely on account of the fact that good Shorthorn bulls have not been used. I know that some of the Texas breeders have made a great success in that line, for

Copyright photo by Erwin E. Smith

A GOOD CATCH—MATADOR RANGE.

Copyright photo by Erwin E. Smith

HOBBLING AN OUTLAW.

instance, Mr. Burnett of Fort Worth. His herd, bred in that way, certainly is a very fine one, and has been brough to that point through careful breeding. On the other hand, I have had lots of experience in breeding the Hereford bull on the Shorthorn cow, and I like the result.''

Joseph Scott.—Another leader in the early line of progress through the use of white-faced bulls upon the open range was Joseph Scott. Born in Ireland from Scotch parentage, a man of enterprise, high intelligence and thoroughly upright in all his dealings, he operated largely in Montana, and later at Halleck, Nev. He first came into prominence as a member of the firm of Scott & Hank, whose old address was Mandel, Wyo. They ranged on the Tongue and Little Powder rivers, their brand being S-H. Joe Scott was not only one of the most expert cattlemen ever identified with western ranching, but he was progressive, and early devoted his attention to Herefords, more especially in the Nevada herd. He was a customer of Mr. C. M. Culbertson and others of the pioneer importers from Herefordshire. He also imported cattle direct from England for the Montana ranch about 1880, and in connection with George Leigh of Aurora, Ill., imported 120 head in 1897, about forty head of which went to the Nevada ranch.

Mr. Scott had a long, eventful and honorable career. He was for several terms President of the Montana Stock Growers' Association, and devoted a great deal of his time to that work. For many years he made his home at Miles City, and from there he went to Spokane, Wash. He underwent all the vicissitudes and ups and downs of the cattle business, and in his later years often said: "I was a millionaire before the winter of 1886-87, and a pauper afterwards." Eventually, however, he left quite an estate. He died and was buried in Italy, and is remembered by all his surviving friends as a man of broad sympathies—one who never tired helping his fellowmen.

B. C. Rhome.—One of the pioneers in purebred Herefords in north Texas was Mr. B. C. Rhome of Fort Worth. He began around 1882 by making purchases of William Powell, who as already stated was probably the first to engage in the trade of supplying purebred Herefords for the Texas range. Along about 1880 Mr. Powell sold quite a number of bulls to various range cattle breeders, many of which went into southern Texas in the region around San Antonio. Mr. Rhome states that shortly after he made his first purchases Mr. G. H. Mathis and G. P. McCampbell of Rockford also bought cattle from Fowler & VanNatta and the T. L. Miller Co. W. S. Ikard of Henrietta began a purebred herd about this time, making selections from the herds about Beecher. According to Mr. Rhome, another early Texas herd was that of F. M. Houts of Decatur, founded upon purchases from Fowler & VanNatta, which included the imported Carwardine bull Wilfred. One of the most important introductions of Herefords into Texas during the late '80's was that of 200 head shipped in by Mr. G. W. Henry of Chicago. They were placed on sale on Mr. Rhome's ranch in charge of William Powell. A good many of these died of the fever, but this was nevertheless the source of a lot of good blood scattered throughout different parts of the Texas range country. In 1888 Mr. Rhome and Mr. Powell formed a partnership, buying some of the Henry cattle and adding to these a lot belonging to Mr. Powell brought in from Beecher. They bought a son of old Fowler and two bulls from Thomas Clark for breeding purposes. In 1890 Rhome & Powell bought the F. M. Houts herd numbering about 50 head. The firm at this time owned about 200 head. On the dissolution of the partnership Mr. Powell established his headquarters at Channing, Tex., at which place he is at this writing still living.

Reynolds Cattle Co.—This is another one of the big Texas cattle companies. It has holdings at the present time of an estimated value of about $2,500,000. Its operations go back to the very beginnings of cattle ranching in the southwest. This company had its first Hereford bulls from T. L. Miller around 1876, the cows at the time being mainly of the ordinary north Texas type. The Reynolds people were among the first to take the Hereford blood into Texas. The company now has about 130,000 acres of broken, hilly, but well watered land in Shackelford and Throckmorton counties on which about 8,000 cattle have been maintained in recent years. It also has 300,000 acres owned and leased in Jeff Davis county, carrying about 12,000 cattle. The company has in previous years, however, run as high as 50,000 cattle at one time.

Mr. W. D. Reynolds of this company states that they regard the Hereford as the best cattle for range use, particularly on short feed in a drouthy country. Their early purchases of Herefords from the north turned out badly on account of lack of knowledge concerning Texas fever and its causes at that time. They have bought persistently, however, from various breeders, besides producing large numbers of bulls from their own herds. They have at different times used Shorthorn bulls, and in the recent past have introduced a few Aberdeen-Angus and Galloways. Like most of their contemporaries they aim to run about four bulls to 100 cows. Bulls of their own breeding are turned in as yearlings, but when purchased from the outside are usually two-year-olds.

The Hereford in California.—Beyond the Sierra Nevada Mountains climatic conditions as relating to cattle-ranching differ materially from those prevailing in the Rocky Mountain region proper. There is, except perhaps in the extreme southern portion of the state of California, a much greater annual rainfall than occurs on this side of the coast ranges, giving a larger supply of succulent feed. The climate is also milder and more equable. Due perhaps to the fact that the natural conditions do not make such severe demands upon range herds, the Hereford has not come into so commanding a position on the Pacific Coast as throughout the more arid regions of the western states. The earliest improved cattle on the coast were undoubtedly Shorthorns, and at one time Devon bulls were quite in evidence.

Not long after the Herefords began attaining popularity all along the line from Texas to Montana, the white-faced blood was introduced into both Oregon and California and subsequently became popular with those who were handling cattle in large numbers. Particularly was this the case with companies running cattle on the ranges of New Mexico, Nevada or Arizona, as well as in the state of California.

An Importation from Australia.—Soon after the demand for Herefords set in among the cattle-owners of the coast it developed that owing to the high price of good purebred "white faces" and the high railway freights in the transportation of such animals Mr. Roland P. Saxe of San Francisco made two importations, comprising fifty-six head, mostly cows and heifers, from Australia. A part of these were for Capt. William Kohl and T. J. Janes of San Mateo. It was found that these cattle could at that time be landed in California cheaper than from the eastern part of the United States. We are without information as to the exact sources from which these cattle were obtained or as to their use, but it is a matter of record that the shipments were made and that forty of the fifty-six cattle so imported were from New Zealand. A portion of these shipments came from New South Wales.

Mr. Jastro's Experience.—No one has been more

active in the introduction and dissemination of Hereford blood in California than Mr. H. A. Jastro of Bakersfield, general superintendent for the Kern County Land Co. Mr. Jastro has maintained since some time in the '80's a herd of purebred Hereford cows, which now numbers probably 500 head and runs on this company's Stockdale range. His aim has been to produce bulls of good quality under natural conditions for use on the extensive Arizona and New Mexico ranches operated by his company. The size of the herd, as well as the purpose for which it is maintained, makes it impractical as well as unnecessary to maintain registration for the cattle produced.

Inasmuch as something like 20,000 calves are branded yearly in connection with the operations of his company, and in view of the long and successful experience of Mr. Jastro with cattle in the southwest his endorsement of the Hereford for range purposes must carry weight. He testifies that they are hardier, will travel farther for water and keep in better condition on short feed than Shorthorns or any other breed except the Devons. "In fact," says Mr. Jastro, "it is my judgment that by carefully supplying the range with bulls of the right stamp the Hereford is really the only breed for range purposes where water is scarce and feed at certain seasons of the year is short. On the other hand, the rancher who has lots of feed and plenty of water will in my judgment find the Shorthorns more profitable."

Mr. Jastro states that while he thinks there is a tendency towards some loss of size where the Hereford cross is repeated during a succession of years his remedy for this is to turn out Shorthorn bulls every third or fourth year, so that cows will get back to their original size. He adds: "Our best success is in crossing a Hereford bull with cows well bred up in Shorthorn blood."

Mr. Mackey, who was at one time manager of the Rancho Del Paso, commonly known as the "Haggin Grant," and famous for its Thoroughbred horses, at one time introduced Hereford cattle which later found their way to Bakersfield and were merged into the Stockdale herd when the grant was sold.

The Chowchilla Herd.—In April, 1882, Mr. John Clay, of Clay, Robinson & Co., purchased the Chowchilla Ranch, comprising 116,000 acres of land and the herd of 12,000 head of cattle, for a Scotch syndicate. Mr. Clay had visited California some years prior to that date. He states that at the time of his first visit the cattle of California were still strong in Spanish blood—a big, bony, stretchy lot, much heavier than Texas cattle of the same ages. These responded rapidly to the Shorthorn cross. Miller & Lux had brought down to their ranches a lot of Oregon cows which were coarse-boned, heavily-built Shorthorn types upon which they had used Devon bulls in the hope of acquiring more quality. About this time Mr. Clay bought two carloads of bulls by Devon sires and from ordinary Oregon cows from this firm. They were put into the "Sev-

enty-one Quarter Circle" herd on the Sweetwater, but the results of their work were swept away in the winter of 1886-87 so that no satisfactory estimate as to the value of the cross could be made.

The first move of the Chowchilla company was the purchase in Illinois of a number of good Shorthorn bulls and heifers for the purpose of replenishing the then limited number of purebred cattle on the ranch and of increasing the herd. The year following Mr. Clay shipped several carloads of extra Shorthorn heifers from the east, which supplied a good foundation for a registered herd.

In the fall of 1884 Mr. Isaac Bird was given full management of the business by Mr. Clay and from that time on the greatest care was given to the breeding of the cattle. In a few years a Shorthorn herd of high quality was established. In due time the blood began to show in the range herd, as all the bulls were purebreds. None but Shorthorns were used until the year 1898, when Mr. Bird purchased several loads of Hereford cows and bulls; and it was at this point that the "hit" of the company's career was made. The crosses by Hereford bulls on Shorthorn cows were extra good, developing into A-1 stock which was always in good flesh, while at times the other cattle on the range were thin. Speaking of the success of this cross Mr. Bird says: "My thirty years' experience in the cattle business has taught me that the best steers I ever raised were those bred from Hereford bulls on Shorthorn cows, the Shorthorn giving them large bone and square rumps."

COWS AND CALVES OF THE KERN CO. (CAL.) LAND COMPANY'S HERD. H. A. JASTRO, SUPERINTENDENT.

In December, 1900, the Chowchilla outfit exhibited at the Chicago International one carload of Shorthorn calves on which the blue ribbon for the southern district was secured. These calves averaged 600 pounds and were purchased by Mr. Judy, of Menard Co., Ill., at $7 per cwt. He in turn fed sixteen head of them for the International of the following year and on these was awarded first prize for fed yearlings. They averaged at that time 1,260 pounds. At this same show the company exhibited one carload of Shorthorn calves and a carload of half-bred Herefords and drew blue ribbons again.

In 1905 the Chowchilla people sold their beef steers to the Western Meat Co. (Swift & Co.), and the tops proving too fat for the San Francisco market two trains of twenty cars each of these three-year-olds were shipped to the Chicago stock yards on the 15th and 20th of April. The first trainload averaged 1,280 pounds and the second shipment 1,200 pounds, with an average shrinkage of 90 pounds per head in transit. Mr. Charles Robinson (of Clay, Robinson & Co.) wrote at the time in regard to these cattle that no one in the yards could believe that they were grass cattle, owing to their being so fat and such early beef. At least 90 per cent of these steers were half-bred Herefords. Owing to the high price of land the Chowchilla Ranch was sold in 1911 to a colonization company, and the remnant of the herd was sold to Miller & Lux in 1912. The cow had to give way to the farmer.

Mr. Henry Miller, of Miller & Lux, was never par-

Copyright photo by Erwin E. Smith
A RANGE BOSS ON THE O. R. RANCH, ARIZONA.

Copyright photo by Erwin E. Smith
THE MARK OF THE "WHITE-FACE." VIEW ON O. R. RANCH, ARIZONA.

ticularly favorable to Herefords, and often said, "A red Shorthorn is good enough for me." Mr. Bird says: "Herefords do as well if not better than any breed we have ever handled, and were I to raise cattle again I would always be partial to the 'white faces.' At this time there is scarcely a herd in California in which you do not see some Herefords, and I feel that within a few years the number will be greatly increased. I like them, for I know what they did for us. And I may state that our herd of 18,000 cattle was considered in its day the best large herd west of the Rockies."

CHAPTER XVIII.

THE RED ROBE OF COURAGE.

The winter of 1886-87 was one of the worst ever experienced on the western ranges. It brought widespread disaster and an almost complete collapse of the cattle business as then conducted on the open range. The result of this great calamity, which brought ruin to many leading operators, was an increased demand for Hereford bulls. Heavy losses had occurred in different localities during preceding years, and in most cases the comparison as to the relative hardiness of the different breeds had from the beginning been altogether favorable to the "white faces." When the supreme test came in the winter of 1886-87, while the Herefords themselves suffered considerable losses, the general consensus of opinion was that they had stood the test in a manner which demonstrated that they were better qualified to endure privations than any other known type. Not only did the range men turn more generally to the use of Hereford bulls, but what was equally important, they began making better provision for the future maintenance of their herds. There was less overcrowding of the ranges, more businesslike methods of administration generally, and in the end better results than had been attained during the wild period of speculation that had previously prevailed.

Hereford Hardiness Hereditary.—The breed that passed through this harsh experience so successfully presents an interesting study in heredity—the persistent transmission of ancestral qualities, even after the lapse of generations. The Hereford of old Herefordshire, the Hereford of a century and a half ago, was bred for the yoke. He was not reared in the lap of luxury. He was not pampered. His was a life of plain living and heavy hauling. No corn and little cake entered into his rations. He tilled the fields of his owner, subsisted mainly on grass, and often worked hard till more than ten years old. Beef-making as a business prior to the time of Tomkins the Younger did not enter specially into the calculation. What this did for the Herefordshire cattle may be read today in those heavy shoulders and broad chests, those legs and muscles that enable them to tramp the range and win their way through storm and stress and drouth and heat and cold, traversing distances that are hopeless to most cattle of other improved breeds, and through it all maintaining fair condition.*

---

*Will C. Barnes, author of "Western Grazing Grounds," speaking of the superior hardiness of the Hereford says: "Range cattle with considerable infusions of Shorthorn blood are never quite so hardy as the old stock, and in the early spring when the heel flies are about they seem to delight in finding the worst bog holes. Once down they often lose all pluck and grit, and where a Hereford would fight her way out to hard ground the cow with the infusion of Shorthorn blood is apt to give up after the first struggle. Even when dragged out by the bog rider she may make no effort to get to her feet, but will lie there and starve, losing herself and her calf to the owner. On the other hand, the longhorn or Hereford when thus dragged out will, if she has a single spark of life left in her, get to her feet some way and chase her rescuer off the range."

Swan's Failure.—A. H. Swan personally went down in 1887, but the Swan company, which was financed in Scotland, survived. No attempt had ever been made to count the cattle until the summer of 1887, when Mr. Finlay Dun, with his famous paint pot and brush, attempted to tally the herd. Summer showers and other causes soon disposed of the patches of paint and the work was given up, as it was found that cattle were being tallied twice over. The company had, however, purchased a lot of land. The spring of 1887 found the corporation possessed of a herd of cattle estimated at 50,000 head, possibly less. In addition the company had about 576,000 acres of land. A large part of this was intermediate sections of railroad land on the Laramie Plains—a high and exposed plateau. Part of this was eventually allowed to go back to the railroad, the original owner. Today, we believe, the company owns about 270,000 acres of railroad land (intermediate sections) and 40,000 acres of land on the Chug, Sybille and Richeau, and in Goshen's Hole. Mr. Dun managed the property during the summer, fall and winter of 1887-1888.

Mr. John Clay was appointed manager of the Swan company on the first of March, 1888, and remained in charge until July, 1896. Mr. Al Bowie succeeded Mr. Clay, and after him came Mr. William Dawson, who resigned in 1912 and was succeeded by Mr. M. R. Johnstone. The company today, so far as actual management is concerned, lies in the hands of an executive committee composed of John Clay, James T. Craig and M. R. Johnson, all practical western men.

The Swan range in the old days extended from Ogallala, Neb., to Fort Steele, Wyo. In a general way the cattle ranged with many others over the whole territory north from the Union Pacific Railroad (taking the points named as the east and west limits) to the Platte River. In round numbers this was a terirtory 200 miles long and 100 miles wide. Gradually this range has been encroached upon. In 1910 most of the cattle were sold. The company had run quite a number of sheep previous to that time, and now it is largely a sheep proposition. In five years' time, at the present rate, the dry-farmers will take all the public lands in Goshen's Hole, on the Chug, Sybille and other streams. What thirty years ago was purely a grazing area in a region considered arid is now being taken up for farming purposes. Up to date the dry-farmer has not yet located on the higher altitudes of the Laramie Plains.

Al Bowie's Testimony.—Mr. Bowie, so long identified with the Swan company, has spent the best part of his life upon the Wyoming range, and is a willing witness in behalf of the value of the Herefords under conditions there prevailing. He says:

"When in 1884 we purchased a large lot of Shorthorns as well as several hundred Herefords we were feeding the Shorthorns all the hay they wanted. In fact, we had to in order to keep them alive, while the Herefords ran to pasture and kept

in much better condition than the Shorthorns did on hay. Since that time I have been much more in favor of the Herefords than Shorthorns. They are more quiet, have better coats of hair, stand the winter better and running on a poor range show a much less loss under same conditions. Furthermore, they cross well if you have good Shorthorn cows, as we had in 1880—cows that came in from Oregon and Washington. In fact, I have never since seen as good ones.

"In-breeding and poor feed have caused some loss of size and weight among the Herefords, at least that has been my experience. They are not naturally as big cattle as the Shorthorns, but they are more blocky and there will be fewer culls in a big herd. They naturally have shorter legs than the Shorthorns, and do much better on short, poor feed.

"The cattlemen and commission men of Chicago will condemn our Herefords in seasons when we have hard winters and poor feed in summers and cattle do not get fat. Then when we have the reverse in seasons and cattle get fat they think the Herefords all right. Where the Hereford shines is in a feedlot. As T. B. Hord used to say, 'Fat is a good color.' He also said, 'Give the Hereford one cross of blood and three crosses of corn and you have beef good enough for any one.' "

Robert Kleberg.—It is given to few men to be afforded the opportunity for such big constructive work as fell to Mr. Robert J. Kleberg, Capt. King's son-in-law, at the decease of the proprietor of the great Santa Gertrudis property in southeast Texas, to which reference has already been made. Few of those who fall heir to such opportunities rise to their full achievement. The development of

ROBERT J. KLEBERG.

the lands and cattle in the hands of Mr. Kleberg, however, constitutes one of the most important chapters in the latter-day history of the state of Texas. However, our story of the Hereford cattle is in itself too long for us to enter into great detail as to the modern history of the King ranch. We must therefore sketch rapidly.

The two half-million-acre ranches mentioned in a previous chapter were subdivided into numerous "small" pastures, ranging in size from 1,000 acres to 50,000 acres each. This was done for the purpose of carrying out certain clearly defined purposes in the introduction of purebreds. The underground rivers were tapped, artesian wells gushed forth their pure waters wherever wanted, the railway finally pierced the great principality, and towns and irrigated farms came into existence where once half-wild cattle and horses roamed the unfenced plains.

Mr. Kleberg was for a number of years a liberal buyer of registered Shorthorn cattle, purchased from the best herds of the middle west, particularly those of Kentucky and Missouri. These of course had to undergo the trying process of becoming acclimated, and losses were frequently so heavy as to be altogether discouraging. Mr. Kleberg was one of the first to undertake the risks attending the introduction of high-priced purebred bulls below the fever line, but his persistence and enterprise were finally rewarded. In due course of time discovery was made that the cattle tick was the cause

MRS. KING'S NEW FIREPROOF RESIDENCE, UNDER CONSTRUCTION AT SANTA GERTRUDIS.

A ROUND-UP OF OLD COWS OVER TEN YEARS OLD.
Cattle being sold and delivered on lower ranch of Mrs. King, Cameron Co., Texas.

of the so-called Texas fever. Indeed, it is asserted that the preliminary proofs in this most important discovery were first furnished at the Santa Gertrudis Ranch. The Department of Agriculture, then under the efficient direction of the late Secretary Rusk, took the matter actively in hand, and by a series of investigations established the truth of what had previously been a mere theory in regard to the mysterious origin of this southern plague. At length the process of immunizing the northern cattle against the disease was scientifically worked out, after which the cattle intended for breeding purposes could be shipped from above the fever line with comparative safety.

It was not until after the cattle on the King ranch had been well graded up with Shorthorn blood that the Hereford was introduced. Bulls, as well as heifers, sired by the most noted prize-winning bulls of the north were bought in large numbers, so that at the present time probably 90 per cent of all the cattle on the upper ranch are grade Hereford-Shorthorns, the others being either purebred Shorthorns or purebred Herefords. Of the latter there are now about 2,000 head, and of the former about 4,000, kept on the upper ranch for the purpose of breeding bulls for the main herd. The best are retained for this purpose and the remainder are sold to be used on ranches in southern Texas and Old Mexico.

Some years ago the Laureles ranch was acquired by the King estate from the Texas Land & Cattle

Co. This was a property of something over 300,000 acres which joined the Santa Gertrudis on the east of the upper ranch. Other lands were added until over 1,000,000 acres were included in the two ranches. Since the construction of the railway through the property, several hundred thousand acres have been subdivided and sold to farmers, and four towns with a population of from 1,000 to 5,000 inhabitants each are now located on what was formerly the ranch proper.

Mr. Kleberg joins with most of the other experienced range men in giving the palm to the Herefords in the matter of maintaining their condition under ordinary range conditions. Nevertheless, he is a great admirer of the Shorthorns, and as above stated still maintains them in large numbers. That they require rather better care than the Herefords, however, in order to secure the best results is freely admitted. It was from the King ranch that Mr. Murdo Mackenzie selected in 1912, 150 heifers and 650 bulls, mostly Herefords, for export to Brazil.

Capt. John Tod.—This veteran Scottish herd manager had his first experience with Herefords between 1883 and 1886 in Wabaunsee and Chase counties in Kansas. From 1887 to 1907 he had charge of the Laureles Ranch of 325,000 acres—the property of the Texas Land & Cattle Co. on the Gulf Coast in Nueces Co., Tex., since sold to the King estate. From 1887 to 1891 Capt. Tod also managed a ranch in the Panhandle.

During a period of some twenty years he handled

annually on an average probably 10,000 head of grade and purebred Herefords and 20,000 Shorthorns, the bulk of the cows being from a Texas foundation. He says that in his experience the Hereford bull is "far and away the best for range purposes." As a rustler he insists that the Hereford is "infinitely better than the Shorthorn, having more vitality", and he testifies that "when Hereford bulls and Shorthorn bulls are turned out in the spring in very large pastures with Shorthorn cows, the bulk of the early calves are 'white faces.'"

Capt. Tod states that though it is generally agreed that Hereford bulls have made their greatest success when the cow herds had previously been more or less improved by the use of Shorthorn blood, it should be explained that in the early days of grading up from common cows, while there were plenty of good Shorthorn bulls used, many of the Hereford bulls resorted to were little better than "scrubs", and were largely grades.

In regard to the assertion that the continued use of one Hereford cross after another upon a herd already well graded up with white-faced blood tends to an ultimate loss of size and weight, Capt. Tod says:

"My observation is that ranchmen have not been persistent enough, have got tired out too soon with the long effort necessary, have shut up the purse strings too tightly, and have not kept on purchasing better and better bulls. My experience is that if this is done the seven-eighths or fifteen-sixteenths grade Hereford cows do not show a loss of weight and size. The Herefords are generally preferred throughout the range country on account of their good constitutions, grazing qualities, prepotency and masculinity. The cows are good nurses, rear their calves as well or better than any other breed, and while doing so keep in better condition. Graziers and feeders have a preference for the steer from a Shorthorn cow, by a Hereford bull, and the nearer to a perfect 'white face' the better they like him.''

The Capitol Syndicate XIT Ranch.—This great ranch was situated in the northwest corner of the Panhandle of Texas, lying in the counties of Dallam, Hartley, Oldham, Deaf Smith, Palmer, Castro, Bailey, Lamb, Hockley and Cochran, and consisted of 3,000,000 acres. The state of Texas retained all of the land lying within its boundaries when it was admitted into the Union. To provide an adequate capitol building at Austin the legislature passed an act in 1879 appropriating 3,000,000 acres of land for disposition in that connection, the same to be selected from the unappropriated state lands by a commissioner appointed to locate none but agricultural or grazing lands. There were at that time 5,000,000 acres from which to make the selection. This was made in 1879 and 1880 and the land was surveyed into leagues, a league comprising 4,428 acres. These 3,000,000 acres lay in a tract averaging about 25 miles wide east and west by about 200 miles north and south, the west border being the line between Texas and New Mexico. The property was offered to a responsible party who would enter into a contract to erect a state capitol according to plans and specifications furnished by the state. This contract was ultimately assigned to a syndicate consisting of U. S. Senator C. B. Farwell, John V. Farwell, the well known wholesale dry goods man of Chicago, Col. Abner Taylor, who at one time represented in Congress the first district of Chicago, and Col. A. C. Babcock of Canton, Ill., all now deceased.

Description of the Property.—Generally speaking this tract of land is a level plain or plateau varying from 2,300 to 4,700 feet in altitude covered by a luxuriant growth of buffalo, mesquite, grama, bluestem, bunch, sage and other grasses. The soil varies from chocolate loam to red sandy loam, with subsoil of practically the same character under which lies a stratum of clay. The Canadian River traverses the tract in an easterly direction through Oldham county and tributary to it the land is rolling or gently undulating. There were a few springs on this tract of land and these and the waters of the Canadian were all that the buffalo and wild animals of the early days could depend upon. Many lake basins are to be found on the plains which contain water for some time after heavy rainfalls. These were entirely inadequate to supply water for the large herds that were put on this tract by the Capitol Syndicate, and it became necessary to bore wells, erect windmills over them and provide drinking troughs and reservoirs. The syndicate bored about 300 such wells which varied in depth from 10 to 400 feet and averaged about 125 feet. Dams were thrown across ravines or draws to conserve the rainfall. The watering facilities were developed in this way sufficiently to take care of 150,000 head of cattle. The water in the bored wells was "freestone" of good quality.

Character of the XIT Cattle.—The ranch was first stocked with cattle during the years of 1885, 1886 and 1887. These were cattle of very indifferent quality—some from near the Gulf of Mexico and some from the country lying tributary to and south of the Texas & Pacific Railway. The herd at its maximum size numbered about 150,000 head. There were seven divisions of the ranch and each of the seven divisions had its territory cut into a number of pastures by barbed wire fences. After the ranch had been separated into divisions, which in turn were divided into several pastures, the work of improving the herd was undertaken—some time about 1889. The sorting of the cattle with a view to following out different lines of breeding was a matter of several years. With such large numbers and such great distances this sorting necessarily took considerable time. But by degrees the cattle which in the judgment of the management seemed most suitable for mating with Hereford bulls were put in certain pastures, others which it was thought promised good results from Aberdeen-Angus crossing were placed in certain other pastures, and those that seemed most in need of the Shorthorn blood were quartered in still other pastures, so that an effort could be made to experiment and improve along the lines of these three distinct breeds.

The scarcity and high price of purebred bulls in

the late '80's and early '90's made their use in any large numbers impossible for this ranch. Many low-grade bulls had been put on the ranch in the late '80's, but each year a better class of sires was procured and after 1892 only purebred bulls were purchased. With cattle of such an ordinary foundation and with grade bulls improvement was necessarily slow. However, after the introduction of purebred bulls, as was the case on the Matador and other great Panhandle ranges, improvement was rapid, so that when the cattle began to be dispersed on account of sales of land in 1901 the herd for one of such large size was of exceptional quality. At the sale of the last cattle in 1912 it was, for all practical beef purposes, a purebred herd.

Purebred Bulls Purchased.—The Hereford bulls to work this improvement were purchased largely from William Powell, Channing, Tex., the Farwell Bros., Montezuma, Ia., and the T. L. Miller Co., Beecher, Ill. The Aberdeen-Angus bulls came from Farwell Bros., Montezuma, Ia., Anderson & Findlay, Lake Forest, Ill., George Farwell, Mt. Morris, Ill., and Arnold Bros., Hansford Co., Tex. The herds of John D. Gillett, Elkhart, Ill., and C. S. Barclay, West Liberty, Ia., furnished the Shorthorns. Besides these a goodly number of bulls were purchased each year from individuals who would undertake a contract of getting together a number of good ones from the leading herds of the various beef breeds of the country. In 1892 the company purchased from the T. L. Miller Co. forty-four bulls and

111 Hereford cows. Some years later it purchased from Mr. Cook, Odeboldt, Ia., a number of registered Hereford females, and in 1892 from Arnold Bros. fifty-five registered Aberdeen-Angus females and a large number of bulls, besides a number of Herefords and Aberdeen-Angus from the Farwell Bros.

With these purebred females of the Aberdeen-Angus and of the Hereford breeds small herds were established into which were introduced bulls from the best herds of the country for the purpose of raising sires for use on the company's ranch. In the later years almost enough bulls for its own use were supplied from this source.

Herefords Predominate.—While Mr. George Findlay of this company was closely identified with Aberdeen-Angus interests, the management made no claim as to superior adaptability for any of the three beef breeds. Mr. A. G. Boyce, the manager under whom most of the early improvements were effected, undertook the work at a time when he was strongly of the belief that the Shorthorns were the best cattle. After a number of years' experience with these breeds it is stated that he ended his career on the ranch in the belief that the Angus were in the first place and the Herefords second, but the manager who followed him, Mr. H. S. Boice, a ranchman of large experience and owner of large herds, was strongly in favor of the Herefords. Of course there were many things to consider in seeking an answer to a question of this kind on a ranch of this character.* Two herds might be separated 200 miles apart on such an enormous range, and one year climatic conditions might be more favorable for one herd than the other, and vice versa the next year. But it was the belief of those who had most to do with the property that either of these breeds properly looked after would do well in the Panhandle.

Views of H. S. Boice.—In this connection the following narration by H. S. Boice, formerly with the XIT syndicate, is of interest. In a letter to the author under date of June 19, 1914, written from Los Angeles, Cal., he says:

"About thirty-five years ago while I was working as a hand on the range in southern Colorado, we had a drouth followed by a very severe winter. In those days range cattle, including bulls, were left to the mercy of the elements. The losses during that winter were simply tremendous. The next spring our round-ups showed very plainly the survival of the fittest in the depleted herds and the Herefords, compared with the other breeds, were conspicuously numerous, and of the bulls that survived the many

---

*Sales of large tracts of land were made by the syndicate in 1901 and 1902 to ranchmen who were beginning to realize that the public domain was fast being settled up and that the day when it would be necessary to own the land in order to control the grass was very near at hand. Several hundred thousand acres of the tract were sold to George W. Littlefield, the well known banker and ranchman of Texas, owner of the LFD brand. Another large tract went to W. E. Halsell, large cattle owner of Indian Territory, another to the Matador Land & Cattle Co., another to T. S. Hutton and E. L. Halsell of Kansas City, another to Rhea Bros. of New Mexico, another to F. D. Wight of Trindad, Colo., and another to W. J. Tod, Maplehill, Kans. A few years later, beginning about 1906, a large influx of northern farmers took place, and a great deal of the land was sold in tracts ranging from 40 acres to a section or several sections. Some idea of the extent of this business may be gathered from the fact that the company has executed and delivered over 2,500 deeds to land from the Capitol Reservation grant, and the lands sold aggregate about 2,000,000 acres.

months of grief, the Herefords were about the only ones left. This experience of course made a lasting impression on me in favor of the Herefords and my varied experience since has confirmed it.

"During the year 1897 the H. S. Boice Cattle Co. was organized and purchased the Beaty Bros. ranch and cattle in southeastern Colorado and southwestern Kansas. We continued the ranch about ten years, when the settlers came in on us and obliged us to close out our cattle. These cattle were fairly well improved. We eliminated all bulls from the herd except the Herefords and soon raised all the sires required from a fine little herd of select cows in which we kept the best bulls that money would buy. We were very particular in selecting the heavy-boned, big-framed bulls for both the small herd and the large one. When it was known that we intended to use Hereford bulls continuously, without crossing with the Shorthorn occasionally, it was often remarked that our cattle would grow smaller until we would be obliged to cross with the Shorthorn. But our experience did not justify the predictions. Our herd grew in numbers until we were branding over 5,000 calves. It became very uniform and attractive in quality and steadily grew heavier in bone, frame and weight and our feeders sold on the range at the top of the market year after year. The fat cows likewise generally topped the market at Kansas City in weight and price.

"I was connected with the Capital Freehold Land & Investment Co. (XIT outfit) as general manager of their cattle interests in the Panhandle of Texas for seven years, closing out the same in the fall of 1912. When I took charge they were branding about 20,000 calves, much the larger number of which were from their Hereford herd. These cattle were of good quality and had been graded up from the unimproved straight Texas cows that were placed on the ranch about 1885. Nothing but purebred Hereford bulls were being used and they were raised from a purebred herd that numbered at one time about 3,500 head. We kept the standard of this purebred herd high by cutting out every year and turning into the large herd everything that showed a lack of quality.

"I wish to emphasize the fact that the purebred Hereford is a hardier, thriftier range animal than the grade Hereford. As a demonstration of the fact I will say in this instance that our purebred herd was handled just the same and had no better treatment than our large graded herd and yet was always in better condition. Both herds had to depend upon the grass and natural shelter of the pastures in which they were located, with simply a wire fence between them. Neither herd was given any additional feed in the winter. Those cattle that became poor and weak were gathered into a smaller pasture and fed cake on the grass.

"During the last six years I have given most of my time to the breeding ranch of Boice, Gates & Johnson, formerly known as the Chiricahua Cattle Co., or CCC outfit, in southern Arizona. It is one of the oldest herds in the state and has been one of the best improved for many years, though we have materially improved it and it is still in the process of improvement. At the beginning of my administration we bought a select purebred herd to raise bulls from and shipped them to our upper ranch where they are located at an elevation of about 6,500 feet. They have never had any feed except the natural grass and browsing and have always been in better condition than our main graded herd adjoining. When we brought this little herd there, the old foreman, who had been in the business all his life, shook his head, knowing the disaster that would follow in turning out purebred cattle to rustle for themselves. But since then he has often stated that it was too bad that all our cattle were not purebred Herefords.

"We are now able to turn into our main herd 150 choice bulls a year from our purebred herd. The three herds to which I have referred, in Colorado, Texas and Arizona, reached the highest standard in quality for range cattle in the several localities by the same method—careful elimination and selection. In following this method the very best bulls, regardless of price, should be obtained for the purebred herds. Everything undesirable in quality should be culled out every year and the young bulls turned into the main herd should be most liberally culled, leaving only the big-boned, big-framed, loose-hided, rangy fellows even though some of them may seem a little coarse. The main or large herd should also be constantly culled. By the way, I think most of the breeders of our finest herds of purebred cattle are not as particular as they should be in culling out from their herds the poorer quality and undesirable animals."

CHAPTER XIX.
PROOF PILED ON PROOF.

As has been already stated the first crosses of the Shorthorn on the Longhorn and other native types had made a marked improvement but, unfortunately for the best interests of a breed which was not solid-colored, the western demand for Shorthorns in the old days persistently prescribed "red-and-all-red" as the only color wanted. The reason was plain. Light or broken-colored bulls left a motley progeny when mated with the black, dun or brindle cows so common in the old Texan stock. Roan is the one distinctive Shorthorn color, the one color never counterfeited by any other breed. White and red-and-white Shorthorns have also always been common.

This range demand for red Shorthorns during the boom days of the business led inevitably to the sacrifice by the Shorthorn breeders of Kentucky and the central west of thousands of their best bulls, and to the retention in many cases of inferior animals of the right color for getting stock available for range purposes—to the palpable injury of the breed. We have only to observe a ring of Shorthorns at any of our leading shows of today, where perhaps two-thirds of all the best animals will be roan or white, to realize what was really lost to the breed through the failure twenty-five years ago to utilize the best material at hand regardless of color. It was a condition, however, and not a theory that confronted the breeders of that time, and they pursued the only course then open to them. They were forced to cater to the range, and therefore it may well be said at this time that the rise of Hereford power in the far west was really the beginning of a great renaissance in the popularity as well as in the merit of the Shorthorn in the older states, the abatement of the range demand for red bulls proving a real blessing in disguise to the old-time favorites. But, to our story.

The Carey Co.—The J. M. Carey & Bro. Co., the owner of the CY brand, at one time ran as many as 40,000 cattle. In recent years on account of their range becoming restricted these men have reduced the number of their cattle to about 6,000 head. They produce enough feed during the growing season to feed their cattle through the winters. Their ranches are well improved and they have about 4,000 acres of irrigated lands. About twenty years ago they purchased from George Morgan, of the Wyoming Hereford Cattle Co., three head of Hereford bulls, paying $1,000 for the trio. These were calves and the buyers did not have much luck with them. At this time their herd was mostly Shorthorns, bred up from Texas stock. In spite of the fact that they did not have much success with the first Hereford bulls purchased they have continued to use them until their entire herd has become as thoroughly marked as if registered.

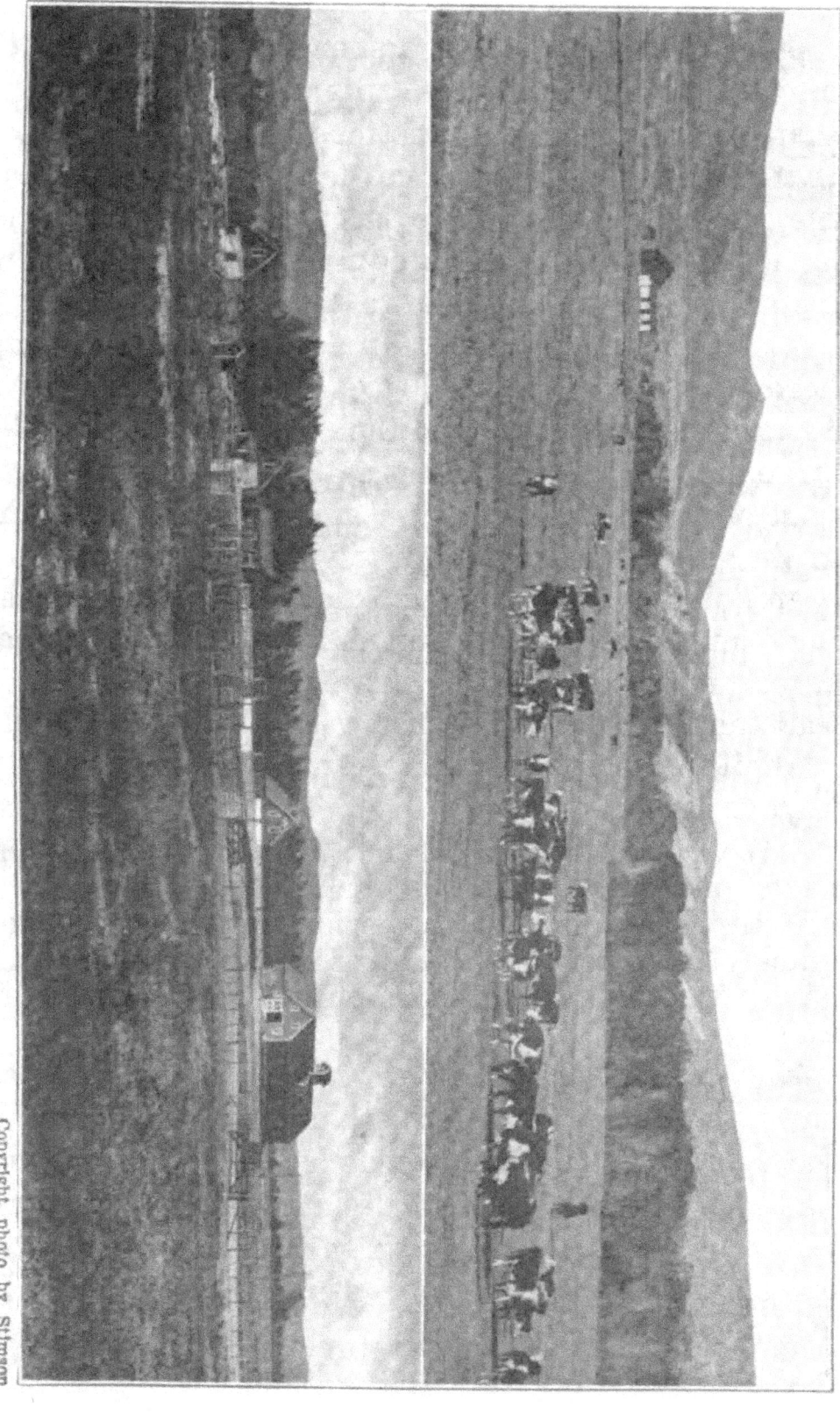

ABOVE: PURE-BRED HEREFORDS ON THE CAREY PROPERTY, CONVERSE CO., WYOMING.
BELOW: VIEW ON THE S. O. RANCH, CAREYHURST, WYOMING.

Copyright photo by Stimson

They corroborate the general testimony to the effect that "white faces" stand the winter better than other cattle and are better rustlers. They report that the weights of their cattle have increased since the early days due to the fact that the Texas blood has been entirely eliminated. Bulls are put in service at the age of eighteen to thirty months, four or five to the hundred cows. In addition to their range herd they have about 600 head of registered cows. From these they produce their own bulls and sell a large number yearly. These cattle are fed the greater part of the year, the calves and yearlings receiving grain as well as hay. For several years they have been following the system known as "hand breeding," but have recently gone back to breeding in pastures on account of the small percentage of calves that they have been getting by the former method.

While adhering to the Herefords for range use the Careys say:

"We have studied the matter for a good many years on small farms, and believe Shorthorn cattle are the best on such places. They mature more quickly. However, the Herefords are hardier and better where it is not possible to give cattle every care."

The LS Cattle.—The late Lucien Scott of Leavenworth, Kans., was an enthusiastic advocate of the Hereford for range purposes. In connection with Mr. W. M. D. Lee he maintained for many years near Tascosa, Tex., in the Panhandle country, an extensive ranch upon which the famous LS brand was developed. As high as 35,000 head of cattle were at one time maintained.

This land, consisting of 180,000 acres, was located in the Panhandle country and in Oldham and Potter counties. In 1888 Mr. Scott and Mr. Lee dissolved partnership, Mr. Scott purchasing Mr. Lee's interest and continuing the ranch under the same name. At Leavenworth, Kans., he had a farm called Ridgewood which he maintained for the breeding of fine cattle of pure Hereford strains. The young bulls from that farm were sent from time to time to the LS Ranch and in that way, and by culling out the poorer of the bulls on the ranch, the herd was gradually improved.

Mr. Scott died in 1893 and left his property to his widow Julia H. Scott. She left the management of the ranch in the hands of her brother Charles N. Whitman, a practical cattleman who was much imbued with the idea of breeding a finer grade of cattle. They gradually acquired more land until they had 204,000 acres, all in fee simple and all fenced in. The Ridgewood herd had in the meantime been enlarged and improved by the constant purchase of good registered animals. An increasing number of the pedigree bulls were sent each year to the LS Ranch either from the farm or by purchases, until in more recent years nothing but purebred bulls were used. In 1897 Mrs. Scott transferred one-half the poperty to Charles N. Whitman and in 1899 she sold him the remaining other half. Late in the year 1899 Mr. Whitman died and his widow, Mrs. Whitman, continued the business until 1907, when she sold the entire property to Edward F. Swift of Chicago. The LS Ranch, recently divided and sold, was a breeding ranch solely, the young steers being sold at one and two years old, when their fine quality with uniform color and white faces commanded at all times the top prices. They were frequently exhibited and a great many prizes were won at the Chicago, Kansas City and Fort Worth shows.

Mrs. Whitman,* now Mrs. F. H. Kreismann of St. Louis, bears this testimony as to the good results following the use of the Hereford bulls in the Tascosa herd:

"As is well known the Hereford blood has always been very prepotent and the greater percentage of calves bred from purebred bulls and off-colored cows will be white-faced. But again, as the quality of the cattle on the ranch became finer and finer they lost some of the hardihood of the Texas rangers, and more care in the way of feed and shelter had to be given them. But this extra outlay and expense was more than made up by the much greater price which the young stock commanded. By selling off every year as many of the off-colored and inferior cows as possible without reducing the herd too much, and keeping for breeding purposes only such young heifers as markedly showed the Hereford strain, the herd in a few years was so much improved and so decidedly 'Hereford' that it became one of the show herds of the Texas Panhandle."

---

*Mrs. Whitman, a woman of high intelligence, had a deep personal interest in cattle-breeding. For a number of years she was frequently seen at the leading cattle sales and shows, often finding upon these occasions a congenial spirit in Mrs. Kate Wilder Cross, widow of C. S. Cross of Sunny Slope Farm, Emporia, Kans.

Speaking of the quality of the bulls bought for service on this ranch, John Gosling writes: "During the time of Mrs. Whitman's ownership I furnished her ranch with as many as eighty bulls in a season, which included in three consecutive years the entire crop by the famous bull Painter, a son of Beau Brummel. Painter was owned by W. W. Gray, Fayette, Mo., and later by Mr. Robert H. Hazlett of Eldorado, in whose hands he finished

The Matador Land and Cattle Co.—This corporation, one of the largest in the west, was organized in 1882, and its holdings are now reputed to be worth around $6,000,000, under the management of John McBain. It began operations by purchasing between 300,0000 and 350,000 acres of land, and from time to time added to this until now it has two divisions, one of about 500,000 acres and the other comprising over 250,000 acres, both in the "breaks" of the plains and mostly rough country. Both ranches are well improved. Besides this the company has on lease about 450,000 acres in Dakota and 150,000 acres in Canada.

The original herd consisted of about 40,000 head of cattle but from year to year it was increased until in 1891, when Mr. Murdo Mackenzie became manager, it had about 65,000 head, the maximum herd during Mr. Mackenzie's administration being about 70,000 head. The cattle originally purchased by the company were of the ordinary type prevailing at that time in southern Texas, and were driven from that section to the Matador range by A. M. Britton and Henry Campbell. Later on the company purchased from 8,000 to 10,000 cattle from Mr. Chisholm in the Pecos Valley. These cattle were to some extent graded up by the infusion of Shorthorn blood, but were not by any means what would be considered good cattle today. Up to 1891 there

---

his career with pronounced successful results. Indeed the sons of Painter had much to do with the prowess of the LS calves when exhibited at the Kansas City Royal ten and twelve years ago, and attracted the attention of Landrigan Bros., then at Eureka, Kans., who leased some 90,000 acres of the LS Ranch."

had been some attempt at improving the herd, but this had been tried by the use of grade bulls. Some of the bulls used were purchased in Kansas, but the majority of them were raised from the herd itself. By this method there must of necessity have been considerable in-breeding, and in any event there could not have been much progress made in the way of grading up from such a foundation by continuing this policy. On taking charge of the property Mr. Mackenzie came to the conclusion that in order to obtain the best results some radical changes had to be made, first, in the kind of bulls used, and second, by culling out all inferior cows undesirable for breeding purposes.

In the old days the Matador sold practically all its steer cattle as two-year-olds to buyers from Dakota, Montana and Wyoming, and the prices realized were not sufficient to pay the cost of production. It was decided that something must be done to raise the grade of the cattle to a point where they would be in demand by the Kansas and Missouri feeders. To attain such a standard not only would the inferior cows have to be culled out, but a different type of bull would have to be used. Yet to carry out this plan required considerable money, and at that time money with cattlemen was not as plentiful as it is today. Few had as yet attempted to use purebred bulls on the open ranges of the Panhandle or in large pastures, and the accepted idea was that if a herd of cows was brought up beyond a certain grade their reproductive quali-

A SECTION OF THE MATADOR HEADQUARTERS.

JUDGE H. H. CAMPBELL, WHO ORIGINALLY LOCATED AND ESTABLISHED THE MATADOR RANCH.

ties on the range would be impaired. Referring to this situation, Mr. Mackenzie says:

"I realized that several important changes must be made to put the company on a paying basis, that instead of selling off our fat cows we should dispose of those undesirable for breeding purposes, and that we must supply the herd with purebred bulls exclusively. It took several years to accomplish this because the purchase price of the bulls, as well as the running expenses of the company, had to be paid from the revenue derived from the annual sales of cattle. Furthermore, I suggested that instead of selling our two-year-old steers to northern buyers we adopt the plan of sending all our steer cattle to northern ranges of our own and holding them there for two years and then sending them to the Chicago market. This policy has been carried out by the company since that time."

Manager Mackenzie Discards Grade Bulls.—At the time that Mr. Mackenzie took hold of the Matador company there were a few Hereford grade bulls being used in the herd, and not first-class ones at that, the balance being Shorthorn grades and a few Aberdeen-Angus. The Hereford was not altogether yet in favor at that date, many claiming that the produce was smaller both in bone and size than the cross by the Shorthorn. This was undoubtedly the truth in certain instances, but was explained, in part at least, by the fact that many of the so-called Herefords in use were only grades and in other cases were very inferior specimens of purebreds. Accordingly it was decided to change the policy—instead of using grades to breed from purebreds only.

The first pedigree Hereford bulls bought for the

purebred herd which was then started at the Matador headquarters were obtained from Fowler & Tod of Maple Hill, Kans., about 1892. About the same time other purebred Hereford and some Shorthorn bulls were purchased in Colorado, Kansas and Missouri for the main Matador herd, which consisted mainly of Texas cows. This resort to purebreds proved successful, cattle of good conformation and first-class quality being obtained. Another interesting fact which developed was that the accepted idea that cattle highly-bred necessarily become unproductive was erroneous; no difference whatever was observable in that particular. As an illustration of this, from 100 purebred cows in a pasture where the cattle had nothing but the grass they gathered a calf crop of 99 per cent was one year recorded, and this statement can be verified from the record of the branding in the company's office at Trinidad.

A Purebred Herd Established.—Mr. Mackenzie states that early in his experience in Texas he found that to supply such a large herd of cows as the Matadors with a sufficient number of purebred bulls would be a very expensive process, so he adopted the policy of buying purebred cows and holding them on the range for the purpose of breeding at least a part of the bulls required. In this he was successful. He says:

"If the Hereford cow is supplied with a sufficient amount of grass she will produce a calf as regularly as the cows on Missouri and Iowa pastures, and at four years old the bull is just as large and of as good conformation and quality as any you will find in the grain-growing states. All that is required is care and the culling out of the cows undesirable for breeding, but not only is this care required on the range, but it is required on the farm as well, if one expects to get the best results. I do not wish the breeders of other purebred cattle to feel that I have any prejudice against the other breeds; all breeds to my mind are good in their own place, but in large pastures where cattle have to hustle for themselves I have no hesitation in stating that my experience has been that the Hereford has it over them all."

The Matador management has had a marked predilection for the Anxiety blood when obtainable. From $100 to $250 was paid for bulls to be used on the range, and as high as $1,000 for bulls for the purebred herd. Ordinarily about 150 purebred bulls were raised each year for use in the herd. These were reared in the usual range way, although usually placed in an extra good pasture and fed a little during the winter months. The bulls are turned out when two years old, and the aim is to run four bulls to every hundred cows.

"In this enlightened age," says Mr. Mackenzie, "everybody knows that Herefords cannot be equaled as range animals, and we have found that the nearer purebred they are the better they do. Many think that the continuous use of Hereford blood makes cattle small and with poor hindquarters. We find this not to be the case where really good northern bulls are used on sufficient range. In order that the Herefords do well during the year round they should have at least 15 acres to every

MURDO MACKENZIE.

head with a plentiful supply of water at distances not greater than four miles between watering places."

Tod of Maple Hill.—Prominent among the successful cattlemen of the trans-Missouri country is W. J. Tod of Maple Hill, Kans. Experienced in all the varying conditions met with upon the range, as well as in the feedlots of the middle west, a familiar figure at all our leading shows and markets, his experience with the Herefords in connection with Mr. Fowler cannot fail to be of interest.

Mr. Tod originally came out from Scotland and was for several years with the Prairie Cattle Co., but subsequently formed a partnership with Mr. George Fowler of the Fowler Packing Co. In the year 1884 they imported from England a few purebred Hereford cows, a few Shorthorn cows, one or two Shorthorn bulls and a large number of Hereford bulls, with a view to using the bulls chiefly on grade cows. During the dull years of 1889, 1890 and 1891 they disposed of these purebreds and a few years later started a herd of grade cows in northern New Mexico using Hereford bulls exclusively. These cows were ordinary Colorado range-bred cows, with the exception of a few practically purebred Herefords. They continued using Hereford bulls in this herd until it was closed out on account of the land being acquired at a price so high that it was thought it could not longer be held profitably. The bulls bought were chiefly of Gudgell & Simpson breeding, and Mr. Tod was careful to buy no bull which was not strong-

W. J. TOD OF MAPLE HILL.

ly "Anxiety-bred." The result in the herd was most pronounced, and the improvement in quality, feeding capabilities, style and size most marked.

"From long experience," says Mr. Tod, "we have found that in a range country the Shorthorns have not the constitution nor the rustling qualities that we get in the Herefords. We had not only a better calf crop with the Herefords, but they stood the winter with less feed in better condition than the Shorthorns, and they are undoubtedly better grazers. Now that baby beef is so much in demand we have found that there is no breed of cattle that equals the Hereford in capability of becoming prime fat as yearlings, but it of course must be clearly understood that we feed in a manner in the rough. None of our cattle are stabled, nor have we sheds. The only shelter we have is the timber. These are the conditions we have to contend with, and we have found the Hereford equal to them."

While the herd in New Mexico was maintained from 1,000 to 1,200 head were branded annually, and about 3,000 steers a year were fed in Kansas. The greater number of these were purchased in the Panhandle and in southern Colorado. The range in New Mexico contained about 60,000 acres of fenced land. No young cattle were sold, everything disposed of being fed off in Kansas, and marketed chiefly in Kansas City, with occasional shipments to Chicago and St. Joseph.

The New Mexico ranch was located in the northern part of the state and the quality of the grama and other grasses was excellent. Some alfalfa hay was used, and the calves had hay and a little cottonseed during the first winter. The cows, however, got only what the range produced. Having a small range it was considered best to keep the bulls only two years, and as the yearling heifers were never bred there was no chance of in-breeding. High-class bulls were bought and turned on the range at two years of age, allowing four to 100 cows. At four years old these were sold to some of the large ranchers in the west, and there was no trouble whatever in disposing of them. Indeed there was recorded a very active inquiry. It is the judgment of Mr. Tod that there are few cattle handled and fed in the rough capable of making as fine-finished, high-class, high-priced beef at one or two years old as well as Herefords.

As proof of the excellence that can be attained by the use of pedigree Hereford bulls with ordinary range cows Fowler & Tod showed at the International Live Stock Exposition, Chicago, in 1905, a carload of calves, a carload of yearlings, a carload of two-year-olds and a carload of three-year-olds. These four carloads of cattle took three first premiums and one second in their respective classes. They were also the champion Herefords, and the three-year-olds were the champion three-year-olds over all breeds. These cattle were all from the same range, and this record was duplicated exactly when four carloads from the same herd were shown at Chicago in 1909.

Some years later the same firm showed a carload

of fat cattle and a carload of feeder cattle from the same herd at the Kansas City Royal Show, and both of these were grand champions in their respective divisions. Inasmuch as these cattle were handled on the range until they were put on feed in Kansas the results demonstrate conclusively what can be done by the careful selection of Hereford bulls to place on good range cows.

The Swensons.—Few brands of Texas cattle are better known or in higher repute than the SMS. The Swenson Bros., of New York City, owners of an enormous property in Jones and five other counties, at one time ran 55,000 head of cattle. That was when they owned the Spur cattle as well as the SMS herd. They have been selling largely of their lands in recent years, but still manage 400,000 acres —250,000 acres of their own and a lease of the 150,000 acres of the Spur property, all under wire fence. At this writing they are probably running about 22,000 cattle exclusive of calves, of which they had in 1913 about 10,000 head.

The first registered Hereford bulls taken to the Swenson Ranch were bought from Fowler & Van-Natta in 1884, and they were bred upon a small herd of cross-bred Hereford and Shorthorn heifers, and the bulls from that herd were in turn bred on the SMS main herd which was made up of native, but well selected, Texas cattle. The first Hereford bulls used on the Spur range were introduced the same year by manager C. L. Goff from O. H. Nelson's.

Speaking of the foundation of the SMS cattle

manager Frank S. Hastings says that the early heifers in the herd were carefully selected native cattle, and these were crossed with bulls from an unregistered herd that had been bred up from crossbred Shorthorn-Hereford heifers mated with registered Fowler & VanNatta bulls. This unregistered herd was probably a little stronger in Shorthorn than Hereford blood originally, but it has been persistently crossed with registered Hereford bulls and now consists of about 1,400 cows which will show an undercurrent of only about 5 per cent Shorthorn. This is distinctly a "white face" herd, and probably no herd in the west carries a wider range of Hereford blood. It has had the service of more than fifty head of imported registered Hereford bulls, it has drawn from the Fowler & VanNatta herd, it has had several drafts from the Armour herd and several shipments from the Gudgell & Simpson, the Dr. James A. Logan and other good herds. In recent years it has had drafts from some of the best Texas herds, added to which there has been a "throw-in" each year from a registered herd of the ranch. These registered bulls after a service of two years are thrown into the main herd and scattered over the various ranches.

With this unregistered herd as a bull basis the native type in the main herd soon disappeared, and when Mr. Hastings took charge in 1902 it was distinctly a high-grade Hereford herd, but with some weeds in it. A pruning process has continued almost to the point of extravagance ever since, until

FRANK S. HASTINGS.

today out of 16,000 breeding cows the nearest approach to an off-color is something spotted showing the result of the Shorthorn-Hereford cross. Figured out in the way of fractions the commercial herd which for ten years has been sending all of its progeny to cornbelt feedlots with good records both there and in the showrings will show within a fraction of 99 99/100 per cent purebred Hereford.

A strong bull tally is a part of the Swenson policy, one aged bull to sixteen cows, and in addition to that all the bulls whether of their own raising or purchased are turned out in their yearling period, though not counted in the breeding complement. Naturally they sire a few calves, but in the main the benefit of turning them out is to acquaint them thoroughly with the range and to get them acclimated and so give them a usefulness that cannot be obtained in any other way.

Yearling heifers are not bred at all. The entire yearling crop is pruned 10 per cent every season and that cut is sent to the block. The Swensons consider that this early pruning before maturity with a careful subsequent culling as development may suggest, leaves a uniformity of type which could not be obtained by waiting until the cows had served their period of usefulness before pruning. The average winter loss on the ranch, or the average loss from all causes for twelve months, is about 5 per cent.

The Swensons bought the entire Spur property, but never mixed the Spur herd with the SMS and later sold the entire Spur head to W. J. Lewis. Asked for a summary of his reasons for preferring Hereford blood for range purposes Mr. Hastings says:

"The Hereford has been the redeemer of the range on account of his hustling ability, capacity for taking care of himself under adverse circumstances and general adaptability to large pasture work."

The largest investment for improvements on the SMS property has been for water, over $100,000, but the greatest expense item has been for the extermination of prairie dogs, the sum of $50,000 having been successfully expended to rid the big pastures of "dogs."

Calves for the Cornbelt..—Beginning about 1904 the Swensons began selling their youngsters to cornbelt feeders, and many a great load of "white faces" bearing their brand has graced the pens of leading shows and markets. In 1912 they delivered 5,000 head to the cornbelt, and not a single buyer was present to receive his cattle. "With the exception of one man who wrote that he thought the calves a little young," says Mr. Hastings, "we had a clean sweep of voluntary letters expressing absolute satisfaction with the cattle. In fact, our business is up to a place now where we cannot half supply the demand for the cattle, and we grade them just as one would grade sacks of granulated sugar, that is, our standard has been established and we are able to sell them by correspondence with universal satisfaction."

Richards & Comstock.—This firm, at one time very prominent in Nebraska, running as high as 70,000 cattle, bought its first Hereford bulls in 1882, a portion of them coming from T. L. Miller. The cow herd at that time consisted of native Montanas. Richards & Comstock were pleased with the Hereford cross, and used for the most part Hereford bulls ever after. It was their opinion that continued crossing from Hereford bulls would not tend to increase the weight, but was apt to decrease it. They admired the Herefords, however, as "hardy, producing good colors and giving the best cross for market purposes."

Richards & Comstock testified that their outlet varied, according to grass and market conditions. Some years ago they only produced feeders, while again they had good beef. For their feeders they found a market in Omaha, for good beef they favored Chicago. They secured their bulls from various breeders in Nebraska, Missouri, Iowa and Illinois, paying for them at different times from $75 to $200 per head. They never attempted to raise the bulls required for the herd. They preferred two-year-olds and allowed twenty to twenty-five cows per bull. In selecting Hereford bulls for breeding on the ranch they always aimed to select those that were inclined to be a little coarse and rough, rather than those that were fine in the bone. Their experience was that they had to constantly be on their guard lest they get their cattle "too fine."

Big Horn Land and Cattle Co.—This company, controlling property valued at approximately $750,-

000, has 18,000 acres of patented land, all under fence. Manager William Marr had his earliest experience with Hereford cattle in North Park, Larimer Co., Colo., beginning in 1880, and in a letter written to the author several years ago testified as follows:

"I have used several hundred Hereford bulls and have at present between thirty and forty, the balance being Shorthorns. I think the Hereford a good cross on big strong rough cows, but no better than a Shorthorn, in fact, I do not think as good on a well bred smooth bunch of cows. It is my experience that Hereford bulls have made their greatest success on the range and elsewhere when the cow herds had been previously more or less improved by the use of Shorthorn blood. The best cross on the Texas cow was the Shorthorn, afterwards the Hereford; the latter seemed to smooth them up and give them a uniform color. It has been my experience that with continuous use of Hereford bulls my cattle got to weigh less, and for the past six or seven years I have been using as many Shorthorn bulls as Herefords, and am getting more size and weight.

"In my neighborhood the Hereford has been pushed more than the Shorthorn by the owners of purebred herds. There were no purebred Shorthorn herds, and the Hereford did well on the coarse western cows. They are good rustlers, and on cows with no particular breeding they would get a calf with a white face. My idea of an animal for the range is a cross between the Shorthorn and Hereford, and if I was starting a herd again I would get Shorthorn cows and Hereford bulls."

The Sparks Herd.—The large and first-class Alamo herd maintained for so many years by the late Governor John Sparks of Reno, Nev., probably supplied more good Hereford bulls to the ranchmen of California than any other one herd. Offshoots from the Alamo were numerous, one of them being the herd of Whitaker & Ray, of Gault, and another the Jacks herd at Salinas. Joseph Marsden of Lovelocks, Nev., maintained a Hereford herd for many years, and upon his retiring from business this was taken over by a company at Newman, Cal., the herd being successfully maintained at a high standard.

From the Sparks herd also was obtained the foundation Herefords for the Fred H. Bixby cattle, running in southern California and Arizona. Mr. Bixby has used both the Hereford and the Shorthorn blood, and believes that a cross of those bloods constitutes the ideal animal. He is partial to the Hereford, however, and gives four reasons for his preference:

"First, the Hereford is the first to fatten; second, the Hereford is a better rustler; third, the Hereford can stand more hardship; fourth, the Hereford as a rule has a better loin."

Continental Land and Cattle Co.—Col. William E. Hughes of Denver, chief owner of the "Mill Iron" cattle, one of the important Texas herds, had his first experience with Herefords about 1895 in Collingsworth Co., Tex., and has used the blood ever since. Some years ago he purchased 500 unregistered Hereford heifers from the Adair herd in the Panhandle of Texas, bred up from the well known Palo Duro foundation of Shorthorn cows mated to Hereford bulls. The Continental company has continued to breed these cattle and their offspring to registered Hereford bulls bought in Missouri, Iowa and Kansas.

During the same time and in the same country the company has bred an equal number of the same class of cows to registered Shorthorn bulls. Col. Hughes gives it as his experience that there is no great difference in the progeny, but "if there is a difference, it is in favor of the Hereford." In 1906 he exhibited both kinds of cattle at the International show, winning first prizes on both in their classes. "These yearlings," he says, " had an equal chance in feed, range and breed." The Shorthorns averaged 1,150 pounds and the Herefords 1,077 pounds, the former bringing $8.35 and the latter $9.75 per cwt. on the December, 1906, market.

While Col. Hughes inclines to the opinion that the Hereford bull is the best ranger and rustler he says that "it is generally understood that Hereford bulls have made their greatest success on the range when the cow herds had previously more or less Shorthorn blood in them. This is my experience." He also adds: "I think, however, if the Hereford breeding is kept up for any great length of time there is a loss in weight and size. The Herefords are inclined to get too peaked and light behind. This is obviated by occasional crossing back to the Shorthorn, getting a square rump. The Herefords are generally preferred in the range country because they are less sluggish than the Shorthorns and are better rustlers and rangers."

The Marcus Daly Outfit.—Manager P. J. Shan-

ON THE WALLUP RANCH, SHERIDAN CO., WYOMING.

THE MONCRIEFFE RANCH, SHERIDAN CO., WYOMING.

non of this Montana property has been using Hereford bulls for the past ten years on the range cows, and considers them the very best available for that purpose. He is of the opinion that the Hereford bulls have made their greatest success on the range when the cow herds had previously been more or less improved by the use of Shorthorn blood, and agrees with the view that "continued use of the Hereford tends to loss of size, particularly in the hindquarters of the animal." To counteract this he intended to change to Shorthorn bulls in the season of 1913 and use them for a few years, then returning to the Hereford. He adds:

"Our main reason for preferring the Herefords is that they are the best rustlers and will make a living and breed well in a poor rough mountain range such as we have here."

The Bell Ranch.—Mr. C. M. O'Donel, manager of the Bell Ranch owned by the Red River Valley Co., in San Miguel Co., N. M., had his first experience with the Herefords in the Texas Panhandle shortly after Goodnight started at Palo Duro. He is another one of those who while admiring the hardiness of the Hereford still has a warm spot in his heart for the Shorthorn. In response to a letter of inquiry from the author, Mr. O'Donel writes:

"While I have never bred purebred Herefords I have used and seen them used on range cows extensively for the last twenty-five years. For the past nine years I have had from 300 to 500 white-faced bulls at a time on this range. Comparing the Hereford with the Shorthorn bull for range purposes, its advantages are: (1) undoubtedly superior thrift under adverse conditions, which I attribute as much to his placid and equable temperament as to his compact and easily nourished frame; (2) his generally superior coat; (3) his popularity with steer buyers, though this is less marked in recent years. His disadvantages are: (1) a want of scale; (2) less breeding activity when young (this latter is, I believe, not generally acknowledged, but I am convinced of its correctness, although it is partly compensated for by the superior condition that the Hereford maintains in consequence of that fact); (3) a weight of horn and lightness and angularity of hindquarter which is not well calculated to remove these same defects from the native cattle of the southwest.

"The popularity of the Hereford on the range is due undoubtedly to his conspicuous, uniform and attractive coloring which proclaims the blood even to the most inexperienced. At the time of the first introduction there was excessive mortality among the Shorthorn bulls, and while the Hereford is generally regarded as having saved the situation, his breed sometimes receives the credit for survival which was really due to more judicious stocking of the ranges. I do not know of any range herd in the front rank as regards quality that has been produced by the use of Hereford bulls alone on the native scrub cow. I am convinced that there is a tendency where one Hereford cross is followed by another for a long period towards some loss in size and weight. This might perhaps be avoided by the careful selection of Herefords of large frame. The obvious remedy for this is the use of a Shorthorn cross. The exclusive use of Shorthorns has its drawbacks also. Nevertheless, I am free to confess that I do not believe that Herefords can ever be dispensed with on the range."

This frank statement from a man who is very partial to Shorthorn blood perhaps carries quite as much weight as some of the more enthusiastic praises bestowed upon the Herefords by their special advocates.

Governor McDonald's Evidence.—Hon. W. C. McDonald, at this writing Governor of the state of New Mexico, in his capacity as manager of two ranches and live stock companies, adds his expression of appreciation of the Herefords as being the best "doers" on short range. He manages properties aggregating perhaps 1,000,000 acres—mostly rolling land along the foot of the mountains, where the grass is principally grama. The ranges are now well equipped with windmills and gas engines, although not much of the land is as yet fenced. The cow herds were originally grade Shorthorns, and the first Hereford bulls came from the Richards stock at Watrous, N. M. Since these bulls have been used the Governor states that the cattle have increased in weight. A few Shorthorns have usually been kept, however, along with the "white faces." In more recent years the bulls have been bought in Kansas, Missouri and Texas.

George W. Baker.—Another New Mexico ranchman, Mr. George W. Baker, of Folsom, who is now running about 600 cows and has had at various times as many as 1,500, operates a 10,000-acre ranch under fence with plenty of water and good corrals. Most of the land is rough and broken, carrying wild grass, mostly grama. He puts up

Copyright photo by Erwin E. Smith
"ON THE TRAIL THAT LED NOT BACKWARD."

Copyright photo by McClure
A PRIZE-WINNING BUNCH OF HEREFORDS ON THE RANGE.

300 tons of alfalfa annually. His first Herefords came from C. A. Stannard, who bought the Sunny Slope herd, originally made famous by the late Charles S. Cross, and began breeding them to cows that were grade Herefords with a Shorthorn cross. He has used Herefords continuously since, and states that his cattle have increased in weight. He believes the Hereford to be the hardiest and most prolific sire, and that his use results in more uniform cattle. All of Mr. Baker's bulls are bought from other herds. After relating his experience, which has been favorable to the Hereford, Mr. Baker states:

"Breeding cattle on the range is a thing of the past in this country. There are a few steers still on the range, and some cattlemen turn their herds out in the summer, but they are closely herded, and range conditions as generally understood do not exist here."

The H. G. Adams XI Ranches.—In Meade and Seward counties in Kansas, and in Beaver Co., Okla., Adams & Robert own about 36,000 acres of deeded land and have some 30,000 acres additional under lease. The property is well improved, is watered from windmills and ponds and is situated about 18 miles from the Cimarron River. The land and cattle represent holdings said to be worth at this time around $500,000. In addition to this Mr. Adams has a 7,000-acre place of his own at Maple Hill, Kans., where he handles and feeds steers exclusively.

Adams & Robert began with unregistered Hereford cows and bought their first pedigree bulls from Gudgell & Simpson, using since only purebred bulls selected from that herd and from Armour's and Hazlett's. They have, therefore, a lot of the Anxiety blood. Their experience coincides with others who testify to the superior hardiness and general adaptability of the Hereford for range uses. They do not breed from bulls of their own production, preferring to keep up fresh infusions from good sources. They graze their young bulls through the summer months, and in the winter give them ground kafir corn, cottonseed meal or cake, and hay. Their young cattle are largely sold to feeders in Kansas, Missouri and Illinois.

John Z. Means.—Something like 250 sections of land in Jeff Davis and Culberson counties on the west side of the Davis Mountains and 350 sections north of Pecos City and lying on both sides of the Pecos River, well equipped for the cattle business, are controlled by Mr. Means. The entire property is valued somewhere around $1,000,000. Replying to queries submitted some time since, Mr. Means wrote the author as follows:

"I own about 15,000 cattle and prefer the Herefords because they are good, thrifty cattle of the type best adapted to this dry country. We have not had anything to discourage us in breeding to Hereford bulls, and the more we see of them the better we like them. While we have never bred any registered bulls, we bought twenty-nine at one time from Gov. John Sparks, and additional ones at different times from northern herds. The first 'white faces' used were from a Mr. Adams of Moffit, Colo., the lot consisting of two registered and thirty or more grade bulls. The herd upon which our first Herefords were crossed was obtained from R. K. Miley of this state. While I have used some Shorthorn bulls I have preferred the Herefords ever since we began using them."

Ike Pryor Prefers Herefords.—One of the best known Texas cattle-growers is Hon. Ike T. Pryor of San Antonio. He bought his first Hereford bulls in 1880 from Towers & Gudgell, and placed them on his ranch in Colorado. Thereafter he followed up this purchase each year with other registered Herefords, principally from Missouri and Kansas, until 1885. Between 1880 and 1884 Pryor Bros. bought large herds of grade Herefords in Colorado, as well as several lots of Shorthorns, and turned them on the open range in the southern part of that state. As already set forth in these pages the hard winter of 1884-85 gave the different breeds of cattle on the open range a thorough test of ability to withstand the hardships incident to open range conditions. Referring to this experience Mr. Pryor says:

"At least 75 per cent of our high-bred Shorthorn cattle died in that disastrous winter, while not over 25 per cent of the Herefords died—all running on the same range. This convinced me that the Hereford was the best animal a stockman could use for range purposes.

"I am the owner today of a large herd of Hereford cows on the Membres River in Grant Co., N. M. This is a mountainous range and I am using on this ranch exclusively registered Hereford bulls, because of the fact that bulls from this breed of

BEEF STEERS IN THE ROUND-UP.

NOON-TIDE AT THE WATER HOLE.

cattle will follow the cows to the top of high mountains, while Shorthorn bulls will remain near the water where feed is usually poor and of course give less service than the Herefords that go out in the mountains with the cattle. I would not think of changing the Herefords for any other breed of cattle. So much for the Hereford as a range animal.

"My observation of this breed of cattle is that you can make them into good beef at any age from six months to a four-year-old. There is no animal superior to the Hereford for making baby beef; in fact, as I said before, it is possible to put him in prime condition at any age up to the time he is fully grown, and this is a strong point in favor of the Hereford, because an animal out of which one can create baby beef at from a year to eighteen months old is the popular type. The farmer can make choice baby beef of a Hereford at from twelve to eighteen months, thereby saving from one to two years' time, whereas the more you feed a Shorthorn the more he grows, and does not seem to take on fat in proportion to the Hereford of the same age. I can, however, cite instances where one cross of a Shorthorn on a Hereford herd has increased their size for range purposes and probably did not decrease their vitality."

George H. Webster, Jr.—The Uracca Ranch, near Cimarron, N. M., is a property of some 80,000 acres of semi-mountainous land divided into summer ranges with an average altitude of 7,000 to 9,000 feet and winter ranges averaging 6,000 feet above sea level. It is mostly in blue grama grass. Steers only are run on this ranch at the present time.

Mr. Webster prefers the Herefords because of their superior constitutions and rustling power, but inclines to the opinion that western range cattle generally are "inclined to grow lighter where Hereford bulls are used exclusively."

James A. Lockhart.—Another admirer of the Hereford for the west is J. A. Lockhart of Colorado, who used the Herefords first in New Mexico from 1888 to 1892 and in Colorado since 1892. He considers the Hereford the best bull to use on the open range in an arid country where drouth and short grass prevail, as in New Mexico and parts of Colorado. Mr. Lockhart's firm had 15,000 stock cattle at one time in New Mexico. Only Texas and New Mexico native cow herds were maintained, carrying but little Shorthorn blood.

Mr. Lockhart says that range cattle crossed repeatedly with Hereford bulls "gradually grow smaller and with less vigorous constitutions, the remedy being to cross with large-boned Shorthorn bulls or other good cattle." Like most of his brother ranchmen, however, he expresses a decided preference for the Herefords, "because they are better rustlers and stand grief (short grass, scarcity of water and long distance traveling to obtain the same) better than the Shorthorns."

The C. B. Company.—Mr. Julian M. Bassett, manager of the C. B. Live Stock Co., operating in Crosby Co., Tex., states that his people began using Hereford bulls about 1900, their first purebreds being obtained from K. B. Armour. The cows at that time were mixed Hereford and Shorthorn, and bulls of both breeds have been used since. Mr. Bassett states that the average weight of the cattle is about 100 pounds more at four years old now than was the case twelve years ago. The company closed out its cow herd last year, but in the light of the experience of the management it is believed that if good Hereford bulls are bought, and close breeding avoided, the cattle will not deteriorate in size. Mr. Bassett corroborates the testimony of others that Hereford bulls are better rustlers and easier kept.

"Look for Bone, All You Can Get."—Such is the laconic and eminently sound advice of James Callan of Menard Co., Tex., in the selection of Herefords for use on the range. In giving us his experience he employs the language quoted in the course of an admonition as follows:

"Avoid cheap bulls. Disregard showring decisions. Look for bone, all you can get, and then it will be fine enough in the offspring raised under range conditions."

The Callan company has a property valued (including cattle) at around $750,000. The two ranches comprise 60,000 acres of live-oak country, and the remainder is open. The company runs a main herd of 3,000 head and bought its first registered Hereford bulls in 1895. These were largely of Tom Clark breeding. The Callan she stock at that date carried both Hereford and Shorthorn blood, the latter predominating. White-faced bulls have been used ever since and the Shorthorns have been "cleaned up" entirely, the result being "more uniformity of type and color and thriftier animals."

Mr. Callan reports weights as 20 per cent heavier

FROM THE SAND HILLS—A PRIZE LOAD WAS DRAWN FROM THIS BUNCH.

MOUNTAIN LAKE AT TOP OF THE CONTINENTAL DIVIDE—A. E. DE RICQLES AT "THE PARTING OF THE WATERS."

now than with the old-time natives, but "not heavier than the Shorthorn cross." In his experience the Herefords are "far the hardiest on the range."

Wallis Huidekoper.—Another large operator on the northern range is Mr. Wallis Huidekoper, whose North Dakota and Montana experiences lead him to place a high estimate upon the Hereford. Mr. Huidekoper is running about $100,000 worth of cattle on a $350,000 range in Sweetwater Co., Mont., which comprises some 30,000 acres under fence, well irrigated and with modern equipment. His grazing is a good quality of buffalo grass on the hills and flats. Wild hay and alfalfa are put up for winter feeding.

Mr. Huidekoper bought his first Hereford bulls in 1900, his cow herd at that time consisting of half-blood range-bred Shorthorns. He placed four successive crosses of Hereford bulls upon this foundation, and says that the first cross produced the best beef and the heaviest. Each succeeding cross resulted in a neater type with less size. Mr. Huidekoper has also used Galloway bulls upon Shorthorn cows with good satisfaction, and uses Shorthorn bulls on Shorthorn cows to keep up a supply of that blood. He nevertheless joins with a large majority of all leading western cattle-growers in regarding the Hereford's as the most valuable of all blood elements for range purposes. He says: "They have the heart, they are great rustlers and they hold their flesh well in adversity. If you will watch a large bunch of mixed cattle leave the brush after a winter storm you will see the 'white faces' in the lead."

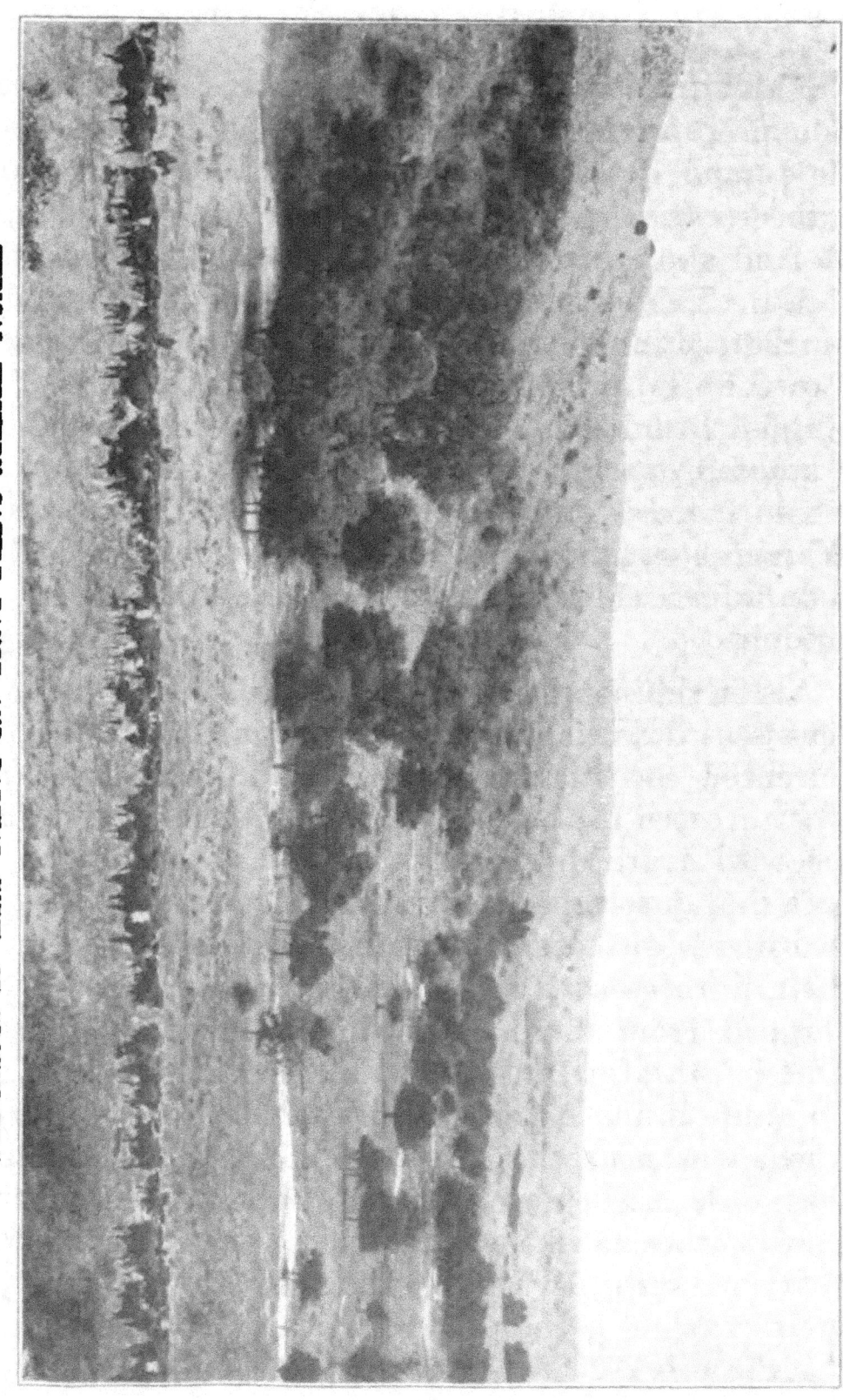
TYPICAL WESTERN CATTLE RANGE AND PARTIAL VIEW OF "ROUND-UP."

Making Good in Old Mexico.—E. K. Warren & Son, proprietors of the U— brand, own land and cattle roughly valued at around $2,750,000, including three ranches covering approximately 900,000 acres of land and carrying on an average 25,000 head of cattle. Their Ojitos Ranch, which is situated in northern Chihuahua in Old Mexico, was formerly owned by Lord Delaval Beresford, a brother of the English Admiral, Lord Charles Beresford, and is a noted property, all under fence with fine improvements, good springs and windmills. The Messrs. Warren bought it in 1909. They also own the Palatada Ranch which joins the Ojitos, and it is also well equipped.

At the time of the purchase of the Ojitos the Warrens sent down 250 Hereford bulls and 350 head of purebred cows from their ranch at Bovina, Tex. They crossed the bulls upon cows purchased in Mexico with gratifying results, as is evidenced by the fact that for three years past they have sold their two-year-old steers at Amarillo at $44 per head, and their three-year-olds at $62.50, the latter being shipped from their Mule Shoe Ranch at Bovina. This bunch of cattle went to the sugar beet mills in Colorado in the fall of 1913 at $62.50 for the threes, which was near the top price for plains cattle at that date. They are said to make a wonderful growth when taken on the plains as yearlings. The Warrens ship all their steers and 75 per cent of their yearling heifers to the plains each year, keeping the other 25 per cent on the Ojitos Ranch for breeding. The three-year-old heifers which they had spayed brought $50 per head at Bovina in the autumn of 1913, and the two-year-olds $40 per head.

The Warren firm owns also the 250,000-acre Alamo Hueco Ranch in New Mexico, which joins the International line and lies exactly 16 miles north of the north line of fence of the Ojitos in Old Mexico. They first began operations at Bovina in 1902, with the purchase of 40,000 acres of land, and later on added 45,000 acres more. This is what is known as the Mule Shoe Ranch. This land was used for grazing purposes only until last year, when the townsite of Muleshoe and 83,000 acres of land were sold.

Manager C. K. Warren says:

"We commenced the purchase of Hereford bulls the first year we went into the business. In our experience this is the only breed of cattle for ranch purposes, especially when cattle are handled in large herds, as they are the best rustlers, have thicker hides, carry the most uniform flesh, stand cold better, produce a larger percentage of calf crop and it takes less feed to carry them than with other cattle. Still we have been putting in every other year a few Shorthorns with good results. They have a little more bone and the cows give more milk for the calves.

"We have now a herd of purebred Herefords in Michigan from which we are raising and shipping our registered bulls to a purebred herd, not registered, in Mexico. Our purebred herd in Mexico numbers about 800 and from this we are raising bulls that are used for breeding purposes both in Mexico and New Mexico. We have gone out of the breeding proposition on the plains at Bovina, and are simply handling our yearlings from that point. We take the bulls away from the herd Nov. 1 each year, and put them back June 15. On good pasture all will winter strong and fat without grain. We brand approximately 5,000 head of calves each year."

Scale Retained Through Selection.—Just as the interests of the Shorthorn cattle, as bred in the older states, have frequently been sacrificed through excess of zeal on the part of their advocates, so the widespread popularity of the Hereford in the west has not been without its unfavorable effect in certain cases upon the character of the "white faces" produced under range conditions. So eager were herd managers to establish thoroughly the type that had proved so useful in their business that lack of care in selection of bulls led not infrequently to deterioration in size. The best registered bulls were, except during certain years of depression, comparatively high in price. Furthermore, some rangemen while keen judges of values of cattle en bloc were not formerly expert in the matter of what constituted the cardinal points to be sought in an individual purebred animal. In too many cases the red coat and the white face seemed to be about the only qualifications required. Bone, size, constitution and correct conformation were not always appreciated or demanded. To this rule, however, there were of course many exceptions, but the owners of registered herds in the older states usually outbid rangemen for the best individual cattle. As has always

Copyright photo by McClure
WHITE-FACES ON A NORTHERN RANGE.

Copyright photo by Erwin E. Smith
IN ARIZONA.

been the case with Shorthorns, the refuse of the pedigree Herefords, not to mention large numbers of grades, went to the range along with a certain percentage of good ones. Then came the often careless piling of Hereford upon Hereford, with more or less disregard of individual excellence, within the same pastures and with few infusions of fresh blood. At this time, however, there is a very general appreciation of the fact that by care in the selection of big type bulls resort to other blood may be avoided.

In-Breeding from Poor Material Fatal.—In-and-in-breeding, or close breeding, is the greatest potential power for good in the whole realm of animal breeding, but it is attended by good results only where the animals closely mated are of a robust and desirable individual character. The doubling of the blood of inferior or weak specimens of any breed is the shortest of all cuts to absolute ruin. It is apt to intensify faults even faster than it fixes good qualities. Happily, some ranchmen have been wise enough to diagnose this situation and avoid the pitfalls into which certain of their contemporaries fell. Such criticism as has been passed upon some of the range-bred Herefords as lacking in size and quality can in almost every case be traced not to any inherent defect in the breed, but to the application of unscientific methods in the handling of the blood.

In some cases where loss of size resulted resort has been had to a cross back to the Shorthorn. Therefore, the latter breed has in recent years been regaining some of the ground it was forced to yield during the days of the overcrowding of the ranges and the appalling losses following severe winters. A good Shorthorn cross undoubtedly tends to restore bone, scale and stretch to herds that have lost in weight, but it is a somewhat costly remedy and many owners of big herds of "white faces" hesitate about incurring the expense and taking the chance of disturbing an established course. These men are finding that by the exercise of care and judgment, more particularly in the matter of bone and scale, they can maintain a high standard of merit through the use of good Hereford bulls of the right stamp without admixture of other blood.

The Open Range Gone.—The open range is now virtually a thing of the past. The fencing of the land and the water has put the big outfits out of business save in cases where they absolutely control large and well watered tracts by purchase or lease. The dry-farmers and the home-seekers have ushered in another era in the evolution of the west, and other types of cattle will now undoubtedly become more numerous in that region than they have been in its recent past. At the same time there can be little danger of the Hereford ever losing popularity in any land where the conversion of grass into beef is an important business.

CHAPTER XX.

THE CREST OF ANOTHER WAVE.

The latter day records of the Hereford in the central west may fairly be dated from the great Hereford association show and sale held at the Kansas City Stock Yards late in the autumn of 1899, as referred to at the conclusion of a previous chapter. The members of the executive committee of the association at that time were Charles Gudgell, Thomas Clark and H. H. Clough. A special advisory committee as to the details of the proposed show consisted of F. A. Nave, John Sparks and James A. Funkhouser. A big sale was also planned, and Tom Smith, C. A. Stannard and T. F. B. Sotham were named as a committee to handle it. The whole affair was a monumental success, no less than 541 animals being listed in the official catalog. The prizelist was so extensive and was supplemented by so many special prizes that it is impossible to make detailed mention of the scores of beautifully fitted prize-winners. Hereford quality and enthusiasm was here at top notch.

Twenty-five years of American breeding had brought the average merit of the show cattle up to the best standards set by the English-bred winners of the earlier western fairs. Refinement of head and horn and improved hindquarters and thighs were everywhere in evidence. The west was Hereford-mad, and Kansas City was the "white-face" capital. The whole countryside in that territory was wild over the wide-backed, rich-fleshed, furry-haired, low-legged, American-bred Herefords which here presented an amazing collection of well-nigh perfect specimens of the breeder's and feeder's art.

Dale and Armour Rose.—Mr. K. B. Armour of the Armour Packing Co., a leading Kansas City business man now keenly interested in purebred Herefords, gave a $400 challenge cup for the best bull of any age on exhibition. This was captured by Mr. Frank Nave, Attica, Ind., with Dale 66481. Dale was bred by Clem Graves of Bunker Hill, Ind., being sired by Columbus 51875 and out of Rose Blossom, a cow bred by Thomas Smith of Beecher, Ill., from Clark's Peerless Wilton. The second dam was the imported cow Blossom, bred by John Price and owned at one time by A. C. Reed of Chicago, who had a farm near Beecher. Blossom was by Auctioneer, a son of Horace 2d. Columbus was bred by G. W. Harness, Jr., of Galveston, Ind., and was sired by Earl of Shadeland 41st (by Garfield) out of Tom Clark's Pet (by imp. Prince Edward 7001 of Carwardine's breeding). Here, then, was rich fruit from the great Earl & Stuart importation.

Dale was not a bull of as much refinement as many of the others produced in the west about this period, but his feeder, James Price, had not allowed his charge to go hungry. Dale had Garfield's strength

of constitution and stood up nobly to the test imposed. He had put on flesh about as thickly as a compactly fashioned bovine carcass ever carries, and shared with the heifer Armour Rose the adoration of the Hereford-worshipping multitudes that thronged this sensational ringside during the most memorable week of American Hereford history up to that date.*

Armour Rose 75086 was a very perfect yearling heifer that had been given by Mr. Armour to the promoters of the Kansas City Coliseum or Convention Hall, a large structure projected in the public interest, to be raffled off for the benefit of the building fund. As the citizens were all enthusiastic over the project, thousands of tickets were disposed of

---

*The occasional references made in these pages to various herdsmen prominently identified with Hereford breeding in the west should include some mention of another man whom the author has always held in high esteem.

George Waters, one of the best of the old-time herdsmen, has had a long and successful experience. He was born on Jan. 4, 1850, at Barton-in-Fabis near Nottingham, England, in the vale of the River Trent. This valley is noted for rich pastures that have not known the plow for many generations. Shorthorn cattle and Leicester sheep and their grades, and wonderful droves of fat bullocks and fat sheep have roamed those pastures. Waters' father was a butcher, and when George was ten years of age rented a farm, carrying on the butchering trade as well.

On Feb. 3, 1870, George left Liverpool for America on the S. S. Nestorian of the Allan Line, landing at Portland, Me., and proceeding direct to Montreal. Here he made a stay of one day and night, taking while there a sleigh ride out to the Victoria Bridge and crossing the St. Lawrence River where men were taking out ice 3 feet in thickness. Some change that, thought the young man, from the green pastures of the Midland counties of old England! Leaving Montreal he went to Guelph. He walked out to the F. W. Stone farm, Moreton Lodge. Henry Arkell, a native of Gloustershire, England, was then foreman and manager. George applied for work, and Arkell hired him for one month. When part of the month had passed he was engaged for the year. He worked here nearly four years. Waters has always regarded Arkell as the best manager he ever met in charge of a pedigree stock-breeding farm.

From Stone's, George Waters went to Buffalo, N. Y., for one season, but in the spring of 1875 returned to Canada to handle John R. Craig's Shorthorns at Burnhamthrope, near Toronto. In the fall of that year Craig made a public sale of Bates cattle in Toronto the day following a notable Shorthorn convention, and Waters led into the ring 38 head of cows, heifers, bulls and calves

DON CARLOS 33734, AS DRAWN BY BURK.

LAMPLIGHTER 51834, AS DRAWN BY BURK.

and the heifer was drawn by a lady resident from whom Mr. Armour brought her back for $1,000. She later went through the sale to Gov. Sparks at $2,500, as we have already mentioned. She was got by Beau Brummel Jr. 65073, of Gudgell & Simpson breeding, a son of Beau Brummel 51817 out of Petunia 6th by Don Juan 11069. Her dam was by Kansas Lad 36932, he by Beau Real out of Bertha by Torro. Beau Brummel Jr. was by the Don Carlos bull Beau Brummel out of a cow by Don Juan by Anxiety 4th, so that Armour Rose was another one of the many proofs now coming forward of the efficacy of the in-bred Anxiety blood of that period.

Other Notable Winners.—Dale's chief competitors at Kansas City were VanNatta's Christopher, Sotham's Thickset, Scott & March's Hesiod 29th, W.

---

that sold for $52,600. He then took charge of A. A. Crane's Herefords at Osco, Ill. Following this he was with F. P. Crane's Herefords at Independence and Kansas City, Mo., for something like three years. He was also herdsman for Robert Otley, an old-time Shorthorn breeder, for a short time at Kewanee, Ill. He also fed for J. H. Spears at Tallula, Ill., at the time of his closing-out sale in Dexter Park, Chicago, in the spring of 1877, going next to Minneapolis to handle Col. W. S. King's Shorthorns at Lyndale Farm.

In 1880 Waters commenced work for T. L. Miller, Beecher, Ill., handling first his purebred flock of Cotswold sheep, but in the following May Mr. Miller decided to place his Hereford show steers and some young bulls on exhibition at the Union Stock Yards in Chicago and George was chosen to handle them. These were kept in the back part of a livery barn on Halsted St., opposite the Transit House until Miller built his stable on Forty-first St. At the close of the Fat Stock Show the steer Conqueror was sold to "Billy" Smith of Detroit. George still thinks Conqueror was the best steer alive and dead he ever saw. Upon his return to Beecher, Mr. Miller having sold some yearling heifers to W. S. VanNatta, Waters made an engagement at Fowler and fed and showed the first Herefords brought out from the Hickory Grove herd—a yearling herd, two heifer calves and the imported bull Tregrehan.

For many years past George has been in business on his own account. After farming and stock-raising at Windom, Minn., for twenty years, on account of the illness of his wife he sold out and is now located on the south shore of beautiful Lake Pulaski amongst the butternuts, oaks, hard maples, elms and basswoods. Here the grey and red squirrels play and scamper through the trees, and he lives again in memory his boyish days in the Valley of the Trent.

H. Curtice's Beau Donald, Mrs. Whitman's imp. Randolph, Murray Boocock's imp. Salisbury and other celebrities. A son of Dale called Perfection, that was also destined to fame, won the senior bull calf prize.

Christopher drew second to Dale and was a favorite with many for premier place on account of his superior breed character. He was sired by Eureka 58549, a bull of Culbertson's breeding sired by Kansas Lad out of The Grove Maid 22d by Grove 3d. The dam of Christopher was Columbia, by the show bull Washington out of Miss Beau Real 3d by Beau Real. It will thus be seen that Mr. VanNatta's great bull was rich in the blood of good Herefords and had two lines to Beau Real. The sum of $5,000 was refused for him during the week, and one of his calves, the bull Aaron, was bought at auction a few days later at $1,950. Sotham's Thickset, for which $5,000 had been declined earlier in the season, was third. Hesiod 29th came next and Beau Donald fifth. Beau Donald was in his sixth year. He had been bought by Mr. Curtice from H. B. Watts of Fayette, Mo., and was a rare stamp of a good sire—full of character and quality. He was by Beau Brummel out of Donna by Anxiety 4th, and proved one of the greatest sires in Hereford history.

The female classes at this show were real revelations. From the aged cows, including Nave's Dolly 5th—first and champion—down to the junior calves the bloom and beauty of the matrons and heifers supplied ample proof that in the hands of the en-

CHRISTOPHER 44565, BRED BY WM. S. VANNATTA & SON.

IMP. MARCH ON 76035, FROM PALMER'S DRAWING.

thusiastic western breeders of that day the good material that had been transferred so lavishly from English to American pastures in the preceding years had been wisely utilized. Mr. Stannard was coming strong at Sunny Slope and had second prize in cows on the big Lady Matchless 2d by the Don Carlos bull Pride of the Clan. VanNatta's Clodia, by Cherry Boy out of Clover 4th by Parmelee's Anxiety 4th, was third. Sotham came next with Lady Charming, by Corrector out of Cherry 24th by Cedric, of which cow we wrote at the time that she possesses "beautiful character, the sweet head and clean throttle that are characteristic of the Weavergrace cattle, and that furry coat of yellow-red hair, as mellow-looking as a ripened peach, that has come to be the trade-mark of the Correctors."

Tom Clark's massive Everest, by Lars out of Jessie Clark 2d by Anxiety 3d, led the two-year-olds, followed by Sotham's brown-eyed beauty Benison, Clark's Winona, also by Lars, and Gudgell & Simpson's great heifer Mischievous by Lamplighter. Nave's Carnation, by the Shadeland bull Acrobat out of Erica 51st by Garfield, won in yearlings. The same exhibitor got first in senior heifer calves with Theressa by Dale.

Nave also won the grand herd prize, with Sotham second, Clark third and Gudgell & Simpson fourth. The victorious Indiana cattle again scored in the young herd competition. The produce-of-cow prize fell to Clark and the get-of-bull award went to Sotham's Correctors.

Excess Fat vs. Real Bloom.—"The Breeders' Gazette" special for "herd shown in best bloom" was sent by the Shorthorn judges who decided the contest to Mr. Nave's heavily conditioned cattle with Dale at their head—an award which was commented upon at the time by the author of this volume in the following language:

"The prize offered by the publishers of 'The Gazette' for herds shown in the best bloom was awarded by agreement of Messrs. Leonard and Dustin to the lots shown by Messrs. Nave, Sotham and Clark in the order named. In view of the fact that 'The Gazette' has persistently deprecated the awarding of prizes in breeding rings to cattle burdened with excessive fat, and as this prize was offered largely with a view towards encouraging those who do not approve of extreme obesity in these contests, it had been generally believed by those who clearly understood our purpose that the herd headed by Dale would be regarded as rather too rich to receive this recognition. The meat of our definition of showyard bloom was to be found in the clause which stipulated that there must be 'ample evidence of the fact that the bringing out of the beast in suitable showyard form has probably not threatened its physical well being.' Without questioning the right of Mr. Nave's thoroughly trained cattle to win in the regular competitions under existing standards of showyard judging, we do not believe that any experienced breeder will contend that breeding cattle can be brought to such ripeness for the block without threatening seriously their future usefulness.

"We have scarcely read the history of cattle-breeding correctly, however, if we accept a herd of cattle in that condition as presenting what a breeder should regard as bloom. We respectfully refer those who hold to the contrary to the history of Warlaby. The annals of cattle-breeding afford no more striking example of the deadly effects of excessive fat to breeding stock than is afforded by the decline of merit and fertility of the celebrated Booth Shorthorn herd that once dominated the showyards of all Britain. To those who have given this matter special study there could be but little question that Mr. Sotham's Herefords were forwarded in a more practical working condition. We are aware that there is some difference of opinion, and a lot of misunderstanding as to what really constitutes bloom in the showring. There is evidently a considerable lack of information upon that subject in cattle-breeding circles. If, therefore, the offering of these prizes by 'The Gazette' at this show shall have served to call attention to this highly important subject, and shall direct the thoughts of breeders and exhibitors towards a study of the matter of putting a reasonable limit upon feeding for breeding shows, they will have served the purpose for which they were provided. The question of who won and who lost in this particular case is wholly swallowed up in the broader considerations involved in the main proposition.''

More Money for Shows.—At the annual meeting of the Hereford association at Chicago on Nov. 21, 1899, Mr. Sotham presided in the absence of President K. B. Armour and was elected President for the succeeding year. The report of the executive committee was a glowing one. The tide of prosperity was flowing high. The sum of $15,000 was appropriated to be offered as prizes for Herefords at the shows of 1900. The Kansas City event was to

JESSAMINE, BRED AND SHOWN BY THOS. CLARK.—Drawn by Throop.

HESIOD 2D 40670, BRED BY GEO. W. HENRY AND USED BY JAS. A. FUNKHOUSER—Drawing by Palmer.

be repeated, $2,000 was matched against a like amount to be offered by the Minnesota State Fair management, and $5,000 was set aside for a proposed new show at Chicago.

The International Projected.—For some years leading stockmen had hoped that a truly national, and indeed international, exhibition might be established at some central point in the middle west. There came into this field about this time a combination of circumstances that resulted in a realization of this dream.

Arthur G. Leonard, a man of action, a man who deservedly had the confidence of his superior officer, John A. Spoor, was at the time General Manager of the Chicago Union Stock Yards. He had at his side as his general agent at this date William E. Skinner, a man of vision who had the confidence of American stockmen. The time, the place, the men and the money were now in conjunction. A general meeting was called at Chicago for Nov. 24, 1899, and was well attended by representatives of the various stock yard and stock-breeding interests. The International Live Stock Exposition Association was formed, and the first week of December, 1900, fixed as the date for the initial show. The committee on rules, regulations and classification consisted of Alvin H. Sanders, chairman; T. F. B. Sotham, representing cattle breeders; A. J. Lovejoy, representing swine breeders; G. Howard Davison, representing sheep breeders; R. B. Ogilvie, representing horse breeders, and D. O. Lively, representing fat stock owners.

Dawn of the Twentieth Century.—The year 1900 came in with the general enthusiasm unabated, albeit an element of speculation had been creeping into the trade that was causing some anxiety to those who prefer conservative growth and moderate prices to so-called "booms."

Out in Nebraska William Humphrey was running 1,100 head of registered Hereford cattle, the herd being under the management of Capt. E. C. Scarlett. Down at Albany, Mo., Charles G. Comstock had built up at his Grandview Farm "the largest Hereford breeding establishment east of the Missouri River." Gudgell & Simpson, Funkhouser, Sotham, Cornish & Patten, K. B. Armour, Benton Gabbert & Son, Dr. Jas. E. Logan, O. Harris, H. C. Taylor & Son, N. Kirtley, C. B. Smith, "Hamp" Watts, John B. Bell, Miller & Balch, T. H. Pugh, W. J. Boney & Son and many others were vying with one another in upholding the colors of the Hereford in the state of Missouri. Mr. Armour was bringing over 100 head more from England. Kansas City was the center of activity, for it was there that the corn states and the range met most frequently in the course of the transaction of the business of transferring Herefords to the ranchmen now clamoring for white-faced bulls.

A Hereford-Shorthorn Alliance.—The executive committee of the Hereford association met in Chicago early in January, 1900, to plan the year's show-yard campaign. The directors of the Shorthorn association were in session at the same time. The latter had been so impressed by the big Hereford demonstration at Kansas City on the preceding November that they decided to hold a Shorthorn show at the same place during the fall of 1900. This led to a conference with the Hereford committeemen which resulted in an agreement to hold both shows at the same time and place. This action was significant of the fact that the Shorthorn breeders not only realized the necessity for more aggressive methods in presenting the claims of their cattle for public consideration, but that they had full confidence in the ability of the Shorthorn exhibitors to show cattle of a modern type that would not suffer by close comparison with the best Herefords. Both associations likewise agreed to get behind their respective classes at the new Chicago International.

Spring Sales.—About 200 head of cattle were put up at auction at Kansas City on March 1, 1900, and brought an average of $331.80. These were from the herds of Stannard, Sotham, VanNatta & Son and Scott & March. The feature of this series was the offering of the show bull Thickset, generally regarded as the best Sotham had ever bred, and he fell to the bidding of William Humphrey at $5,100. The Stanton Breeding Farm of Nebraska took another son of Corrector, Grandee, at $1,500, and Sir Comewell, another good son of the same bull, was taken by Humphrey at $1,000. Sotham's lot of 50 head averaged $454.70. On March 20 and 21, 1900, Gudgell & Simpson and H. H. Clough sold 99 head of cattle at Independence, Mo., for an average of

BEAU BRUMMEL 51817, AS DRAWN BY BURK.

COLUMBUS 51875, AS DRAWN BY HILLS.

$259.15. A lot of good blood went out from this sale to the western ranges, Col. Torrey, John Scharbauer and other western men being free buyers.

On April 25 and 26 Armour, Funkhouser, Dr. Logan and John Sparks sold 115 head at Kansas City at an average of $283.50. Cattle were bought here for several leading range outfits including the Adair and Whitman herds. George Tamblyn gave $1,000 for the imported cow Prudence.

Death of T. L. Miller.—Although Mr. Miller's[*] extraordinary personal activities in behalf of the Herefords ceased around 1885, he nevertheless continued to take a deep interest in their success. In his latter years he spent much of his time at De Funiak Springs, Fla., at which place he died on March 15, 1900. His remains were brought to Chicago for burial, and he sleeps the long sleep in the sylvan shades of Graceland. He had lived to see his faith in the white-faced cattle shared by a majority of all the cattle-owners on the western range and a tidal wave of prosperity sweeping over the Hereford-breeding business of the cornbelt.

At a meeting of the American Hereford Association held at Chicago in December, 1901, a resolution of respect to his memory was unanimously adopted, in the course of which the following language was employed:

---

[*]In this connection it should be noted that Mr. T. E. Miller, son of the great western promoter of the breed, was for a number of years identified with his father's cattle-breeding operations at Beecher, maintaining at one time a herd of his own. He is now engaged in business in the city of Chicago, and the author hereby acknowledges his indebtedness to him for information supplied in connection with various transactions.

"We freely and unhesitatingly accord to him the position of originator and leader in the propaganda of the Hereford in America; and to his enterprise and courage is largely due the position the Herefords have attained in this country."

Dale Sold for $7,500.—High-water mark up to that date in the way of prices was recorded at Chicago on April 17 and 18, 1900, when Mr. F. A. Nave sold 96 head of Herefords, including his show herd, at the record-breaking average of $671. The champion Dale was taken by his breeder, Clem Graves of Bunker Hill, Ind., at $7,500. O. Harris paid $3,000 each for the young cows Theressa and Russett, $1,000 for the heifer calf Sister Theressa, and $1,400 for the imported bull Bruce. William Humphrey got the imported bull Viscount Rupert at $3,100. Tom Clark paid $1,300 for Perfection. Graves gave $2,600 for imp. Lady Help, $2,100 for Dolly 5th, and $1,600 for Carnation. J. C. Adams, Moweaqua, Ill., took Melley May at $1,000.

The day following the conclusion of this sensational event Tom C. Ponting, Moweaqua, Ill., sold 61 head at his farm for an average of $243, the young Corrector cow Blendress bringing $1,010 from Jesse Adams of Moweaqua.

Kansas City's Dual Show of 1900.—The Hereford-Shorthorn show at Kansas City under the joint management of the two organizations was a huge success. At the Hereford show of 1899 Shorthorn breeders of distinction had been called to place the prizes, but on this occasion resort was had to talent within the ranks. The committee to award the class

prizes proper consisted of the veteran importer and breeder William Powell, W. A. Morgan of Kansas and W. M. Atkinson of New Mexico.

The show was even greater than that of the year before, surpassing in quality anything yet seen in the Hereford section of any American show. In fact, the English Royal has probably seen no better show cattle than the season of 1900 developed in the middle west.

In the bull classes there was a fine specimen of latter-day British breeding presented by "Tom" Sotham, who was a great student and close analyst of Hereford pedigrees. His able and intelligent herd manager, Edward J. Taylor,* had spent the summer

---

*Edward J. Taylor was born at Stansbatch, Herefordshire, in 1866. His father, John Taylor, collected and sucessfully bred a very useful herd of Herefords and one of the best flocks of Shropshire sheep in the country, and as chairman of the Kington Stud Co. did much to improve the class of Shire horses in that section. John Taylor had assisted in the building up of the famous herd of S. Robinson of Lynhales and he personally selected all the foundation females of the afterwards noted herd of R. Green of The Whittern. While in quest of these, and also at home, young Edward had the benefit of his father's advice, and sound judgment, accompanying him to such noted sales as Chadnor Court, The Leen, Stocktonbury, etc. In 1876 the father removed from Stansbatch to Elsdon, a farm of some 400 acres adjoining Lynhales and owned by Mr. Robinson, where he remained until failing health compelled him to relinquish business.

"Ed" came to America in 1888, accompanying a small but select importation of heifers for Merrill & Fifield of Bay City, Mich., and remaining in charge of their herd between three and four years. He afterwards became associated with the Rockland herd of H. H. Clough, Elyria, O. In the spring of 1893 he was commissioned by Mr. Clough to return to England and import a bull and two females to augment his already formidable herd for the World's Fair at Chicago. Speaking of this event Mr. Taylor says:

"I shall never forget the beam on Mr. Clough's countenance as he sized up Ancient Briton when I led him off the boat onto the dock in New York. He said, 'Ed, he's all right!' It was a proud day for both of us when he landed as champion at the Columbian."

After Mr. Clough's dispersion sale Taylor went to Troy, Pa., and fitted a herd for George O. Holcomb, showing them successfully on the eastern circuit. Mr. Goodwin, of "The Breeder's Gazette," commenting on the Hereford exhibit at the New York State Fair, said, in part:

"A few years ago we remarked in a report of this fair that

of 1898 in his native land, and acting under instructions selected and shipped out to Weavergrace the young bull Improver 94020, of Arkwright's breeding, for which $1,500 was paid. He was sired by the Royal champion Red Cross, and was descended all-around from long lines of prize-winners. The bull was brought into competition at the Kansas City show with the best products of American breeding, and while much admired for his forward finish and his width and depth, he suffered somewhat by comparison with the best American bulls when it came to a rear-end examination. Nevertheless he had met on the state fair circuit and defeated such redoubtable champions as Dale, Christopher and Dandy Rex 71689, Gudgell & Simpson's great son of Lamplighter. But upon this occasion Dandy Rex won, with Improver second, Dale third and Christopher fourth.

The sensational young bull of the year was the yearling Perfection 92891, a son of Dale, bred by Frank Nave and sold to Thomas Clark, by whom he was exhibited at this Kansas City show. Benton Gabbert, who bred a lot of good Herefords, including Columbus, the sire of Dale, had second here on Columbus 17th.

Mr. George O. Holcomb needed to introduce his Herefords to corn. He made judicious purchases at Shadeland, but they lacked the finish afforded by feed when set before the public. He has thoroughly learned the lesson of showyard fitting, and his herd as seen on this occasion was one of the best-fitted which has ever come under our review, and would rank well up in any competition."

In the fall of 1894 Taylor assumed the management of the Weavergrace herd of T. F. B. Sotham. He remained at Weavergrace until 1902 when he purchased his present farm home at Fremont, Mich., where he still maintains a small herd of his favorites.

The cows were headed by VanNatta's Columbine after a hard battle against such marvels as Dolly 5th, Benison, Dolly 2d, Everest, Mischievous and Lady Charming. In heifers Gudgell & Simpson turned out a most extraordinary lot, in all of which the Anxiety blood was double-distilled. Such specimens as Blanche 13th, the two-year-old winner, and Modesty, a senior yearling winner, both by Beau Brummel, such heifers as Sophronisba and Dorana 3d by Lamplighter, and above all such a wonder as the junior yearling winner Mischief Maker, by Militant out of Mischievous, and Cleopatra by same sire, have never been surpassed in one year by any one establishment. When to these are added the heifer calves Honora 2d, Miss Caprice, Donna Ada, Bright Duchess 32d, Silver Lining 5th and Gipsy Lady, all prize-takers in this phenomenal exhibition, little more need be said of the success attending the concentration of Anxiety blood by this firm. Mischievous and Mischief Maker were declared best cow-and-calf in the show. The special for best cow-and-two-calves was won by the same pair with Miss Caprice added. The special for best pair of yearling bulls fell to the same herd on Patrolman and Donald Dhu, and the ribbon for best pair of yearling heifers went to Mischief Maker and Modesty.

Fall Sales of 1900.—During the Minnesota State Fair in September 117 head were sold at auction for an average of $188, the 53 females averaging $208. During the Kansas City show in October 185 head were sold at an average of $320, Mr. Gabbert's Columbus 17th going to Frank Rockefeller for $5,050. C. B. Smith paid $1,025 for the Armour yearling heifer Saint Justina. On Nov. 9 the Elmendorf herd was closed out at Omaha, the 66 head offered commanding an average of $207. On Dec. 11 and 12 at Kansas City Messrs. Armour and Funkhouser disposed of 106 head at an average of $350, Frank Rockefeller giving $1,125 for imp. Busybody and $1,025 for Beau Real's Maid. At Chicago while the initial International was in progress 95 head sold for an average of $419, Moffatt Bros. paying $3,500 for VanNatta's March On 13th, C. A. Jamison $3,150 for Clem Graves' Dolly 5th and J. C. Adams $2,800 for Lady Help.

The First International.—The Kansas City show, reinforced by contributions from other herds in the States and Canada, was repeated at the formal opening of the International Live Stock Exposition at Chicago the first week in December, 1900.

Dandy Rex headed the aged bulls again, with Dale, Improver and Christopher following in the order named. Dale was made senior champion, however, later in the week. C. G. Comstock's Gentry Lars, son of Clark's old champion Lars, headed the two-year-olds, and O. Harris of Harris, Mo., had second on Goodenough by Benjamin Wilton. Perfection again led the yearlings, and Sotham's Thickflesh, by Corrector, was best senior bull calf.

Columbine again beat Dolly 5th in aged cows, although the latter was subsequently made senior female champion. Harris forged to the front in two-year-olds with the Benjamin Wilton heifer Betty 2d. This grand heifer had been first at Hamline and many thought she should have beaten Blanche 13th at Kansas City. She was certainly a popular winner at the International. Modesty held down the senior yearlings, and Mischief Maker turned the same trick among the juniors. Lady Dewdrop, from the Harris stalls, was best senior heifer calf.

Gudgell & Simpson won the grand herd prize over Dale and his harem. The Anxieties also drew the young herd trophy. Sotham's Correctors won the get-of-sire contest.

The Big Trade of 1901.—Sotham opened the successful sale season of 1901 by selling 50 head at Kansas City on Jan. 21 at an average of $423.50. the 26 females bringing an average of $477. Clem Graves paid $1,080 for the Corrector heifer Happiness. On the succeeding day Mr. Humphrey sold 70 head from his Riverside herd in Nebraska at an average of $344.50, Mr. Benton Gabbert giving $1,275 for the cow Erica 78th. At a combination sale held at same place on Jan. 23 Clem Graves sold 19 head at an average of $584.20, Mr. J. C. Adams taking the cow Columbia at $1,000, and the heifers Columbia 2d and Carnation at $1,325 and $3,700 respectively. At these sales near 200 head brought an average of $380.

On Feb. 19 and 20 K. B. Armour and James A. Funkhouser sold 104 head at Kansas City at an average of $257. On Feb. 26 and 27 a combination sale was held at Kansas City by Gudgell & Simpson, C. A. Stannard, Scott & March and W. S. VanNatta at which 202 head averaged $294.30, Gudgell & Simpson topping the sale with an average of $383.50 on 45 head. Mr. VanNatta bought the cow Cleopatra at $1,010 and B. E. Keyt took the bull Pretorian at $1,000. On May 21 at a combination sale at Chicago N. W. Bowen of Indiana bid off Dolly 2d and her heifer calf at $5,000, and Belle of Maplewood 3d at $1,900, both exposed by John Hooker. The average on 98 head was $343.

Among the important private transactions in the spring of 1901 was the purchase in England by Mr. Frank Nave of the four-year-old prize-winning bull Protector at $6,000 and his importation to Indiana. Protector was bred by Allen Hughes of Wintercott and was a rich-fleshed deep-bodied bull got by Albion (15027) out of a cow by Rudolph. Capt. "Ned" Scarlett, in charge of the Riverside Ranch, Ashland, Neb., sold to C. A. Jamison of Illinois the imported bull Diplomat and a large number of females. Diplomat met with an accident, however, and lived but one year thereafter.

During the first five months of 1901 nearly 10,000 registered Herefords changed hands at public and private sale. About 1,000 of these went into Texas alone, and some 2,000 head were taken by Wyoming, Montana, Colorado, New Mexico, Utah, Nevada and Oklahoma.

In October 135 head from various herds sold at Kansas City for an average of $253.25, William

Humphrey paying $1,005 for Mr. Armour's imp. Southington and C. B. Stoll of Hamburg, Ia., the same price for Beau Donald 37th. During the first week of December 96 head sold at Chicago at an average of $380, J. C. Adams taking out Harris' show cow Betty 2d at $4,500, the Stanton Farm being runner-up.

Tom Ponting Closes Out.—An important private transaction of the year 1901 was the sale of something over 200 head of cattle by Tom Ponting, Moweaqua, Ill., to William Humphrey, Ashland, Neb., at $35,000. This practically marked the close of Mr. Ponting's career as a breeder of pedigree "white faces." While he had never made any particular effort to force himself or his herd into the limelight, he nevertheless contribtued in a very practical way for a long series of years to the successful extension of Hereford breeding throughout the western states. Mr. Ponting was born in England in 1824, came out to the States in 1847, and engaged in the Hereford business in 1878. He made his first importation in 1882, buying several head at the Carwardine sale. He at one time imported three head of the old gray sort from J. G. Haynes of Monmouthshire. At this writing (1914) Mr. Ponting is still living at the ripe old age of ninety years.

Death of K. B. Armour.—On June 27, 1901, Kirkland Brooks Armour, one of the strongest supporters the Hereford interests had in the west, passed away while yet in the prime of a busy and eminently useful life. His first introduction to Herefords was through the gift of a fine collection of purebred cows made by his uncle, the late P. D. Armour of Chicago. The latter had bought a very valuable group of cows and heifers, full of Grove 3d and Lord Wilton blood, from his friend Mr. C. M. Culbertson, Newman, Ill., intending them as an attraction for a country place owned by P. D. Armour, Jr. This young man showed no special fondness for the cattle, however, and on this account they were shipped to Kansas City to the Excelsior Farm of K. B. Armour. Here they met with adequate appreciation, and with the general revival of interest in cattle-breeding Mr. Armour resolved to materially enlarge and strengthen the herd. He became a heavy buyer of high-class breeding animals from nearly all of the leading herds of the United States, and later on began a series of importations from Herefordshire, England, that culminated in the shipment of nearly 300 head which landed in Baltimore during the summer of 1901. In this work he had the active personal assistance of two of his most trusted employes, Mr. William Cummings and Mr. Frank Hastings.

Kirk B. Armour's brother, Charles W. Armour, succeeded to his Hereford interests and for a long series of years continued to maintain a large herd near Kansas City. On Dec. 10 and 11, 1901, the Armour estate and Mr. Funkhouser made a sale at Kansas City, at which 110 head averaged $338.

Important Contests of 1901.—Interest in the big shows of 1901 centered largely in the competition for premier place among the aged bulls. Sotham's Improver was sent forward in considerably higher condition than he showed during the previous year and made his first appearance at the Iowa State Fair, Des Moines, where he received the blue ribbon with limited competition. The Minnesota show at Hamline was in those days one of the most important events of the year in Hereford circles, and here the imported bull had to meet Gudgell & Simpson's Dandy Rex. Victory rested in this first encounter with Dandy Rex, and in the class competition at Kansas City later on this verdict met with the approval of William S. VanNatta and Thomas Mortimer as judges. Later in the week, however, at the same show the senior bull championship was sent to Prince Rupert 79539, a son of the now famous Beau Donald, exhibited by W. H. Curtice of Kentucky. The Prince was brought forward in high condition, showed the characteristic good Anxiety head and horn, and had a lot of scale and the extraordinary loin that has now come to be looked for in all good specimens of the Gudgell & Simpson breeding. He had stood second to Dandy Rex in the class judging, Improver being third and the Armour entry, imp. Southington, fourth. Curiously enough when the Armour special trophy for best bull of any age came to be awarded Dandy Rex was preferred. The committee which had sent the senior championship to Prince Rupert consisted of Thomas Mortimer and William H. Giltner. The Armour trophy was awarded by Mr. Mortimer and

William VanNatta, the former returning to his first love after having forsaken him an hour before for Prince Rupert.

At the Chicago International a few weeks later Improver was first and Dandy Rex second, the imported bull Protector and Prince Rupert being turned down to fourth and fifth places respectively. This judging was done by Mr. T. J. Wornall, at that time a leading Missouri Shorthorn breeder, and William Cummings of the Armour management. The first-prize two-year-old at the International was Clark's Perfection. He had not been shown at Kansas City, and was presented in such capital form, that he ultimately received at this show the senior bull championship.

At Kansas City O. Harris had first prize and senior female championship on Betty 2d. Mischievous had stood second to her in class. Miss Caprice was junior female champion, having been first among senior yearlings. Modesty, by Beau Brummel, was the first-prize two-year-old at Kansas City. At this same show Gudgell & Simpson had first-prize aged herd, while Harris showed the first-prize young herd and also the first-prize calf herd. Sotham's Correctors were again the winners in the get-of-bull class. Betty 2d repeated her Kansas City winnings at the International. Golden Lassie, by Corrector, was placed ahead of Modesty, Theressa and Mischief Maker in the two-year-olds, and Miss Caprice led the senior yearling heifers. Harris won first prize in both the aged and young herd competitions, and Sotham had his customary place in the get-of-bull contest.

Perfection Brings $9,000.—Early in January, 1902, Thomas Clark offered 58 head of cattle at auction at the Chicago sale pavilion, the star attraction being the show bull Perfection. This proved to be one of the sensational episodes of this period, a spirited contest for the possession of the noted son of Dale between Thomas Mortimer and Gilbert H. Hoxie resulting in the sale of the bull to the latter at the previously unheard-of price for a Hereford bull of $9,000. At this sale it was announced that Dale had been sold privately to Mr. Jesse Adams of Moweaqua, Ill., for $10,000. Mr. Clark's entire lot upon this occasion averaged $497, although the great sum given for Perfection was the only extraordinary figure registered.*

---

*Speaking of Perfection reminds us of "Bert" Fluck. The number of young Englishmen who came out to the States during the period of active importations was large, and many notable successes have been achieved by them. Their stories are always interesting, and in most cases inspiring. In these notes we have taken delight in reciting a few representative narratives of success achieved by young men who came out with nothing but pluck, a natural aptitude for the cattle business, and an inherited attachment for good animals.

Here is the story of "Bert" Fluck, cousin to Tom Clark and Harry Fluck. Let him tell it in his own way:

"My first experience with Hereford cattle dates back to the year 1882, when I was a boy of ten years at home with my father, the late Henry Fluck of Meer Court Farm, Kingstone, Herefordshire, England. Father kept a small heard of twenty breeding cows and always kept the best of sires. It was always my delight to be with him while he was feeding and caring for them as that seemed to be my chosen occupation, which I continued to follow. As I grew older father put more confidence in me. At the age of fifteen years I had complete charge of my father's herd and all herd records, which I considered quite an honor.

"In the summer of 1888 'Uncle John' Lewis, who was then manager of the Shadeland herd, owned by the late Earl & Stuart of Lafayette, Ind., came back to England on a visit. He was staying at my home and it being my duty to show him the herd he became deeply interested in me. He said to father: 'That is the kind of a boy we need in America. He can get a position at any time; you had better let him go back with me.' 'John,

The day following the Clark sale 74 head offered by various breeders at the same place brought an average of $227. At Kansas City on the 14th and 15th of January in a combination sale of cattle consigned from 23 different herds 171 head averaged $227.70.

Sotham's "Criterion" Sale.—On Jan. 28-30, 1902, Sotham held what he called his "criterion" combination sale at Kansas City, upon which occasion 184 head of cattle sold at an average of $341.70. Mr. Sotham's own consignment, consisting of 51 head,

---

I can't spare him', said father. 'Uncle John' said, 'Henry, he can do more for himself in America than he can here in England.' Father said, 'If he wants to go he can', thinking at the same time I lacked the sand to start out. However, I met 'Uncle John' in Hereford and we talked the matter over, which looked bright to me, so I booked my passage to America on the Cunard steamship Servia, which at that time was a very fast boat.

"We set sail July 26, 1888, from Liverpool, England, and landed in New York, Aug. 5, 1888. From there we took the train for Lafayette, Ind. After arriving there we went out to Shadeland Farm, where the herd was kept. It was a sight to behold. The herd was then at its best; the bull Earl of Shadeland 22d was a marvel. After staying at Shadeland a few days 'Uncle' said, 'I am going up to Beecher to see Tom Clark. You had better come along.' Tom being my cousin had visited us in England a few years previous when he made his large importation. I was quite young at that time but remembered him well. To Beecher, Ill., we went. There I found another Hereford herd equal to the Shadeland herd, headed by Anxiety 3d and Peerless Wilton 12774. After visiting there a few days 'Uncle' said, 'You better stay with Tom', which I did and made it my home for six years, then returning to England on a visit. Upon arriving back in America I accepted a position as herdsman with the Hugh Paul Galloway herd of Heron Lake, Minn., under the charge of David M. Fyffe, where I remained until Mr. Edward Paul dispersed the herd. David Fyffe informing me that there was nothing to do except farm work, which at that time I did not care to do, I accepted my old position with Tom Clark, where I remained until he sold his farm and dispersed his herd. After the cattle had all gone it became somewhat lonesome for me and I then accepted a position with the late G. H. Hoxie as manager of his Thorn Creek Herd, at Thornton, Ill., where I again had charge of my old chum Perfection 92891, staying with him four years. I then moved back to Beecher on a farm which I had bought, and there engaged in raising hogs and feeding steers for the Chicago market. Selling my farm at Beecher, I purchased one at Grant Park, Ill., where I still carry on the cattle-feeding business. I hope to engage in the breeding of pure-bred Herefords when my son is old enough to take the responsibility off my shoulders to some extent, as I wish him to follow in my footsteps."

averaged $384.30, the top price being $3,995 offered by Mr. S. H. Godman, representing the Wabash Stock Farm Co. of Indiana, for the young bull Goodcross, sired by imp. Improver out of the famous old matron Grove Maid 22d by The Grove 3d, grandam Mr. Culbertson's celebrated Royal champion Prettyface by old Anxiety. The Corrector bull Bequeather was taken by Carruthers Bros. of Ryan, Ia., at $1,100. Mr. Clem Graves sold 8 head at an average of $1,077.50, the top being $2,300 paid by Carruthers Bros. for the Corrector cow Happiness. Jesse Adams took Bright Duchess 32d at $1,200 and O. Harris bought Madrona by Earl of Shadeland 22d at $1,050. Nine head offered by S. H. Godman, Wabash, Ind., averaged $586.65, the lot being topped by the Cherry Boy cow Park Blanche going at $1,080 to J. Hartley, Fairmount, Ind. F. A. Nave's 10 head averaged $336.50. The Egger Hereford Cattle Co., Appleton City, Mo., sent 40 head through the ring at an average of $215; Jesse Adams of Moweaqua, Ill., 12 head at an average of $317; Geo. P. Henry, Goodenow, Ill., 12 head at an average of $285; C. B. Smith, Fayette, Mo., 10 head at an average of $249; Dan W. Black, Lyndon, O., 5 head at an average of $229; Makin Bros., 9 head at an average of $197; E. B. Keyt, Newton, Ind., 4 head at an average of $212.50 and Geo. H. Adams, Crestone, Colo., 6 head at an average of $186.

Changes in Hereford Headquarters.—The office of the American Hereford Breeders' Association which had for so many years been at Kansas City,

Mo., was removed in 1902 to Chicago. The management of the Chicago Union Stock Yards, in addition to financing the newly established International show, had erected a substantial structure known as the Pedigree Record Building in which quarters free of rent were offered to the various national herd book associations. While there was some opposition to this removal of the Hereford record office the transfer was made, nevertheless. The office remained in Chicago for several years, but it was finally decided to re-establish headquarters at Kansas City, at which point the herd book is still published.*

March On 6th and Queenly.—There was a wealth of new material seen on the show circuits of 1902, the heroes and heroines of the immediate past giving way in all directions to fresher candidates for honors. The first clash was at Des Moines with Tom Clark in the judge's box. A new king had arisen among the bulls. His name was March On 6th 96537, bred by the VanNattas and brought forward by Will Willis from the Funkhouser stalls. He was a son of imp. March On, of the memorable Cross importation, out of Jewel Fowler by Fowler. Wide, deep and wrapped in thick mellow flesh he

---

*In this connection portraits are presented of Mr. Charles R. Thomas and Mr. R. J. Kinzer, the former long-time Secretary of the American Hereford Breeders' Association, and the latter the present holder of that important office. Mr. Thomas served the association for a great many years, not only handling the heavy work of the office during the frequent periods of heavy registration, but having charge of the association's interest in connection with the holding of a great number of special Hereford exhibits at different shows, as well as the conduct of numerous combination sales under the auspices of the national organization. Mr. Thomas also visited England and South America in the interest of the association.

took rank at once as one of the best American-bred show bulls of his day. VanNatta's Marmaduke, by the old champion Christopher out of a Cherry Boy Dam, a bull of pronounced substance, stood second. In the two-year-old class John Letham, manager for George P. Henry, won with Prime Lad 108911, a bull which even then gave promise of the greatness that was to come his way. Another showyard model that was to win her way to future championships was the two-year-old heifer Queenly, bred by Steward & Hutcheon and now owned by Messrs. Van-Natta. She topped her class and later was adjudged best female of any age. March On 6th was champion over all bulls.

At Hamline the following week, under Ed Taylor's judgment, March On 6th was again at the head of his class, but in the bull championship the wondeful character and quality of Prime Lad brought Mr. Henry that high honor. This grand young bull was sired by Kansas Lad Jr. out of Primrose, a cow bred by Arthur Turner and imported by K. B. Armour. Gudgell & Simpson* won the blue ribbons on both senior and junior yearling bulls with Belis-

---

*George Shand was born near Huntley, Aberdeenshire, Scotland, in 1845, left Scotland in 1882 and came to Canada, where he lived for three years. He came to Gudgell & Simpson at Independence, Mo., in March, 1885, when Anxiety 4th was in his prime. He left Gudgell & Simpson in 1896 and went to work for Charles B. Dustin in Illinois, staying there until the Dustin Shorthorn herd was sold in 1900. He came back to Independence in 1900 and worked for J. M. Curtice eighteen months. At the end of that time he went to work for Alexander Fraser as foreman and herdsman of a Shorthorn herd and stayed there for thirteen years, or until Mr. Fraser's death, when the herd was dispersed. He then went to work for W. C. Thompson at Plano, Ill., with a herd of Shorthorns, and stayed there until the fall of 1913, when he returned to Independence to make his home with his son-in-law, George Hendry, who succeeded him as head cattleman on the Gudgell & Simpson farms.

arcus 126243, by Militant out of a Don Carlos dam, and Bright Donald by Donald Dhu, and had first on senior bull calves for Rex Premier by the champion Dandy Rex. Harris took first in cows with Russett over Modesty. Queenly was first in two-year-olds over Miss Caprice, as well as female champion.

Beau Donalds to the Front.—While this was going on in the west W. H. Curtice of Kentucky, F. L. Studebaker, Warren, Ind., C. A. Jamison, Peoria, Ill., and G. W. Harness, Galveston, Ind., were putting up a good show east of the river. At Columbus, O., Mr. Curtice appeared with fourteen entries of which twelve were the get of Beau Donald—all young things of real quality. He had Prince Rupert out again to head the senior bulls. The Beau Donald youngsters, however, were the real attraction of the show, and with them the young herd, the get-of-bull and the produce-of-cow prizes were won, Beau Donalds 39th, 41st and 54th and Belle Donalds 27th, 55th, 56th and 59th specially honored. Belle Donald 59th was made champion female under two years old, all breeds competing, and the Curtice herd won grand championship of the yard over the Hanna Shorthorns and the Bradfute Aberdeen-Angus.

At the Illinois State Fair O. Harris won first in aged bulls with Beau Donald 5th over C. A. Jamison's Arlington by Earl of Shadeland 22d and the same owner's Sailor by Acrobat. Mr. Curtice met heavier metal here, however, in the young herd competition and had to accept second to the Harris entries.

Clem Graves' $1,000 Average.—During the Indiana State Fair of 1902 Clem Graves made a sale of 43 head of cattle which resulted in the extraordinary average of $1,007. This sensational figure was reached largely through the fact that the bull Crusader 86596 was run up to $10,000, and knocked off to Ed. Hawkins of Earl Park, Ind. Dolly 2d was taken by the same bidder at $7,000, and Cosmo, the dam of Crusader, at $3,000. Crusader was a richly-bred, low-legged, wide-bodied bull with a grand front and had just been made champion in strong competition.

These prices were so startling that they created a veritable sensation in the American cattle-breeding world, and when some time later it developed that Mr. Graves had taken the cattle back there was some doubt created in the public mind as to the genuineness of the transaction at the sale ring at Indianapolis. This being the case, the author has requested Mr. Graves to make public a plain statement of the real facts in the case, and in compliance he has furnished the following:

"Crusader, sire Cherry Ben, full brother to Columbus, dam Cosmo by Cherry Boy, was at the head of my herd when I sold the Dale Stock Farm to A. C. Huxley. I engaged Col. David Wallace to act as manager of my dispersion sale held Tuesday of the state fair, Sept. 16, 1902. There were fifty-four cattle listed and the sale expense was $103 on each lot. I believed that the class of cattle I had to offer merited this outlay, and the interest in this sale was such that I was honored by the presence of nearly every Hereford breeder of prominence, and many of the Shorthorn and Angus breeders, as well.

"There were several bidders on Crusader. Among them I recall S. J. Peabody, Gilbert Hoxie, S. L. Wright and James R. Henry, who later in the sale purchased Dale Wilton. Ed. S. Hawkins and C. E. Amsden were the contending bidders up to $10,000, when he was sold to Mr. Hawkins. I learned after the sale that Mr. Amsden, then recorder of Shelby county and an ardent Hereford enthusiast, thinking that Crusader would likely sell at a high figure, had interested a number of Hereford breeders in his section of the state to join him in the attempt to secure the bull, and that one of the bankers at Shelbyville came to the sale with them to make the settlement should they succeed in buying him. If Crusader had been sold to Mr. Amsden the deal would have been closed with cash.

"Mr. Hawkins bought Cosmo, the dam of Crusader, with Amy Dale at foot and bred to Dale, for $3,000 and several other cattle, his total purchase amounting to $17,520. I had sold him cattle in a breeders' sale at Chicago in the spring of 1902, and he had promptly settled with his check. I had visited his home, a palatial residence situated on a farm of 6,000 acres of Benton county's richest land, stocked with Thoroughbred horses, Hereford cattle, and a large number of feeding cattle. Col. Wallace made the settlement for the sale and when he informed me that Mr. Hawkins desired time on a part of his purchase I had no reluctance in accepting his note. In May, 1903, Mr. Hawkins made it known to his creditors who held cattle paper that he was financially embarrassed and invited them to meet in conference at Earl Park. We found that the real estate belonged to his mother, the live stock was mortgaged, Mr. Hawkins was broken in health, and unable to

CRUSADER 86596, BRED BY CLEM GRAVES.

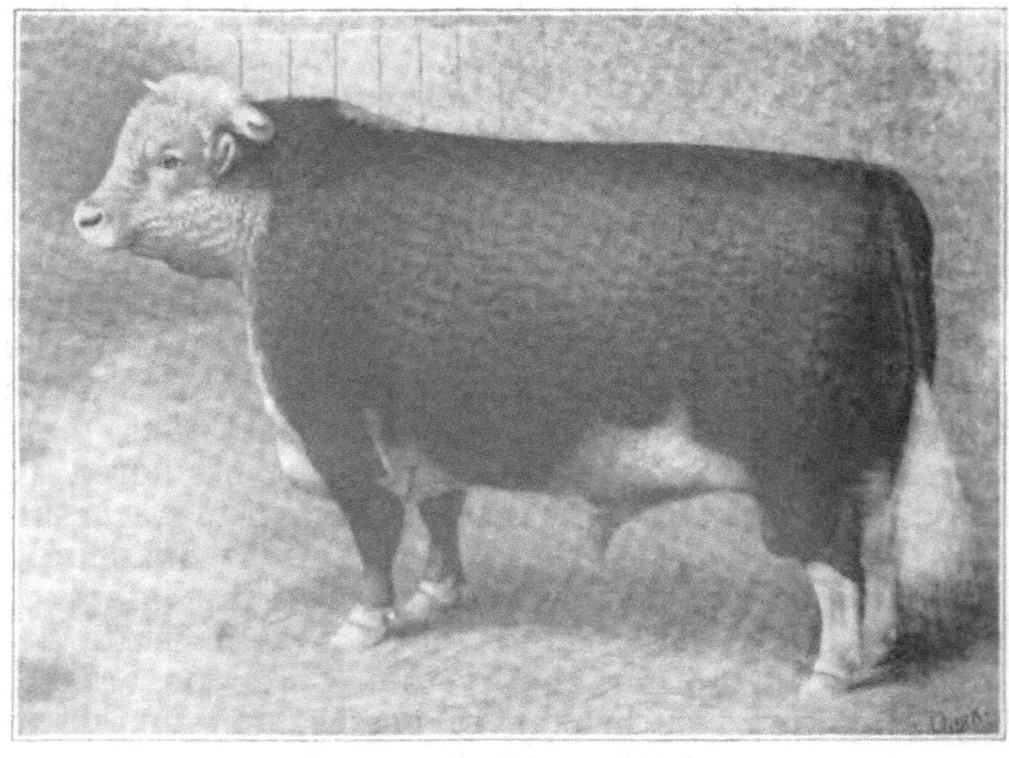

DALE 66481, THE $10,000 CHAMPION—BRED BY CLEM GRAVES.

supervise his business; in short everything was apparently going wrong.

"The cattle had received but little care or feed. 'Cruse' had been left out doors to sleep on refuse straw where the bush of his tail had frozen fast and been pulled out. The cattle were in a deplorable condition. We were in council several days before we could plan a course of action. We then decided to pay this mortgage. I paid $3,225 of it. We then listed the cattle and selected, each man in his turn from this list, until his claim should be satisfied. I selected Crusader, Cosmo and seven other cows for my claim. I kept the bull till December, 1906, when I sold him to Hon. George Chandler, Baker City, Ore.

"Crusader was pronounced by capable judges of Herefords to be the best front-ended bull they ever passed on. I never saw a bull that carried his head so well; in fact he was a remarkable specimen of bovine beauty. Crusader was first and champion bull at the Pan-American. He was not exhibited at the World's Fair at St. Louis, but he met and defeated both the senior and junior champions of that show, Prime Lad at Indianapolis and Mapleton at the Virginia State Fair.

"I am glad to make this statement in order to clear up the doubt as to Crusader selling at all, since he was returned to me. He did sell for $10,000 and was well worth it. If Mr. Hawkins had not become ill, and had not lost heavily in the race-horse business, I am confident that he would have finished paying out on all the cattle."

Broadening the Kansas City Show.—The fourth show since the Hereford association launched the first at Kansas City in 1899, was participated in by eight different breeders' organizations. The American Royal, as it has been called in recent years, was now fairly on its feet.

The Hereford exhibit was as impressive as ever. March On 6th was the senior bull champion and Benton Gabbert produced the two-year-old winner in Columbus 33d, a bull of unusual scale and exceptional substance. Bright Donald was junior bull champion, and Queenly the champion cow.

On Dec. 8 and 9 at Kansas City Benton Gabbert and Dr. Logan sold 76 head for an average of $227. At this sale G. E. Reynolds gave $1,000 for Hesiod's Best.*

Various breeders consigned cattle to a combination sale at Wabash, Ind., on Dec. 18, at which 63 head sold for an average of $225.70.

A New International Champion.—Mr. George Leigh had brought over from Herefordshire one of the biggest white-faced bulls of his time, Britisher, an English showyard favorite bred by Allen Hughes from Albion. He was entered at the Chicago International of 1902, where he not only headed the senior class by defeating his half-brother imp. Protector, Beau Donald 5th, Columbus 17th, and other good ones, but received the senior and grand cham-

---

*Reference has been made to the fact that old Imp. Hesiod, the sire of Hesiod 2d, had a bad temper. It took two men with ropes and staffs to safely present him in the showyard at an old-time Chicago exhibition. Speaking of this John Gosling is our authority for the statement that the fighting spirit in this fine bull was undoubtedly brought to the surface by the flopping of the long white smocks frequently worn by the old English herdsmen. On windy days the flopping of a smock or of an overcoat has been known to develop, for some occult reason, the combativeness of bulls. Mr. Gosling gives it as his opinion, however, that "the width between a bull's eyes has more to do with a bull's disposition than the flopping of a woman's petticoat or a smock." Once the fighting spirit is aroused, however, the staff usually has to be brought into requisition as a measure of safety.

pionship prize as well. He was shown at a weight of nearly 2,800 pounds, and was described at the time as "a bull of magnificent masculinity, most massive and imposing in appearance, with head, horn and crest of pleasing character, a brisket that hangs close to the ground, a tremendous spread of rib, showing the most tablelike back and loin of any bull of the breed that has yet fallen under our observation." His hindquarters were long and heavy, so bulging that his tail hung over them on that slant that was so noticeable in old Grove 3d. His flesh had begun to slip, however, under the strain of repeated fittings and the long voyage to America, and he was counted fortunate in going through this show with such signal honors.

Second to Britisher at the Chicago show was Frank Rockefeller's Columbus 17th, bred by Benton Gabbert and sired by Columbus, the sire of Dale. He sold for $5,050 at Kansas City. He was a great-ribbed bull carrying a lot of flesh on stout, well set legs. Prime Lad had a walk-over in two-year-olds. Harris led the senior yearlings with Goodenough 2d. Bright Donald was best junior yearling. Mr. Henry's Disturber by Beau Donald 3d was best senior bull calf, and Harry Fluck* had a flash win-

---

*H. J. Fluck, cousin to Thomas Clark, was born in the city of Hereford, England, on May 21, 1862. His father was a merchant in that city, born at Dinedor's Court, Herefordshire. At the age of two years Harry was taken to the country by his uncle and aunt, Mr. and Mrs. Henry Fluck of Upper House, Didley, St. Deouvrix, Herefordshire, where he was brought up by them. Always imbued with a fondness for fine stock of all kinds, his environments during his boyhood days added zeal to his ambition, for his uncle, who was one of the best all-around stockmen of his time in Herefordshire, possessed one of the good herds of "white faces" at that period. He took no interest,

ner in junior bull calves in Star Wilton by Peerless Wilton 39th.

Russett, Modesty, Betty 2d and Mischievous—all old friends—were ranked in the order named. Queenly presided by common consent among the two-year-olds, Lucile 2d of the Harris herd was at the top of the senior yearlings, Steward & Hutcheon's burly Madge came forward in junior yearlings, and the same firm had the honor of drawing the blue in the senior heifer calves with Beau's Queen by Beau Brummel.

A Beau Brummel-Fowler Nick.—Queenly's wonderful wealth of flesh carried her through this show

---

however, in recording cattle, simply keeping the herd on a rent-paying basis by selling steers and culling out females for the block. Mr. Fluck is somewhat proud of the fact that he is of the third generation of the family of Flucks who were closely identified with the raising of good Hereford cattle. His grandfather, Thomas Fluck, owned one of the good herds of Herefordshire, and produced the foundation sire of our latter-day Anxieties, the bull Dinedor 132 (395). The celebrated Walford 47 was a grandson of Dinedor.

Harry became somewhat dissatisfied with his prospects at home and after reading and talking to others about the allurements of other lands determined to leave his mother country and seek his fortune in America. He landed in September, 1880. In the fall of 1881 he became connected with the Culbertson herd, and his initial attempt for showyard honors was in 1883 with the steer Roan Boy and others. Speaking of this Mr. Fluck says: "This was only a preliminary show, and it took six men to lead Roan Boy into the ring. He was exceedingly nervous. He was not considered worthy by the judiciary at that time, but the next year I sprung quite a surprise on the boys. The unexpected happened. Roan Boy appeared as one of the best models of a beef steer that ever graced a show arena, winning every prize from class prize to grand championship, also winning the gold medal presented by 'The Breeder's Gazette,' which under the rules had to be won two years in succession or any three years. We again captured the much coveted prize with a steer named Dysart in 1885 and won it finally in 1886 with the steer Bowdoin. These were three outstanding good steers."

Mr. Fluck severed his connection with Mr. Culbertson in 1886 to take a more lucrative position with G. W. Henry of Rossland Park Farm, Ashkum, Ill. During his connection with that herd he participated in the invasion of the Shorthorn strongholds of Kentucky with a show herd of Herefords as detailed elsewhere in this volume. Mr. Henry's show bulls were Caractacus and Prince Edward. There was also in the herd such other bulls as Hesiod, Anxiety 2d and Lord Derby. Some of the best females were Edwina, Annie Laurie, Lady Pet and Miss Stewart. Under Fluck's management Mr. Henry also had the winning steers Long-season of 1902 without a slip. She was senior and also grand champion over females of all ages at the International, gaining the honor over the junior champion Beau's Queen. And thereby hangs a tale: Queenly was by Beau Brummel out of the VanNatta-bred Fowler Queen by old Fowler, and Beau's Queen was out of the same dam and sired by a son of Beau Brummel! And here they stood, sisters in blood, the two lone contestants at the greatest show of the year for the highest honor that could fall to a Hereford female.

The Giltners Buy Britisher.—At Chicago on Jan. 7 and 8, 1903, in a combination sale 90 head were sold for an average of $265, the top price being $3,800 paid by Giltner Bros., Eminence, Ky., for

---

fellow and Sir Edward. Some of the show-bulls at Rossland Park, which Fluck was partly instrumental in producing, were Hesiod 2d, Sitting Bull and Caractacus Wilton. In 1890 Rossland Park was sold and the stock was disposed of by auction.

Mr. Fluck then embarked in business on his own account, buying out the old-established herd of George F. Baker of Oakland Stock Farm, Goodenow, Ill. In 1893 at the world's fair he showed a good yearling steer which took second prize, a bull calf, and the two-year-old Sitting Bull, which won first in class and was finally made champion over all breeds. At St. Louis shortly after, Mr. Fluck sent down Sitting Bull, the calf and others and took blue ribbons. This same year at the fat stock show in Chicago he won a cup offered by "The Breeder's Gazette" for best steer bred and fed by exhibitor with the yearling Percy that weighed 1,610 pounds. Percy was second to Cherry Brandy at the world's fair, but won over him at this show two months later. The next year the show was held at Tattersall's on Wabash Ave., where Percy won this cup again.

Mr. Fluck has shown at every International since its inception, and won a sweepstakes over all breeds three years in succession. He bred and fed the grand champion Peerless Wilton 39th's Defender in 1906, the reserve grand champion Fluck's Expectation in 1904, and champion herd and the get-of-sire in the same year. At the St. Louis exposition in 1904 he was the only Hereford breeder to win a championship over all breeds. This was taken by the steer Fluck's Expectation. Looking back over his career Harry says: "There are two achievements which I am not a little elated over—one to be the first man of the Hereford fraternity to select, feed and show a Hereford steer that was made champion over all breeds, and another to have taken the Herefords into the state of Kentucky in the '80's and won over Shorthorns in strong competition against many of the illustrious breeders of that day."

imp. Britisher.* Tom Clark gave $1,525 for the young bull Perfection Lad by Perfection.

---

*The firm of Giltner Bros., comprising Messrs. W. H., Robert R. and F. C. Giltner, first engaged in the breeding of Hereford cattle in the spring of 1897. The first cattle purchased were an imported bull and ten Shadeland-bred heifers. This bull proved impotent and after a diligent search for a successor Acrobat 68460, sired by Anxiety Monarch and out of a daughter of the celebrated Earl of Shadeland 22d, was purchased at a cost of $1,500. This bull was used extensively and successfully until nine years of age, when he was sold for $3,500 to C. E. Clapp, Berryville, Va. The next herd bull used by Giltner Bros. was Britisher, assisted by imp. Protector 117878, a one-time English champion imported by Frank A. Nave at a reported cost of $6,000. Britisher and Protector were both sired by Albion 76960, a champion and a sire of English champions.

In the meantime the firm purchased in 1898 forty cows and heifers from Wallace Libbey. These cows were sired mostly by Rantin Robin 50603, he by Earl of Shadeland 12th 20109. The heifers were sired by Welldone 68786, a full brother to the famous Sir Bredwell by Sotham's Corrector. They nicked kindly with Acrobat and from the beginning a class of young stock was produced which sold at from $200 to $600 each, which "looked mighty good" to the firm, considering the fact that a two-year-old steer was at that time bringing only from $50 to $60.

From their own herd the brothers retained the bull Acrobat's Beau Donald 157648, a son of Acrobat 68460 and out of a cow by Beau Donald 58966. This bull proved most useful and was not only a prominent prize-winner himself all through the south at the leading state fairs but sired Florence Acrobat 283070, the 1909 junior champion female at the American Royal, also first-prize heifer calf at the Royal in 1908 and all leading state fairs that year.

About 1903 Messrs. Hornsby Bros., neighbors of the Giltner Bros., purchased from Gudgell & Simpson the Beau Brummel bull Beau Roland 102767, to be used on their own herd. They allowed Giltner Bros. the free use of this animal, and the blending of Britisher and Beau Roland blood gave excellent results. From this cross was secured Beau Columbus, which was first as calf, first as yearling, second as two-year-old and first as aged bull at the Kansas City Royal and first and junior champion at the 1909 International. British Corker 283072, first-prize aged bull at Denver for two years, was sired by Britisher and out of a Beau Roland cow. British Highball 267816, a prominent winner, which sold to S. B. Burnet for $1,500, was bred in the same way.

While Giltner Bros. have not gone in extensively for show-yard competition, they have brought out each year a good herd of their own breeding. They have directed their efforts chiefly to supplying the immense field presented by the southern states and the export trade. Five state colleges have purchased breeding stock from Giltner Bros., as have also the governments of Cuba, Porto Rico and Brazil. They have customers in Argentina, Panama, Santo Domingo, Mexico, Hawaii and Canada. The junior member of the firm, Mr. F. C. Giltner, was for six years a director of the American Hereford Breeders' Association, of which oganization he was president from January, 1912, to October, 1913.

CHAPTER XXI.
HISTORY REPEATS.

The high prices of 1900, 1901 and 1902 could not hold. Just ten years after the panic of 1893 another one of those widespread commercial depressions that have so often been registered in our country's progress was setting in and by 1903 liquidation was general. The cattle business suffered in common with all other industries and the Hereford-breeding fraternity did not escape its share of depression. Values declined rapidly, the speculative element liquidated, and during the years immediately succeeding the bargain counter was very much in evidence. As usual in such cases, those who were in a position to purchase good, well bred cattle at low levels in due course of time reaped full reward. The return tide did not set in until about 1909.

Lower Values at Auction.—At Kansas City on Jan. 14 and 15, 1903, Charles W. Armour and Mr. Funkhouser sold 107 head of cattle at an average of $245.30, the highest price reached being $625 for the young bull Onward 9th, a son of March On 6th bought by Murdo Mackenzie for the Matador herd. At Chicago on Jan. 29 in a combination sale 68 head sold for an average of $164, the top being $600 for imp. Princess Royal, bought by Amsden & Sons,

Shelbyville, Ind. At another Chicago sale on Feb. 17 and 18 the 66 head averaged only $131.50. On Feb. 19 at Indianapolis 74 head consigned from various herds reached an average of $169.40, the highest price being $540 paid by F. L. Studebaker, Warren, Ind., for the cow Lucy M 2d, by Shadeland Dean. At Kansas City on Feb. 25 and 26 109 head from such herds as Gudgell & Simpson's, C. A. Stannard's, Scott & March's and F. R. Rockefeller's went under the hammer at an average of $164.25, the best price being $385 paid by S. L. Brock for Annabel 5th, by Militant. Nothing could better illustrate the trend of the market than the fact that good things offered by Gudgell & Simpson and sired by Beau Brummel, Lamplighter and other noted bulls of the Anxiety blood passed through the ring at around $200 per head. And yet much darker days than these had been experienced in the earlier years and greater gloom was in store for 1904. On Feb. 21 and 24 at Des Moines, Ia., in a combination sale 66 head were knocked down at an average of $111.50. Mr. F. A. Nave sold at Attica, Ind., on March 18 50 head at an average of $280.80, the top being $795 paid by W. S. VanNatta & Son for Royal Daisy 2d, the dam of imp. March On, sold in calf to Dale. The famous show cow Benison went to Giltner Bros. at what was called the bargain price of $505. G. H. Hoxie on May 14 sold 38 head at Thornton, Ill., for an average of $300. On May 22 F. L. Studebaker sold 28 head at Wabash, Ind., for an average of $225. On June 16 and 17 T. F.

B. Sotham at Chillicothe, Mo., disposed of some 1,800 head of stockers and feeders—along with 50 head of pedigree Herefords, the total receipts for the two days' sale aggregating near $54,000. The stockers and feeders were sold in lots to suit purchasers and averaged about $24.30 per head. The pedigree cattle averaged $232. The top price of the sale for the registered cattle was $625 for the two-year-old Clem Graves, by Dale.

In August, 1903, Frank Nave sold his $6,000 bull imp. Protector to Mr. T. A. Fletcher, who was for so many years active in the management of the Indiana Blooded Stock Co. The price was not made public.

Fall Sales of 1903.—Prices did not mend as the season advanced. George H. Adams, Crestone, Colo., closed out 107 head at Linwood, Kans., on July 28 and 29 at an average of $163.65. Mr. Adams was the owner of a 100,000-acre ranch in the San Luis Valley in southern Colorado, upon which he carried some 5,000 head of cattle all showing more or less Hereford blood and including at one time 200 head of registered animals.* These had been procured originally from the best sources, such as the closing-out sale of Thomas J. Higgins'

---

*As early as 1872, Mr. Adams began the improvement of his herd—founded by selection of the best native stocks in 1869—by the use of pedigree blood. He paid J. C. Shropshire of Kentucky $1,060 for two Shorthorn bulls and continued the use of this blood for seven years, when having seen some of the Herefords sent to Colorado by T. L. Miller he concluded that the Hereford possessed superior hardiness. In 1879 he purchased 150 Colorado-bred grade Hereford bulls and later he visited the herd of T. L. Miller and other Illinois and Indiana breeders and purchased $8,000 worth of Hereford bulls. From that time Mr. Adams was one of the most spirited and liberal supporters of the Hereford.

stock and the great Sunny Slope offering of 1898, where he bought 21 head at an average cost of over $500 per head, including the imported heifer Luminous at $1,500 and imp. Leominister Daisy 2d at $1,205.

Mr. Adams had in the meantime bought the famous Linwood Farm—so long celebrated as the home of the Scotch-bred Shorthorn herd of the late Senator W. A. Harris—and had placed George Morgan in charge. On account of failing health, however, he decided to give up the handling of the purebreds and they were disposed of on the dates mentioned. Many of these cattle were range-bred and not halter-broken. They were necessarily presented in pasture condition and naturally failed to bring their real value, especially at this period of depression. Luminous sold for $600 and her son Orpheus 2d for $400. Tom Ponting was a buyer of useful cattle for his sons Everett and Wayne. The top of the sale was $770 for the cow Lulu with twin heifer calves at foot. She was taken by Mr. Sotham, who had managed the dispersion.

On Aug. 11 and 12 at Wabash, Ind., various breeders sold 113 head at an average of $146. At Kansas City on Oct. 22 98 head from various Missouri, Kansas and Nebraska herds averaged $163, W. H. Curtice receiving the top price of $600 for Beau Donald 48th. On Nov. 17 and 18 C. W. Armour and J. A. Funkhouser passed 97 head through the auction ring at an average of $155.70. The best price made upon this occasion was $855, which was regarded as a bargain figure for the capital yearling bull Onward 19th, by March On 6th. He was taken by Benton Gabbert. During the International at Chicago 71 head were sold on Dec. 3 at an average of $168.75. The highest mark made here was $1,400 for the two-year-old bull Prairie Donald 139616, offered by the Stanton Breeding Farms, Madison, Neb., and bought by William Reynolds, Lusk, Wyo. Only four females in the entire lot reached the $400 mark. These were all daughters of Beau Donald and bred by W. H. Curtice. At Kansas City on Dec. 10 and 11 C. A. Stannard and Mrs. Kate Wilder Cross sold from Sunny Slope Farm, Emporia, Kans., 100 head for an average of $172.90. Mr. Stannard's yearling show bull Keep On 26th by imp. Keep On was taken by the Messrs. Harris at $600.

One of the regrettable incidents of this general liquidating movement was the enforced closing-out of the Sotham herd at Chillicothe, Mo., the dispersion occurring at the farm on Dec. 15. The show bull Fulfiller went to O. Harris at $1,510. He was sired by Improver, and was a son of the beautiful Benison by Protection, second dam Benita by Corrector. Protection was by Corrector out of a daughter of Royal Grove. The 128 head averaged only $120.65.*

---

*Speaking of this event "The Breeder's Gazette" commented at the time as follows:

"The results of Mr. Sotham's life-work as a breeder of Herefords were scattered on Dec. 15 at sheriff's sale. William Moffatt, Paw Paw, Ill., foreclosed a mortgage which he held on the herd and sold it out. Mr. Sotham had relied on a promise of financial aid which would have enabled him to save the cream of the herd and retain it under his management, but this failed him almost at the last minute, when it was too late to organize a local

An important private transaction of this period was the disposition of George P. Henry's herd at Goodenow, Ill. It went to James R. Henry of Gosport, Ind., who subsequently resold a number of the cattle to Messrs. VanNatta and S. L. Brock, Macon, Mo. Along with the good cattle obtained from this source Mr. Brock secured as manager Mr. John Letham, in whose hands the herd became the fountain-head of many high-class Herefords in the years that followed.

Death of George Morgan.—For some time prior to the Adams dispersion sale Mr. Morgan, the veteran importer and herd manager, had been in poor health, and late in August, 1903, he died in a hospital at Chillicothe, Mo., treatment for a carbuncle having failed to bring relief. Arrangements had been made by Sotham for Morgan to join him in handling Herefords at Weavergrace Farm, but this was not to be.

The name of George Morgan will ever stand conspicuous among those playing large parts in the introduction of Hereford cattle in the western

---

company to buy the best of the cattle and hence all have been scattered.

"Mr. Sotham has faced some misfortunes in his life, but it may readily be believed that the bitterest of them was when he stood in the salering and lent all possible aid to the forced dispersion of the herd which had been the pride of his heart. Without his assistance buyers were chary of taking hold, but when he entered the ring and guaranteed the transfer of all animals sold and worked earnestly in the interests of the sale the bidding became spirited and an average of around $120 was reached, by young and old, big and little. This is an excellent showing under all the circumstances. Only a tithe of the real value is usually reached at sheriff's sales, and considering the number of old cows in the herd which had been retained on account of demonstrated greatness as producers, and the condition of the cattle, sold without fitting or preparation, the result is better than had been expected. From the cattle and farm implements a total of $17,200 was realized."

states. His relations to various important transactions have already been set forth. He was generally regarded as a keen judge of a good animal, and personally selected in Herefordshire some of the greatest cattle transferred to American soil during the period of extensive importations. In the course of his long career in the business he naturally acquired a great store of information concerning the breed on both sides the water. His facility of expression, his aggressive personality and his keen sense of humor made him the life of almost any company of congenial spirits in which he might be found. The author regrets that he has been unable to procure a photograph of Mr. Morgan for reproduction in this volume along with other notables of his time. However, his work is his own best memorial and title to appreciation at the hands of posterity.*

Prime Lad and Beau Donald 5th.—There was a hard-fought battle at the fairs of 1903 between the coming and the going champions. Prime Lad, younger and fresher and admirably representing the old warrior Beau Real, was hammering hard on the shield of Beau Donald 5th. In the hands of the Messrs. Van-Natta the Lad was slowly but surely making-up into

---

*It was sometimes difficult to tell whether Morgan was talking in jest or earnest. While haranguing a crowd of cattlemen one night in the early days upon the merits of the Hereford for western range purposes he made a statement substantially as follows, which of course created much amusement:

"I'll tell you 'ow it is: You see the 'ereford is something like the buffalo; 'e 'as a 'eavy 'ead and 'orn, is deep through the shoulders and chest, and bein' light be'ind 'e climbs the 'ills fine."

As present-day breeders have long since given the typical Hereford two ends as well as a middle, Morgan's buffalo exaggeration may now be treated, as he intended it at the time, as a joke—a good specimen of the ordinary play of his nimble wit.

a wonderful specimen of the breed—evenly balanced and full of character and quality. The old Beau, with his ponderous hind-quarters and extraordinary expanse of loin, impressed yet again the improvement being wrought in America in respect to rear-end finish. In the preliminary competition at Sedalia, under a Shorthorn judgment, he had been preferred by Mr. Wornall to Steward & Hutcheon's Beaumont, and by Wiley Fall at Des Moines he was set above Prime Lad. At Hamline with W. A. McHenry, of Aberdeen-Angus fame, and D. Y. Robertson, manager of Dan Hanna's Shorthorns, on the bench Beau Donald 5th defeated both bulls, but at Indianapolis N. H. Gentry ordered Prime Lad to the front—a rating which stood for the remainder of the season, being confirmed at the Kentucky State Fair at Owensboro, at Springfield, at Kansas City and at Chicago.

Three Great Groups.—There were at least three overwhelming demonstrations of the prowess of American breeders on the circuits of 1903—the Beau Donalds, which herdsman Hendry continued to send forward with never-failing quality, the Benjamin Wiltons, with which Overton Harris made such a "hit" during this period, and the get of March On 6th, now coming from the Funkhouser herd and showing outstanding character. At Sedalia Funkhouser had the senior bull championship over Beau Donald 5th with Onward 4th, and the junior bull championship on Onward 8th, besides the prize for best four get of the same

BEAU DONALD 5TH 86142, AS DRAWN BY BURK.

PRIME LAD 106911 AT THREE YEARS, AS DRAWN BY THROOP.

sire. At subsequent shows, indeed for a series of years, the depth and width of the Keep Ons and the March Ons attracted fresh attention to the value of the imported blood which John Steward and Harry Yeld had brought out for Mr. Cross in 1898.

The Kansas City and International shows of 1903 were so rich in toppy youngsters that it is impos-

ONWARD 4TH AND HIS TRAINER WILL WILLIS.

sible to enter into details here. Harris, Curtice, Funkhouser, the VanNattas, Stannard, Gabbert, Steward & Hutcheon, Gudgell & Simpson, C. G. Comstock, J. M. Curtice, the Stanton Farm, the Makins, the Steeles, C. N. Moore, Dr. Logan, C. W. Armour, Robert Hazlett, S. L. Brock, and others were now producing show stock as good, if not better than had ever before been seen.

At Kansas City the champion bull was Mr. Funkhouser's Onward 4th, both his sire and dam—Onward 6th and Dewdrop—having been champions before him. VanNatta's Rosalie by March On had the female championship.

At the final round-up at Chicago Prime Lad, Beau Donald 39th, Prairie Donald (by Beau Donald 7th), Right Lad and Benjamin Wiltons 10th and 16th led the bull classes. In a memorable show of cows and heifers the blue ribbons rested with VanNatta's Lorna Doone (by Christopher), Harris' Lucile 2d (by Benjamin Wilton), VanNatta's Rosalie (by March On), and Harris' Amelia, Arminta and Miss Donald 5th. The Beau Donalds had both first and third in the get-of-sire class, with the Benjamin Wiltons coming in between.

Death of Benjamin Wilton.—The great showyard events of this era contain many references to the splendid character of the sons and daughters of the bull Benjamin Wilton, exhibited by Overton Harris. The bull was bred by Cornish & Patten, Osborn, Mo., and had been owned at one time by John E. Stone, Harris, Mo. He was an in-bred Anxiety, having been sired by Wilton Anxiety 41810, he by Tom Clark's Peerless Wilton, out of one of Gudgell & Simpson's Anxiety 4th cows. He was not a big bull, weighing about 2,100 pounds in breeding condition, and was specially distinguished for his extraordinary good temper and docility. He sired in his time about 250 calves, including Betty 2d, the champion female of 1901, that sold with calf at side for $4,500. He was killed by a stroke of lightning in the spring of 1903. But a few days previous Mr. Jesse Adams of Moweaqua, Ill., had closed a deal with Mr. Harris for some ten head of Benjamin Wilton heifers at an even $10,000. Mr. Harris is said to have sold over $42,000 worth of calves sired by this bull within the space of four years.

Death of Dale.—On Oct. 18, 1903, Dale, the champion show bull, died at Woodland Farm, the property of his owner, Jesse Adams, Moweaqua, Ill. Mr. Adams had paid the great sum of $10,000 for the bull in 1901. Dale was one of the many valuable legacies left to the Hereford breed in America as a result of the famous old-time importation of Earl & Stuart. He carried a double cross of Garfield combined with the blood of Peerless and Prince Edward, and through his son Perfection passed on to the Hereford breeders of the United States a factor of demonstrated value. His dam, Rose Blossom, once changed hands at $5,000. She lived to be fifteen years old, was the mother of Columbia the dam of Disturber, and died two months after the decease of Dale, the property of Mr. G. M. Naber of Naberlea Farm, Wabash, Ind.

The Sale Season of 1904.—Breeders had now settled down to an acceptance of a situation which did not promise exceptional prices. There was grim determination all along the line, however, to hold on to that which was good and await the return of

PROTECTOR 117878, BY ALBION 15027, DAM BY RUDOLPH—BRED BY ALLEN HUGHES. IMPORTED BY F. A. NAVE AND SOLD TO GILTNER BROS.

ALBION 76960, IMPORTED BY C. A. JAMISON.

better times. In February of this year Giltner Bros. offered 53 head of registered cattle at auction at Auburn, Ala. This was one of the first attempts of the kind in that section and resulted in the gratifying average of $213 per head. On March 2 Mr. C. A. Jamison sold 136 head of cattle at Hamlet, Ind., but the market would not at this time take so many at strong prices. The average was $124, the top being $1,575 paid for the imported bull Albany by M. E. L. Williams, Peoria, Ill.* Events of this spring in the middle west were large offerings of range-bred Hereford calves at auction. On March 17 the SMS outfit sold about 600 head at C. C. Judy's farm, Tallula, Ill., at an average of from $20 to $30. In May the LS management sold 500 head at Mr. Imboden's, Decatur, Ill., at an average of around $20 per head.

There were no important incidents at the fall sales of this year. Fifty-six head, consigned from various herds, sold at Kansas City on Oct. 21 at an average of $186.25. The top was $1,330 paid by S. L. Stand-

---

*Mr. Jamison lived at Peoria, Ill., and began his Hereford herd in the spring of 1899 by the purchase of 99 animals of breeding age and about 30 calves, all of which were placed upon his large farm near Hamlet, Ind. This purchase included the Corrector bull Well Done 66786. A little later 45 two-year-old heifers were bought and with these came another Corrector bull, Sir Comewell 68776, and Reginald 64067 by Mr. VanNatta's Hengler. Other purchases were made from Mr. Nave, Mr. Armour and other prominent breeders. Mr. Jamison also used the Shadeland bull Sailor 93037 by Acrobat. Subsequently he bought Imp. Diplomat 81547, but he met with an accident which caused his death soon afterwards. Immediately after this event, after consultation with Capt. Scarlett, Mr. Jamison decided to cable an offer for the unbeaten two-year-old bull Albany 132876. The deal was closed at a reported price of $6,000. Albany was bred by Allen Hughes, and was landed in New York along with the two fine heifers, Lady Barbara and Princess Royal, both of William Tudge's breeding. Mr. Jamison's operations were on a very extensive scale, and during the winter of 1902 it was stated that although he had sold 229 head during the two years immediately preceeding he still owned over 400 head of Herefords.

ish for Mr. Funkhouser's Onward 4th. Luce & Moxley took Curtice's Prince Rupert 8th at $850.

The St. Louis World's Fair of 1904.—While business depression continued to restrain activity in the trade there was no let-up in the enthusiasm and interest of the leading producers of high-class Herefords. There was held at St. Louis in the autumn of 1904 a great exposition commemorating the purchase by the United States from France of the so-called Louisiana Territory. A live stock department commensurate with the importance and dignity of the occasion was projected and carried out to a successful consummation, Hon. F. D. Coburn, the veteran Secretary of the Kansas State Board of Agriculture, holding the helm. The various national organizations of breeders of pedigree live stock appropriated money for special prizes and cooperated in making the event a monumental success.

There was a comparatively light display of the "white faces" at the early fairs of 1904. The big guns were being held in reserve, in many instances, for the great exposition contest, which was naturally the outstanding event of the year. As this show ranks with the Chicago Columbian of 1893 in point of historic interest to American cattle breeders, the full account of the Hereford exhibit as presented at the time by "The Breeder's Gazette" is appended:*

---

*This account of one of the greatest Hereford battles of modern times is presented not only for the historical value of the prizelist itself, but because of its descriptions and criticisms, reflecting as they do the standard by which Hereford show cattle were judged at that time. It is from the pen of William R. Goodwin, the present managing editor of "The Breeder's Gazette," who for a quarter-century past has been reviewing the leading live stock shows of the United States. His work has dealt with

"No class of breeders has made more systematic, thorough and painstaking preparations for this world's fair cattle show than the men who handle the 'white faces.' Their reward has been great, for it was a sensational display of the excellencies of the breed. From Kentucky to Nevada they came trooping at the call, determined to make plain the fact of Hereford early maturity and bloom in this world's arena. And admirably did they succeed. From start to finish, from the moment that a Hereford first entered the forum until the last white-faced baby romped from the ring, it was a succession of classes of astonishing strength, with a most notable absence of inferior animals.

"Never had the feed-bucket been more carefully handled, never had the tonsorial art on the bovine coat of hair, brought to such perfection by Hereford herdsmen, been more strikingly illustrated. It was a beautiful display of Hereford strength and must have exerted a powerful impression on even the most careless of observers. The list of exhibitors who participated in this event follows:

"James A. Funkhouser, Plattsburg, Mo.; W. S. VanNatta & Son, Fowler, Ind.; S. W. Anderson, Blaker Mills, W. Va.; Gudgell & Simpson, Independence, Mo.; O. Harris, Harris, Mo.; W. H. Curtice, Eminence, Ky.; Benton Gabbert & Son, Dearborn, Mo.; Egger Hereford Cattle Co., Appleton City, Mo.; C. N. Moore, Lees Summit, Mo.; J. S. Lancaster & Sons, Liberty, Mo.; Fritz & Shea, Blakesburg, Ia.; A. R. Firkins, Worcester, England; Walter B. Waddell, Lexington, Mo.; S. L. Brock, Macon, Mo.; H. J. Fluck, Goodenow, Ill.; C. G. Comstock,

---

all the important types of cattle, horses, sheep and swine known to contemporary stock-breeding and for comprehensive grasp of detail, breadth of field covered, facility of expression and fairness of treatment throughout by common consent it has never been equalled in the whole realm of live stock criticism.

Albany, Mo.; Steele Bros., Belvoir, Kans.; A. R. Haven, Greenfield, Ill.; Steward & Hutcheon, Bolckow, Mo.; John Sparks, Reno, Nev.; Eagle Farm of Indiana; Dette Bros., Brinktown, Mo.; Carter & Curtner of Indiana; R. S. Burcham, Windsor, Mo.; Mrs. K. W. Cross, Emporia, Kans.; J. Condell, Eldorado, Kans., and H. D. Martin, Shelbyville, Ky.

"Thomas Clark, Beecher, Ill., was nominated and confirmed as judge of Herefords, but later C. A. Stannard, Emporia, Kans., was appointed to work with him.

"**Aged Bulls.**—1, Prime Lad; sire, Kansas Lad Jr.; W. S. VanNatta & Son. 2, Bright Donald; sire, Donald Dhu; Gudgell & Simpson. 3, Onward 4th; sire, March On 6th; James A. Funkhouser. 4, Fulfiller; sire, Improver; O. Harris. 5, Beau March On; sire, March On 5th; C. N. Moore. 6, Actor 26th; sire, Actor 3d; S. W. Anderson. 7, Beau Donald 39th; sire, Beau Donald; W. H. Curtice.

"The bulls on the prizelist are familiar figures, or well known by name. The repeated trial of strength between Prime Lad and Onward 4th here resulted in another triumph for the former, the March On 6th bull getting a hard setback to third place. Prime Lad had been handled for a couple of years with especial reference to the St. Louis competition and he fulfilled all hopes and expectations. This level-lined shapely quality-sort bull with his large and drooping horn, his neat bone, his bulging buttocks, furnishes many of the elements which popularly inhere in a champion. He is not of the bulkiest pattern, nor is he on the small side. He has matured a little since last season, but has suffered a slipping of flesh from the shoulders, while the 'band' back of the crops is a trifle more pronounced. He is a bull of flesh and finish, balanced at both ends, and strong in the middle except at the point noted. Compared to the low-legged Onward 4th both Prime Lad and Bright Donald looked a bit off the ground, but that is on account of the remarkable brevity of the underpinning of the March On 6th bull. It would perhaps be asking too much for a bull of his weight, bulk and sappiness to hold his back altogether level, and the remarkably high carriage of head adds to the impression of slackness of top. He is extremely wide and rotund in his turning, magnificent in his head, horn and crest, and great in buttocks and twist. Bright Donald is a little different type, a very straight-lined yellow-red, with strong heavy well-fashioned hind-quarters, a pleasing head and horn and a great weight of mellow flesh, but with a roughness of shoulder that stands against him. Of Fulfiller it may be said that he has not fulfilled his early promise. In his youth he was easily the best bull produced at Weavergrace. He is extremely low and broad and wealthily fleshed, but he does not carry a level topline and his hips are somewhat too wide. Beau March On is a neat-boned bull, of excellent top and plenty of finish, but his underline shows some suggestion of pinch. Actor 26th carries his heavy weight close to the ground and his ribs are well sprung and covered, but he lacks the smoothness of outline carried by Beau Donald 39th. This bull continues to present the bullet-like style of architecture, with tremendous heavy hind end, and his place at the bottom of the list was much of a surprise.

"We depart in this instance from the rule limiting comment to prize-winners to note the presence in the ring of the English champion bull Happy Christmas. He is a bull of great bulk and attractive in his fashioning forward, but evidently he has seen better days and at this time was not in fit form to cope with such a company.

The Two-Year-Olds.—1, Defender; sire, Perfection; C. G. Comstock. 2, Keep On 26th; sire, Keep On; O. Harris. 3, Donald March On; sire, March On; W. S. VanNatta & Son. 4, Prince Rupert 8th; sire, Prince Rupert; W. H. Curtice. 5, Right Lad; sire, Kansas Lad, Jr.; S. L. Brock. 6, Romulus; sire, Militant; Gudgell & Simpson. 7, Marmaduke 5th; sire, Marmaduke; S. W. Anderson.

"This company called forward some bulls of a lot of size for the age, and a few of them were a trifle too up-standing. Brevity of legs and weight in small superficies are cardinal points of excellence in the Herefords, and are too highly prized and too deeply ingrained in the breed to be lost. Defender has been growing in popularity as he has been strengthening in form. He is perhaps on the large side, yet not coarse nor wanting type. He has a commanding presence, a good horn, splendid crest, great width of top and smoothness of turning, and is particularly heavy and well finished in the hind-quarters. Keep On 26th is quite on the other type, presenting the no-legged breadth and bulk proposition in rather spectacular fashion. He has a beautifully carved countenance, but a surplusage of brisket. His ribs are widely sprung, his loins deeply packed, his rounds very full, but there is a little unevenness on the top of his hind-quarters. Donald March On is off the same fashioning block in large degree, holding his widespread frame close to the ground and claiming attention to the remarkable development of thighs and rounds, but a little heavy in his lower lines. Prince Rupert 8th reverts again to the type at the head of the class, possessing stretch, but low of leg; his foreribs could arch a trifle more, but beneath his yellow-red coat he carries a tremendous weight of firm flesh. Right Lad needs to come nearer the ground; he has the head of a feeder, a grand top and a particularly well filled chine and impressive crest. Romulus returns to the blocky sort—a bull of good head and horn and attractive presence.

"**Senior Yearlings.**—1, Onward 18th; sire, March On 6th; James A. Funkhouser. 2, Benjamin Wilton 10th; sire Benjamin Wilton; O. Harris. 3, Beau Donald 58th; sire, Beau Donald; W. H. Curtice. 4, Beau Donald 66th; sire, Beau Donald; W. H. Curtice. 5, Princeps 8th; sire, Princeps; Steele Bros. 6, Rare Lad; sire, Kansas Lad Jr.; S. L. Brock. 7, Leader; sire, Beaumont; Steward & Hutcheon.

"One of the most stubborn contests of the week quickly developed as Onward 18th and Benjamin Wilton 10th assumed positions on the firing line. The latter is a bit the taller, and shows more growth of frame, with clean throttle, a roomy middlepiece that is arched over in strong fashion to carry its weight of beef, loins that are the glory of the Hereford in its best estate, and hips neatly covered. Onward 18th is somewhat wanting the scale of his rival, is well rounded on the rib, but a trifle narrow over the shoulders, remarkable in the full-fashioning of his hind-quarters, surpassing the others in this respect, and showing a blockier conformation throughout. A bull's head, smooth shoulders, level strong back, full flanks, and much heft of hind-quarters characterize Beau Donald 58th. Beau Donald 66th is a soggy sort, neat-boned, great-crested, wide of chest, finely ribbed and plump in covering of loins. Such finish at the tailhead as is carried by Princeps 8th is highly desirable, especially when it is joined with his rotund turn of top, neat hips and nice style; a little more fullness of heart-girth would improve this bull. Rare Lad carries lots of sub-

stance, is well conditioned and presents a width of front and covering of forerib that are pleasing. Leader is a strong-framed chap, with well distributed mellow flesh and loin of attractive fullness.

"**Junior Yearlings.**—1, Beau President; sire, Beau Brummel; Gudgell & Simpson. 2, Onward 23d; sire, March On 6th; James A. Funkhouser. 3, Meteor; sire, Hesiod 17th; Benton Gabbert & Son. 4, Hidrotic Alamo; sire, McCord; John Sparks. 5, Actor 30th; sire, Actor 3d; S. W. Anderson. 6, Marchette; sire, March On 5th; S. L. Brock. 7, World's Fair Winner; sire, Sotham; Dette Bros.

"Brevity of underpinning, levelness of top, and blockiness of build doubtless stood Beau President well in hand when he faced his competitors in this lot. The steaks that can be cut from his swelling rounds would weigh out heavily over the butcher's block. Onward 23d has more stretch than the majority of the March On 6th progeny, the same masculine character, and rotund buttocks. Meteor is a dark-colored red, very round and bullet-like, carrying a lot of flesh in smooth form, but a trifle narrow in his head. Hidrotic Alamo charmed with the beautiful smoothness of his outlines, his neat-laid shoulders and well covered hips uniting to present a bull of much evenness. Actor 30th is a wide-topped bull of good depth, carrying his bulk on hind legs that are nicely modeled. Marchette is one of the egg-like kind with no waste of leg, and nicely ornamented in horn. Length, style, good back and nice quality are present in World's Fair Winner, but he had a mighty narrow margin in which to make good his name.

"**Senior Bull Calves.**—1, Mapleton; sire, Beaumont; Steward & Hutcheon. 2, Good Enough 4th; sire, Good Enough 3d; O. Harris. 3, Distributor; sire, Disturber, S. L. Brock. 4, Bold Rex; sire, Dandy

Rex; Gudgell & Simpson. 5, Actor 35th; sire, Actor 3d; S. W. Anderson. 6, Haven's Protector; sire, Protector; A. R. Haven. 7, Advance; sire, Onward 4th; W. S. VanNatta & Son.

"There were nuts to crack among the senior calves. More than a score assembled and the round finally resolved itself into a consideration of the conflicting claims of Mapleton and Good Enough 4th. In this Mapleton we have champion stuff. He has already attained the title of junior champion, and if the fates are kind to him higher honor seems in sight in the future. He is good enough to be discussed negatively. That is to say, if the tail were a bit more neatly set on the rump the man who would throw stones at him would endeavor to pick flaws in a diamond of the first water. In his form, his substance and his finish Mapleton is 18-karat gold. But it required the services of T. J. Wornall as referee to land him in premier position, so hard did Good Enough 4th push him. The latter is a youngster of wonderful ripeness, with ample style and finish at tailhead, deep flanks and a furry coat. Mr. Wornall agreed with Mr. Clark in sending Mapleton to the top. Barring a little unevenness at the tailhead Distributor is like cast from a bullet mold. Rounds, loins, ribs and head are capital. The cylindrical form is nicely illustrated in Bold Rex, barring a little flatness of forerib. A big end, a good head and compact well finished form are shown by Actor 35th. Haven's Protector is a shade darker than the average in coat, of good growth and mellow in his flesh. There is a lot of growth to Advance, and he shows fair depth of body.

"**Junior Bull Calves.**—1, Sagamore; sire, Bright Donald; Gudgell & Simpson. 2, Good Enough 10th; sire, Good Enough 3d; O. Harris. 3, Onward 30th; sire March On 6th; James A. Funkhouser. 4, Beau

Donald 75th; sire, Beau Donald; W. H. Curtice. 5, Mapleton Beau; sire, Beaumont; Steward & Hutcheon. 6, Don Irving; sire, Henry Irving; W. S. VanNatta & Son. 7, Onward 31st; sire, March On 6th; James A. Funkhouser.

"A genuinely good one fore and aft is Sagamore, a calf of true lines, set right af the ground, level of top and bottom, and with hind-quarters that carry the twist well down to the hocks and fill the lower rounds. Good Enough 10th is a calf of more growth, furry-coated and very sappy, a veal ready for the block but not so strong in the back as Sagamore. Onward 30th and Onward 31st were on the list, the former a little dark in coat, of blocky build and on short legs, the latter a lighter red of choice quality, not so wide as his companion but sweeter and like an apple in his smoothness. Donald 75th is one of those lathe-turned chaps that has the mellowness of a ripe peach. Mapleton Beau is a rather wee one, well lined out and nice in his quality, while Don Irving has scale and length and a real round build over the back with well finished quarters. This was a capital lot of youngsters and likely contains some names that will be widely known to fame in the future.

"**Aged Cows.**—1, Lorna Doone; sire, Christopher; W. S. VanNatta & Son. 2, Belle Donald 44th; sire, Beau Donald; W. H. Curtice. 3, Priscilla 5th; sire, Lamplighter; Gudgell & Simpson. 4, Romaine; sire, March On 6th; James A. Funkhouser. 5, Dorinda; sire, Beau Brummel; Fritz & Shea. 6, Modesty 3d; sire, Beau Brummel; Gudgell & Simpson. 7, Belle Donald 59th; sire, Beau Donald; W. H. Curtice.

"A score of cows, rather variant in size and condition, some in the height of showyard condition and others with sucking calves at foot, made up a company that commanded a large degree of the time of the judges who finally referred the allotment of third prize to N. H. Gentry, being unable to agree. Five cows were selected and sent over to one side the arena and there the scrutiny was prolonged. Lorna Doone, thus far the winner of the year, assumed her accustomed place. She had stout friends for higher favor than she received last year, but now all seem ready to do her reverence. Her kindly countenance is ornamented with a horn that is a bit old-fashioned in its turning but not the least homely; the carcass proposition finds its best exemplification in her broad bosom, widespread frame and generous expanse of top. Some little weakness develops on an examination of the thighs, but the cow in her massiveness and trueness to type readily takes high rank. Belle Donald 44th is entitled to consideration among the best of them. She is of fine size, marked quality, broad-bosomed, deep-middled, with back hooped like a barrel and hind-quarters well filled. By vote of referee Gentry Priscilla 5th was set above Romaine. The former is a very short-legged cow, looking a bit small in her present company, with very sweet feminine head and drooping horns, straight topline and thick flesh. Romaine is another wide-out block, standing right at the ground, with broad turn to the ribs which are literally rolled and padded in smooth flesh. She shows a beautiful face and if she were equally well finished behind would make more trouble for her competitors. Dorinda is a cow of beautiful symmetry, wanting just a little closer carriage to the ground. She may be a bit light about the neck but few of them are so well proportioned, so grandly finished in the hind-quarters and so chock full of quality. Modesty 3d hugs the ground closely, the head is pretty and the horns curving, the veins are full, the top wide and the thighs well fleshed. It is a little singular that full sisters should find place on this prizelist, and yet such is the fact. Belle Donald 59th comes from the mating that produced the second-prize cow in this company. She is a year younger, carries much scale, a fine back and great hind-quarters.

"**The Two-Year-Olds.**—1, Amelia; sire, Premier; O. Harris. 2, Heliotrope; sire, Princeps; Steele Bros. 3, Twila; sire, March On 6th; James A. Funkhouser. 4, Domestic; sire, Princeps; Steele Bros. 5, Belle Donald 60th; sire, Beau Donald; W. H. Curtice. 6, Cleo March On; sire, March On; W. S. VanNatta & Son. 7, Belle Donald 61st; sire, Beau Donald; W. H. Curtice.

"A marvel of a middle is carried by Amelia. Her head is delightful in its femininity, her spread of back is tremendous, its covering deep, but she fails a little at the tail. Heliotrope is a bit higher on the leg, a heifer of ample scale, grand top, neat brisket, well finished hind end and strong thighs. Twila is darker in coat and nearer the ground, with big middle, a rib that is arched to carry weight, a well finished head and horn, but not so neat in the hooks or the rump as Heliotrope. Domestic is a bit smaller than her stable-mate Heliotrope, more compact in form, of admirable outlines, plenty of mellow flesh and a drooping horn. Belle Donald 60th and Belle Donald 61st are an impressive pair, both of size and substance, the former smooth-fronted and square-finished behind, broad of loin but a trifle light of thigh, the latter dark in coat, broad of loin and likewise wide of hips, and heavily fleshed. Cleo March On is a heifer of remarkable width of carcass and depth, one of the largest-middled heifers that the breed has shown us.

"**Senior Yearlings.**—1, Arminta 4th; sire, Premier; O. Harris. 2, Iva 4th; sire, Benjamin Wilton; O. Harris. 3, Ravilla; sire, Hesiod 2d; James A. Funkhouser. 4, Belle Donald 74th; sire, Beau Donald; W.

H. Curtice. 5, Princess May 3d; sire, Princeps; Steele Bros. 6, Belle of Whitebreast; sire, Dandy Rex; Fritz & Shea. 7, Capitola 20th; sire, Martinet; Gudgell & Simpson.

"The leading position in this class was assumed by a heifer in the highest condition. Arminta 4th is a remarkable block, with pleasing countenance, and a smoothness that is carried uniformly through her make-up. It is such width of body, cover of flesh in the high-priced parts and neatness that give heifers call on such honors. Iva 4th is larger than her companion, with the neckvein of a feeder, fine head, well developed thighs, and a big middle, but she is not so neat about the hips. In head Ravilla is hardly the equal of the pair that stood above her; she has enough size, the rib is well let down, the flanks are full and the flesh is ample. Belle Donald 74th is a great hind-ended heifer, but wants swelling out a little in the forerib. She stands on abbreviated legs and her flesh in its mellowness is pleasing. A little more strength of neck would improve Princess May 3d, which joins to a hind-quarter of exceptional excellence a well ribbed top. Belle of Whitebreast shows a little darker in the coat than her companions on the list; she has a levelness of form that is taking and her flesh is well distributed. Capitola 20th is low enough but scarcely carries the width of some of those ahead of her. The head is pretty and is ornamented with a set of incurved horns.

"**Junior Yearlings.**—1, Miss Donald 5th; sire, Beau Donald 5th; O. Harris. 2, Kathleen; sire, March On 6th; James A. Funkhouser. 3, Mayflower; sire, Beaumont; Steward & Hutcheon. 4, Belle Donald 77th; sire, Beau Donald; W. H. Curtice. 5, Belle Donald 76th; sire, Beau Donald; W. H. Curtice. 6, Lady March On; sire, March On; W. S.

VanNatta & Son. 7, Miss Donald 3d; sire, Beau Donald 5th; O. Harris.

"Much wrestling produced the prizelist here. It would have been easier were 't'other dear charmer away,' but she was not. She was right there and silently but none the less forcibly demanding consideration. For the blue ribbon wearer it may be said that she is a nugget that finds her most noticeable weakness at the rumps. The forerib is phenomenal, the face pleasing, the veins full, the flanks well let down, and the back turned out of a mold including the hips, which are imbedded almost 'out of feel.' Kathleen carries a little more depth of rib, some greater size, but hardly the width. She is finished about the head and neatly put together in the shoulders. Mayflower was at a double disadvantage, she was short of age and a suit of hair. The hand found plenty of flesh to make an interesting fight with those above her, but a curly coat gives an advantage to the eye that is with difficulty counteracted by the hand. She lacks the depth of her competitors, but is sweet, level on top and rounding in her outlines, neatly finished at the tail and full in neck-veins. Belle Donald 77th stands nearer the ground than the others, is firm to the touch, but not neat at the tailhead. Her half sister Belle Donald 76th is a broad-faced big-framed heifer with even more weight of flesh. Lady March On is a deep and roomy heifer, with a little unevenness about the hooks. Miss Donald 3d is a wide-topped one and carries plenty of flesh where it should be laid on.

"**Senior Heifer Calves.**—1, Purple Leaf 2d; sire, Good Enough 3d; O. Harris. 2, Beaumont's Queen; sire, Beaumont; Steward & Hutcheon. 3, Miss Donald 6th; sire, Beau Donald 5th; O. Harris. 4, Dawn; sire, March On 6th; James A. Funkhouser. 5,

Blanche 28th; sire, Paladin; Gudgell & Simpson. 6, Dorinne 19th; sire, Dandy Rex; Gudgell & Simpson. 7, Onward's Elsie; sire, Onward 4th; W. S. Van-Natta & Son.

"Purple Leaf 2d could win on her coat, if a prize were offered for a furry robe. But she has more substantial claims to consideration in her width and depth and true proportions. Finish at both ends is ample. Beaumont's Queen is smoothly fashioned, drops her flank to the limit, and shows sweet character. Miss Donald 6th is remarkably mature of form, considerably larger in her middlepiece than the one above her, finely coated and attractive in her countenance. Mellowness sticks out all over Dawn; that yellow-red coat seems commonly to cover a wealth of flesh even and springy to the hand, and it is so in this case. Blanche 28th is of the same hue; she is well spread in hind-quarters and wide enough forward, showing a well rounded chine and excellent forerib. Dorinne 19th is a bit tall in comparison, but well let down in her ribs and padded in her loins. Onward's Elsie is a somewhat better type, showing a lot of breadth and ample cover.

"**Junior Heifer Calves.**—1, Miss Donald 17th; sire, Beau Donald 5th; O. Harris. 2, Belle 17th; sire, Paladin; Steward & Hutcheon. 3, Miss Donald 18th; sire, Beau Donald 5th; O. Harris. 4, Estella; sire, Princeps; Steele Bros. 5, Bonita; sire, Lord Saxon; Mrs. Kate W. Cross. 6, Evangeline; sire, Peerless Wilton 39th; H. J. Fluck. 7, Regina; sire, Hesiod 85th; James A. Funkhouser.

"Miss Donald 17th and Miss Donald 18th were separated on the list by Belle 17th. The first named is sweet and sappy and well grown; her companion holds her width evenly, is neatly finished in her quarters and exceptionally good in her body. Belle

17th is big-ended, wide-backed and carries a bit more scale. Estella is full-thighed and good in middle, but not so smooth in her shoulders or her hips. Bonita was one of the best grown and sappiest in the company, an even calf in make-up with plenty of depth of rib. A little nugget is Evangeline and Regina is on the same order.

"**Get of Sire (Four).**—1, Curtice on Beau Donald. 2, Funkhouser on March On 6th. 3, Steele Bros. on Princeps. 4, Curtice on Beau Donald. 5, Harris on Beau Donald 5th. 6, Steward & Hutcheon on Beaumont. 7, Gudgell & Simpson on Beau Brummel.

"**Produce of Cow (Two).**—1, Curtice on Minnie H. 2, Funkhouser on Keepsake. 3, Curtice on Sophia. 4, Funkhouser on Dewdrop. 5, Harris on Iva. 6, Steele Bros. on Lady May 3d. 7, Fritz & Shea on Dorinda.

"**Championships.**—Senior champion bull, W. S. VanNatta & Son's Prime Lad; reserve senior champion, Harris' Defender. Junior champion bull, Steward & Hutcheon's Mapleton; reserve junior champion, Gudgell & Simpson's Beau President.

"Senior champion cow, VanNatta & Son's Lorna Doone; reserve senior champion, O. Harris' Amelia. Junior champion female, Harris' Arminta 4th; reserve junior champion, Harris' Miss Donald 5th.

"Grand champion bull, Prime Lad; reserve grand champion, Mapleton.

"Grand champion female, Lorna Doone; reserve grand champion, Arminta 4th.

"**Group Prizes.**—Aged herds—1, Harris. 2, Funkhouser. 3, Curtice. 4, VanNatta & Son. 5, Gudgell & Simpson. 6, Steele Bros. 7, Anderson. Aged herds (females bred by exhibitor)—1, Harris. 2, Funkhouser. 3, VanNatta & Son. 4, Gudgell & Simpson. 5, Steele Bros. 6, Anderson.

"Young herds—1, Harris. 2, Funkhouser. 3,

Gudgell & Simpson. 4, Curtice. 5, Steele Bros. 6, Steward & Hutcheon. 7, Harris. Young herds (females bred by exhibitor)—1, Harris. 2, Funkhouser. 3, Gudgell & Simpson. 4, Curtice. 5, Steele Bros. 6, Steward & Hutcheon. 7, Harris.''

A VanNatta Triumph.—The winning of both the male and the female championships at this crowning event in American showyard history with Prime Lad and Lorna Doone as well as at the Chicago International in December, was a fitting climax to the work of William S. VanNatta. For a quarter of a century he had been a steadfast follower of Hereford fortunes, through adversity as well as through seasons of prosperity. Never carried away by his own successes beyond the line of safety, never losing faith in the ultimate place of good "white faces" in the American cattle trade, working away along practical lines at all times, even when pedigree cattle were going at beef prices, his hand never wearied, his heart never wavered in the course of his work in behalf of Hereford cattle in the United States. Now ably seconded by his son, Frank, he not only had the satisfaction of receiving the highest honors of the year, but was secure in the knowledge that his work would be faithfully and intelligently carried forward.

Death of Gov. Simpson.—As the sands of 1903 ran out the life of one of the greatest of all the Hereford pioneers was slowly ebbing, and on Jan. 5, 1904, Thomas Alexander Simpson passed over the great divide at the ripe age of 82 years. In the years 1880, 1881 and 1882 he had selected for importation by the firm of Gudgell & Simpson no less than 500 head of Hereford and Aberdeen-Angus cattle, including Anxiety 4th. He was a striking figure in any company, tall and imposing, with strength of character stamped in every feature, yet dignified, gentle and modest in his bearing. He was one of the keenest judges of cattle this country has produced. No man was more tenacious of his opinion or less aggressive in its expression. He was broadminded and tolerant of the judgment of others, while clear and firm in his own convictions. He was a man of such few words and so deprecatingly modest in his manner that first impressions needed revision. When it came to trading, English and Scotch breeders found concealed behind this quiet demeanor a judgment and a shrewdness which they had little expected to encounter. It was diamond cut diamond. No man left behind him in Britain a greater reputation for keen judgment of animal and price than did Gov. Simpson. It was a wonderful lot of cattle he selected of both breeds and no little of the history of the Hereford and Aberdeen-Angus in America originated in the operations under this master breeder's mind. Gov. Simpson's name stands high on the honor roll of the great American improvers of beef cattle stocks.

Dull Days.—There is little to be said about the status of the trade in 1905. The business of extending the use of purebred bulls on the western ranges was going on as usual, but there was an almost entire absence of speculative spirit. As a matter of

FRANK VANNATTA.

fact, it is during just such dull periods, however, that foundations for future successes are always laid by men possessed of foresight and sufficient capital to make selections of good animals when subnormal prices prevail. The business was entering a quiet stage, but a situation which after all was conducive to real progress. Still waters always run deepest.

On April 25 A. C. Huxley of Bunker Hill, Ind., sold 40 head at $170.50, R. C. Cain of the Hoosier state giving $1,000 for the bull Merry Dale. The usual fall combination sale at Kansas City resulted in an average of $215.25 for 57 head. E. R. Morgan of Kansas gave $1,400 for the Funkhouser bull Onward 18th. At the same place in December Messrs. Armour and Funkhouser had a fairly good sale of 75 head averaging $200.45, at which Messrs. J. & B. Miller of Iowa gave $1,105 for the Gabbert-bred bull Columbus 60th.

"Individual Merit by Inheritance."—This Sotham slogan found marvelous exemplification at the Kansas City Royal and the Chicago International of 1905. At the former show Mr. Comstock had first-prize and senior championship with Defender, son of Perfection, the great son of Dale. Defender had narrowly missed the championship honors in 1904, and at Kansas City a year later could not be denied the place. Another son of Perfection that was to cut a great figure appeared at this same show—the first-prize senior yearling shown by Mr. Huxley, Perfection Fairfax. In the bull calves the senior and junior classes were both headed by sons of

Prime Lad, both shown by Messrs. VanNatta. The junior champion, Prime Lad 16th, represented the mating of Prime Lad with Lorna Doone. These shows also were full of demonstrations of the wonderful manner in which the Anxiety blood was coming on as shown by Curtice, Gudgell & Simpson and others. March On 6th's get were again seen throughout the prizelist, as were also Gabbert's young things of the Columbus blood. At the International Defender was not shown, and Cargill & McMillan led the aged bull class and also had the senior champion in Fulfiller. The Prime Lads led in junior yearling bulls, senior bull calves and junior bull calves. In the cow class Cargill & McMillan's Heliotrope beat the pair of Belle Donalds as well as Lorna Doone.

The sale season of 1906 opened with a public sale by Gudgell & Simpson at Kansas City on Feb. 7, at which 57 head registered an average of $230.50. Early in March Mr. Nave sold 62 head at Attica, Ind., at an average of $208.

Autumn Sales of 1906.—On Oct. 24, 1906, the herd of Mr. Jesse Adams of Moweaqua, Ill., was sold, the 36 head which remained at that time bringing an average of $230. The top price was $1,000 paid by J. W. Altman of Dubuque, Ia., for the bull Perfection 3d by Dale. The next highest price was $500 by Cargill & McMillan for Lady Real by Perfection 3d. While these values seem ridiculously low compared with some of the great prices that had been paid by Mr. Adams for his foundation stock, he

HELIOTROPE, CARGILL & McMILLAN'S SENIOR CHAMPION COW AT THE INTERNATIONAL OF 1905.

Photo by Hildebrand

demonstrated to those in attendance by actual figures that even if the prices had been high for some of his original purchases, without exception every cow had produced offspring, from the sale of which the cost was paid and a profit beside. Mr. Adams is the man who gave the bull Disturber 139989 to the American Hereford cattle-breeding fraternity. That alone justifies a large share of the time and money he had lavished upon his herd.

On Nov. 7 John Steward's cattle were sold at Bolckow, Mo., the 46 head averaging $198.60 and the yearling bull Parsifal bringing $650. On Oct. 12 at Kansas City 47 head averaged $193.65, O. Harris paying the top price of $575 for the Funkhouser-bred bull Onward 31st. The combination sale at Chicago on Dec. 12 resulted in an average of $167 on 51 head.

Cargill & McMillan's Great Success.—Without undertaking any detailed review of the fall fairs of 1906—for we must now hasten our story—it may be stated that Messrs. Cargill & McMillan reaped the highest showyard honors of the year. At Kansas City their best bull, Princeps 4th, was not shown, as it was decided to hold him back for the International. Mr. Steward's Beau Mystic headed the senior bulls, but Cargill & McMillan won in both two-year-olds and senior yearlings. The young bulls Onward 46th and 54th, both by March On 6th and of Funkhouser breeding, were blue ribbon winners here. Mr. Hoxie had the best senior bull calf at Kansas City in Peerless Perfection 10th, sired by

Perfection, but at Chicago Cargill & McMillan's Fulfiller 3d was placed above him. This firm was first on both aged and young herds at Kansas City and Chicago. Princeps 4th was senior champion bull at Chicago. Bonnie Brae 3d was junior champion at Chicago and Princeps 4th grand champion. Heliotrope was senior and grand champion female. It was a proud day for "Jimmy" Price.*

---

*James Price was born in Herefordshire, England, on the first of November, 1869, and worked on his father's farm until 1890, when he started for America to fight his own battles. He had never been away from home before and did not know a person in America. He landed in Chicago on the first of July, 1890, with a few letters of introduction, among others one to Tom Clark. He proceeded at once to Beecher determined to take the first job he could get, as he had by this time just $10 left. When he arrived at Beecher he met Mr. Clark, who gave him employment. He worked on the farm that summer and winter, being paid $14 a month. About that time Harry Fluck bought out the Baker herd and rented the farm, and Price went to assist in the care of the cattle.

In the spring of 1894 the young man returned to Mr. Clark, with whom he then remained for three years. Lars was the champion bull of those days. Speaking of him Price says: "I will never forget the day Lars defeated Ancient Briton. It was on the little LaCrosse, Wis., fair grounds the week before the Minnesota State Fair. Prof. Craig did the judging, and Lars held first place after that for three successive years." These were also the days of the famous sisters Juvenile, Jessamine and Juno. In 1896 Mr. Clark made a show at Madison Square Garden, and from there Price left for a trip home. He returned on the first of April, 1897, and hired to John Lewis of Shadeland, but as he was anxious to feed for show he remained at Shadeland but about ten months. At that time Frank Nave was making his start. He bought at Shadeland a few show animals and engaged Price. This was in March, 1898. Nave had gathered a few show animals together, including Dale, Atoka, Dolly 5th and Carnation.

Referring to the conditions at Attica at the time, Price says: "Things were pretty unhandy the first summer, but in the fall Mr. Nave built a barn, the best one I ever worked in. We made a fair showing the first year, winning the aged herd everywhere, including the world's fair at Omaha, but the next year we got busy and brought out what I always considered the best aged herd I ever fed, winning everywhere. We also had the two famous calves Perfection and Theressa. Perfection was the only bull that ever beat Dale while I handled them. That occurred at Indiana when we were showing for grand champion bull over all breeds and Perfection won. He was only a calf at the time. The judges were George Allen, David McKay and Mr. Pickrell, three good men. In the spring of 1900 Mr. Nave decided to sell out, and his sale the 17th of April in Chicago was a recordbreaker, including, I think, the best lot of cattle I ever saw go through a sale."

Overton Harris was then making his start in showing and now engaged Price. He had bought a few cattle at the Nave sale. Only a small show was made that year, but in 1901 Harris came out with a string hard to beat, winning with three herds everywhere he exhibited. The calf herd that year, Price claims, was the best he ever fitted. It was headed by Goodenough 10th and included the two famous "American beauties," Lucile and Troublesome. In 1902 and 1903 Harris also

Passing of Funkhouser, Steward and Scarlett.— The year 1906 was marked by the passing of several men who had exerted large influence upon Hereford cattle breeding in the west.

In the decease of James A. Funkhouser the state of Missouri lost one of her most successful breeders. He was one of those who contributed substantially to the improvement which most American breeders believed had taken place in the western states since the Herefords were first largely introduced from England. Mr. Funkhouser was born in 1846. He was a feeder and shipper of live stock, and had a small herd of Shorthorns until 1881. While attending the fair at St. Louis in that year he saw Herefords for the first time, and they impressed him so

---

made a good showing. Competition was hard, but he got a good share of the premiums. In 1904 at St. Louis he was "loaded up" with good things, and made a great record.

About this time Mr. W. S. Cargill, a lumberman at LaCrosse, Wis., was about to launch into Herefords. He had bought a few cattle, but wanted tops, so he made a trip to the Harris farms and wanted to buy the show herd providing Price would go with them. Harris was willing to sell at a price, but the price was stiff. However Mr. Cargill was game and bought fifteen-head, which Price was to pick out. After the International of that year the selections were made and "Jim" transferred the scene of his labors to LaCrosse. When he landed there he found the weather severe, but everything necessary to take care of a show herd had been provided. The first show made from the Cargill & McMillan herd was at Sedalia, and from there the herd went to Des Moines, where it won the grand champion herd prize over all breeds with Fulfiller and four first-prize females. The same year they won the Armour cup at Kansas City and repeated the following year, which reduced it to possession. In 1906, 1907, 1908 and 1909 the herd still held its own. In the fall of 1908 Price became interested in the herd, Mr. McMillan dropping out, and the firm name was changed to Cargill & Price.

In 1909, having become tired of the show business, Cargill & Price decided to sell out the show herd. This event took place in April, 1910, and involved considerable sacrifice. Price then took a rest, going back to the old home in Herefordshire for three months. On his return he purchased Mr. Cargill's interest in the cattle, rented the farm and started on his own account. "It was a little up-hill work at first," writes the veteran showman, "but I am thankful to say that I have now a nice little herd of Herefords and a 160-acre farm to put them on, and all clear. I owe what success I have made all to the Herefords, and strict attention to business. The best animals I ever fed were the bulls Dale, Princeps 4th, Bonnie Brae 3d and Lars; the best females, Heliotrope, Arminta 4th, Amelia, Betty 2d, Dolly 5th and Miss Donald 17th."

favorably that he decided to give them a trial. In May, 1882, he purchased from T. L. Miller, Beecher, Ill., a pair of calves—the bull Invincible and heifer Landscape Maid. His first experience in breeding was to produce cross-bred steers, using Shorthorn cows. He was so well pleased that he then bought a few Hereford females. His cross-bred steers were shown successfully in Chicago and Kansas City and sold at good prices for Christmas beef. A few years later he began showing breeding cattle. He was specially happy in his choice of breeding bulls, both Hesiod 2d and March On 6th proving showyard champions, as well as great sires. For the former $5,000 was refused, and an offer of $7,500 for the latter is said to have been declined. Much of the success met by the herd in the showring was due to the careful handling and fitting of Will Willis.

In the death of John Steward Missouri suffered another real loss. He was cut down in the very prime of an eminently useful life. Steward was one of the most sane, most conscientious, most dependable men ever identified with American Herefords. In partnership with Mr. Hutcheon he was engaged in building up a herd that was distinguished for true Hereford character and quality.

In November, 1906, Capt. E. C. Scarlett was added to the list of the notable dead. He was a man of good birth and education, coming from the old family estate of Nordan Hall near Leominster, Herefordshire. He was one of the many natives of that county who came to the west when the Herefords were being actively introduced, and at different times was identified with the management of several large establishments. He was at one time attached to the Iowa Hereford Cattle Co., and was for six years in charge of the Riverside Ranch at Ashland, Neb. Subsequently he was for three years head of the Belton Hereford Cattle Co., Belton, Mo., and later identified with Mr. Cook at Brookmont Farm. He died at Odebolt, Ia., on Nov. 18.

The Hoxie and Other Sales of 1907.—The most notable sale of the year 1907 was that of Gilbert Hoxie at Thornton, Ill., on June 13, when 42 head brought an average of $308. The bull Perfection, then in his eighth year, went to Clem Graves at $3,900.* A. W. Jenkins of Texas took Prime Lad 5th at $1,100. Mr. Nave sold 70 head in March at an average of $229.50, the top being $1,975 given by S. R. Waters of Missouri for the show cow Nutbrown 9th. In February the Avery-Hines Co. sold at St. Louis 63 head at an average price of $127, the Jamison bull Albany, that had cost $6,000 in England and now eight years old, selling to Edmunds, Shade & Co. of Iowa at $1,000. In May Mrs. Cross closed out her cattle at Emporia, 64 head bringing an average of $129. In April Giltner Bros. sold 43 cattle at Nashville at an average of $178.80, and on May 2 Gudgell & Simpson realized $149.75 on 47 head at Kansas City.

---

*Although Mr. Graves bought Perfection at this sale for $3,900, Mr. Hoxie had not closed out all of his females and persuaded Mr. Graves to permit Perfection to remain for a time in his service. Some months later Tom Clark negotiated the sale of the entire herd, including Perfection, to W. H. Curtice.

Giltner's imported bull Protector died in November, 1907, at the age of ten years.

Perfection Fairfax Arrives.—First it was Dale, then Perfection, and now the latter's son Perfection Fairfax, bred by Gilbert Hoxie and sold to Mr. Huxley, who had the satisfaction of winning first with the "future great" sire at Kansas City. In addition to that honor Perfection Fairfax annexed at the Chicago International the senior bull championship, which at Kansas City had been sent to Prime Lad 9th. He was shown at a weight of about 2,340 pounds. The VanNattas were first in two-year-old bulls at both shows with Prime Lad 9th. Cargill & Price were successful in carrying off the aged herd prize at both shows, and at Kansas City also had the young herd prize. At Chicago Mr. Brock had first on young herds, the get of Disturber now beginning to come forward with great promise.

The Bargain Counter of 1908.—This was a hopelessly discouraging year so far as sales were concerned, but the fall shows developed strength. Prices, in fact, were at the lowest ebb of this depression, a total of 936 head being passed through the sale ring at a general average of $116.15, as against an average of $123.70 for 1,358 head sold during the preceding year. Mr. Nave sold 98 head at Attica, Ind., for an average of $124. At a combination sale held at Kansas City early in the year 170 cattle were distributed at the absurd price of $88 per head. The general appreciation, however, of the Anxiety blood was well demonstrated even in those dark days by the fact that two Gudgell & Simpson bulls made on this same occasion $400 and $430 respectively.* It was at the very bottom of this extreme depression that Murray Boocock of Virginia closed out at Chicago the herd which he had founded some years before at Castalia. The cattle were offered in ordinary field condition and were passed through the ring at practically beef prices. The imported bull Salisbury, that had been bought at the Sunny Slope sale, went under the hammer in his twelve-year-old form at $150. At Kansas City during the October show 52 head sold for an average of $173, W. J. Tod of Maple Hill taking a number of bulls for range purposes. During the Chicago International 49 head sold for an average of $160.

It was during this period that an effort was made to obtain a share of the Argentine business in Herefords. Secretary Charles R. Thomas of the Hereford association had been sent to South America to exploit the claims of American Herefords and some trial shipments were made. The expenses connected with these experimental efforts, however, were so heavy that not much headway was made in opening up a market in that country.

In December, 1908, W. H. Curtice bought Perfec-

---

*As indicating how dull the cattle trade can be at times, Mr. Charles Gudgell states that some years ago he wrote a number of letters to various breeders offering and recommending the stock bull Don Quixote, then four years old, just in his prime and a sire of proved worth at the price of $100. He was half-brother to the celebrated Don Carlos, having the same mother, Dowager 6th. But after Mr. Gudgell waited several weeks and received no response the bull was shipped to Kansas City and sold for beef, where at a weight of 2,200 pounds he netted more than the price asked.

tion, then in his tenth year, along with 26 of the best Hoxie cows and heifers. The bull had been successfully shown from calfhood and when let down after his career was ended retained his smoothness of flesh in remarkable fashion. The results of his use upon the Beau Donald herd appear in recent showyard history.

A Famous Heifer Class.—The exhibits of Herefords at both Kansas City and Chicago in the fall of 1908 were altogether remarkable, contrasting strangely with the profound depression that had been experienced by the trade during the year. It afforded ample evidence, however, of the fact that the fortunes of the breed were still in the hands of men who believed in the future of the "white faces." The quality of the exhibits this year was indeed superb, a fine illustration of this fact being the senior heifer calf class at Kansas City, where there were twenty-eight entries, every one of real showyard character. The society distributed ten extra prizes in addition to the eight regular ones, and by way of good measure added a reserve prize. The winner of the lowest place had been a second-prize calf at a strong state fair that year. It was, as a matter of fact, a real record-breaking display of Hereford excellence—a sensational demonstration of the continued success of American breeders in developing cattle of the very highest type and all the more impressive because brought forward at a time when the immediate financial rewards of pedigree breeding were, to say the least, unsatisfactory.

Harris had first in aged bulls with Onward 31st, son of March On 6th, a bull weighing around 2,400 pounds. Mr. Nave's Prime Star Grove, a very smooth low-standing bull, stood next and above Prime Lad 9th. In two-year-olds Cornish & Patten were first on Beau Carlos after a hard fight with Harris, who was showing Dislodger by Disturber. Brock had first in senior yearlings on Distinction by Disturber. Giltners headed the junior yearlings with British Highball, by imp. Britisher and they also won the senior bull calf prize on Beau Columbus. Warren T. McCray, Kentland, Ind., a man who was later on to make notable Hereford history, participated in this memorable show, and drew first prize in aged cows on Prairie Queen. McCray had acquired Perfection Fairfax during the year, and while the bull did not win in the strong senior class, he stood at the head of the herd with which Mr. McCray finally defeated the VanNattas, Harris, Cargill & McMillan and Giltner, in the order named. VanNattas' Prime Lads had first in the class for get of sire, and McCray's lot by Perfection Fairfax was second.

An Omen of Better Days.—The Hereford show of 1908 gathered strength as it moved forward, and by the time the Chicago International was reached in December it called forth the following high encomium:

"The Hereford breed has just brought to a glorious conclusion the most sensational show season ever written into its history. Marshalling their forces in marvelous array at the opening of the campaign, breeders of the 'white faces' have marched forward conquering and to conquer, sweeping triumphantly through the circuit with banners proudly flying, challenging stockmen of America to witness the spuerlative degree of perfection to which the studious breeders of this grazier's race of cattle have brought their favorites through long years of devoted endeavor. There has been nothing half-hearted about this campaign. It was almost Gideon-like in its generalship. There was no sounding of trumpets before the strife, but when the cohorts were uncovered at Des Moines, the surprise of the attack admirably served to strengthen the battle array which in deliberate fashion must be recorded as the most effective campaign the breed has ever conducted.

"Figures may make more pertinent this line of comment. Five years ago the Hereford contingent numbered only 94 at the International. Even last year the numerical strength of the exhibit had risen only to 183, while the week just closed witnessed entries that reached almost to two hundred and a half. The pulses of this camp have been stirred to unwonted vigor of beat, and the genial glow of a stimulated circulation has been felt throughout the live stock world. When in the presence of these wonderfully-ripened 'white faces' mention is made of this year of drouth and feed shortage one must rub his eyes in bewilderment. Never has fitting been carried so far. Nay, rather should we draw a finer distinction and say that no year has revealed such artistic fitting in the Hereford camp. There is a remarkable freedom from that over-ripe condition which reveals prolonged lingering at the meal tub. For the most part entries are in impressive flesh and most attractive bloom, so that the eyes of the enthusiast may be freely feasted in an inspection of the magnificent arrays of animals of this breed."

The judging on this occasion was done by Robert H. Hazlett of Eldorado, Kans., Boog-Scott of Coleman, Tex., and Thomas Mortimer. In the aged bull class a newcomer, Preceptor, by Princeps 8th and shown by Dale & Wight, was assigned premier position over Prime Lad 9th, Prime Star Grove, Bonnie Brae 3d and Harry Fluck's Ben Bolt in the order named. So good a bull as Perfection Fairfax was here set down to eighth in this notable line of twelve. Bonnie Brae 8th, shown by J. C. Robinson & Son, Evansville, Wis., headed the two-year-olds and was subsequently made senior champion, the junior bull championship falling to Prince Rupert 17th, shown by Luce & Moxley, Shelbyville, Ky. McCray was again first in the aged cows with Prairie Queen, but VanNatta's first-prize two-year-old Prime Lad heifer Margaret was made senior champion female, the junior championship falling to Donalda, Clem Graves' first-prize senior heifer. The VanNattas had both the aged and young herd prizes, Giltner was first in calf herds, Cargill & McMillan showed the first produce-of-cow and the VanNattas were first with their Prime Lads on four progeny of one sire.

This heartening show was the beginning of the end of the doldrums into which the trade in pedigree Herefords had now for several years been wallowing.

CHAPTER XXII.
PROSPERITY REGAINED.

We have now but to deal with the great revival of the comparatively recent past, and our story ends. In this we must be brief for obvious reasons. In the first place, the tale already grows too long. In the second place, present-day readers do not require, nor probably desire, as full information as to the herds and contests of today as is rightly demanded in the case of the events of more remote periods. There is personal knowledge of contemporary affairs. History deals more with the past than the present. Let us, therefore, sketch rapidly.*

An Upward Trend in 1909.—The general average of public valuations on offerings aggregating 1,400 head of registered cattle in 1909 was $127.05. This

---

*We should not pass over the events of 1909 without noting the decease of Tom Smith of Crete, Ill., which removed from the ranks of the Hereford breeding fraternity a man of Scottish birth who had rendered yeoman service in the up-building of the interest in the "white faces" in the United States. He managed a Hereford herd at Grimley, Worcestershire, England, before coming to the States. He first located at Manhattan, Kans. Soon after this he went to Beecher, Ill., as herd manager for T. L. Miller, and later on had charge of the herd of A. C. Reed, which was maintained for some years at Goodenow. Upon leaving this work Mr. Smith established himself on a farm at Crete, where he bred Herefords successfully for many years. He was a thorough cattleman, and while he never figured conspicuously in the sale and show lists, in a quiet way he supplied much good material to the breed, including Rose Blossom, the dam of the $10,000 Dale, and of Columbia, the grandam of Disturber. He was a keen judge of values. At Clark's dispersion sale, for example, he purchased a cow and bull calf for $200 which he afterwards disposed of for $1,400. He was a natural-born trader and many a carload of cattle passed through his hands to the mutual advantage of both buyer and seller. Tom had few superiors as a judge, and his services were in constant demand at the leading shows, confidence in his integrity as well as in his intelligence being general.

was significant, not so much because of the slight appreciation in the general level, but as an evidence that the inevitable reaction was setting in. Early in the year Giltner Bros. sold privately the young show bull British Highball to L. B. Burnet of Fort Worth, Tex., at $1,500. In March Cornish & Patten sold 117 head at Osborn, Mo., for an average of $206.90, the Anaconda Copper outfit of Montana paying $1,080 for Beau Carlos. At Kansas City on March 9 in a combination sale 54 head averaged $160, Mr. Cornish giving $775 for the yearling bull Onward 71st. On March 24 Mr. Nave closed out his cattle, 99 head in all, at an average of $140, D. E. McConnell of Nebraska taking out the Prime Lad bull Queen's Lad at $600.

S. W. Anderson of West Virginia, who had been for many years the leading defender of the faith in the Virginias, dispersed his herd, making an average of $140 on 122 head. In September various Kentucky breeders contributed 35 head to a sale at Louisville that averaged $165.70, R. C. Hardeman of West Virginia paying the top price of $1,025 for Curtice's Beau Donald 87th. In October the combination offering at Kansas City averaged $166.50 on 49 head, at which Mr. Cudahy gave $800 for Prime Lad 38th. At Chicago in December 48 head were disposed of at $191, William Reynolds of Wyoming buying the Giltner show heifer Florence Acrobat at $700 and C. A. Tow of Iowa taking Makin's Principal 6th at $960.

Prime Lads at a Premium.—Enthusiasm again ran

PRIME LAD 9TH 213903, BRED BY W. S. VANNATTA & SON.

high at the shows of 1909. The state fairs revealed overwhelming excellence in the Hereford section, and the annual round-up at Kansas City in October was indeed a battle royal.

The extraordinary career of the VanNatta cattle in the American showyard is essentially one of the dominant notes of this volume. Indeed, such continuity of successful effort finds few parallels in bovine histories, and no record that fails to reflect this fact would present the truth. Prime Lad "The Great" was now dead, but how his sons were marching on! At the Kansas City and Chicago shows of 1909 Prime Lad 9th was first-prize aged bull and headed the first-prize aged herd over the Cargill & Price, McCray, Makin Bros. and J. L. VanNatta group. In the get-of-sire contest four Prime Lads were first, not only at Kansas City but at Chicago, in competition with the get of Perfection Fairfax, Beau Paragon, Princeps 4th, Young Beau Brummel and Prince Rupert 8th. At Chicago Prime Lad 9th not only headed the senior bull class but was senior champion as well. The great Prime Lad cow Margaret headed her class at both big shows, and at Kansas City was both senior and grand champion female. Then there was the two-year-old heifer Iva. Moreover, McCray's Prime Lad 16th youngsters were pulling ribbons in the senior bull calf classes and the best of these, Gay Lad 6th, was one of a pair of calves sold to "Joe" Green for the Charles P. Taft ranch in southeast Texas for $2,500, the transaction taking place at the International. And "there were others." Prime Lad 16th and Folly's Lad were in the money in the aged bulls and Prime Lad 38th was wearing a ribbon among two-year-olds. Then there was that rare heifer Rosette; and all the while John Letham was as proud and happy over it as were the owners. Why not? And the Lads were to be heard from again another day.*

Let it not be forgotten, nevertheless, that wrapped

---

*John Letham was born at Stonehouse, Lanarkshire, Scotland, in March, 1859. He came to this country when twenty-one years old and started from the ship's side with $2.50 in his pocket. Two years later he was employed at Youngstown, O., feeding Shorthorns. From there he went to Princeton, Ill., where he had charge of the Shorthorns and Percherons of the late Dr. W. H. Winter. Thence he went to Lealand, Tenn., among the Sussex of Overton Lea, and here found his first chance to enter the showyards of the west—first in 1887 with Mayfern, a 2,180-pound Sussex cow, again in 1888 with Rosewood, the champion yearling and grand champion carcass steer, following in 1889 by George, winner of "The Breeder's Gazette" challenge shield for best steer bred and fed by exhibitor.

In the spring of 1891 John Letham went to Mr. VanNatta's at Fowler, Ind., where he fed Hickory Nut, winner of "The Breeder's Gazette" shield and grand championship, and also the champion herd of three steers at the Chicago Fat Stock Show. In 1892 he fed Jerry Rusk, champion at Des Moines and reserve at Chicago Emergency Show. He was still with Mr. VanNatta in 1893, the World's Fair year at Chicago, when Miss Beau Real 3d headed the aged cow class at eight years and the beautiful Annabel was first-prize two-year-old and champion Hereford female (the reserve going to Miss Beau Real 3d). While it was not Letham's good fortune to show Cherry Brandy at the Fat Stcok Show that fall, he left him ready for the fray after two years' careful training, although the steer had never been shown up to that time.

From 1893 until 1900 Letham was not in the arena, but on the latter date he came forward with the two sensational calves from Mr. Henry's—The Woods Principal and Prince Edward, champion calf and reserve champion. In 1901 The Principal was grand champion and Prince Edward champion grade of the show. This same year Prime Lad was in his yearling form, all three animals being from one crop of calves. In 1902 Prince Edward was grand champion at Des Moines and stood second to Shamrock at Chicago, while Valiant Lad was champion yearling and reserve grand champion. In 1903 Prime Lad won his way to grand champion honors in his two-year-old form, but the herd was dispersed and a new lot had then to be bred or bought. In 1904 came Little Joe, grand champion at St. Louis, and Rare Lad, junior champion bull at Chicago.

Meantime, in the hands of Mr. VanNatta Prime Lad went on and became the St. Louis Exposition and Chicago International grand champion of 1904. In 1905, while still with Mr. Brock, Letham gained the two-year-old and Hereford bullock championship at the International with the great steer Silver Lad, by Kansas Lad Jr. Letham's skill was well established in the handling of this bullock. Three shows were made with him. He had second as a calf and yearling, and in his two-year-old form at a weight of over 1,700 pounds was the champion white-faced bullock of the year.

In more recent years Mr. Letham has greatly added to his laurels through the great records made by the Disturbers.

up in rich furry robes in these same great classes were stars of the first magnitude from the other great central sources of the time. McCray was coming strong on Perfection Fairfax stock. Jim Price was forcing the fighting with the Princepses, Fulfillers and Bonnie Braes. Harris was crowding all winners with his Beau Donald 5ths, besides supplying Dislodger and Repeater by Disturber. The Giltners, Clem Graves, Dr. Logan, Luce & Moxley and J. H. & J. L. VanNatta were also in the thick of the fray with finely fitted cattle.

It was truly a wonderful display, that show of 1909—bloom everywhere with excessive obesity clearly, and properly, at a discount as compared with the olden days.

More Ground Regained in 1910.—The improvement in prices noted in 1909 continued throughout the following year, 1,214 head of cattle going under the hammer at an average of $144 per head. There were no sales of special note, but the tone throughout was better. On Feb. 22 Mr. McCray sold 47 head at Kentland, Ind., at an average of $175, the top price being $500, reached in two cases, first by the show heifer Donalda and again by the yearling bull Fairfax 18th, by Perfection Fairfax. J. P. Cudahy got the show heifer Agnes at $410. On the 24th and 25th of February at Craig, Neb., Minier Bros. disposed of 76 head at an average of $216.40, the outside figure being the $775 paid for the stock bull Peerless Perfection 22d by Frank Ulrich, who for ten years past had been in charge of the herd and who secured at this sale some of the best breeding cows for the founding of a herd on his own account. Thomas Nelson of Nebraska took the two-year-old Pretty Lad at $625 and also the yearling Defiance at $600. On April 14 at Chicago Cargill & Price disposed of their cattle at what was felt at the time to be a great sacrifice. The 48 head averaged but $148. Bonnie Brae 15th, a capital two-year-old bull, went to A. L. Weston of Colorado at $990 after a contest with O. Harris. Mr. Weston bought freely of the show cattle. The famous cow Miss Filler 2d by Fulfiller went to Harris at $540. C. A. Tow of Norway, Ia., got Princess 2d at $400. At the usual October sale at Kansas City 69 head went through the ring at $166.45. There were no animals of unusual excellence in the offering. On Nov. 15 Overton Harris sold 69 head at Harris, Mo., at an average of $181. While no sensational prices were paid the bidding was steady, no animal falling below the $100 mark and three reaching $500. The top was $570, paid by the Commercial Land Co. of Oklahoma for Rose Maid, then nursing a bull calf by Repeater. The Bessemer Iron Co. of Alabama bought some of the best cattle, including Prime Lad 48th for $500. J. F. Gulick of Missouri took Harris' Prince 90th, a winner at the shows of 1909, at $500. Frank Nave was also among the bidders upon this occasion. At the customary combination sale during the Chicago International 41 head brought an average of $197.50. This occasion was notable for the fact that the firm of E. N. Casares of Buenos Aires, Argentina, paid the top price of $600 for each of two bulls, Mr. Robinson's Bonnie Lad 11th and Mr. Brock's Discounter.

Another Richmond in the Field.—The Prime Lads, Beau Donalds and other popular favorites were still very much in evidence at the shows in the autumn of 1910, but tangible evidence was now presented of the appearance of an important new factor in western Hereford breeding operations. We refer to the get of the bull Disturber, bred by Jesse Adams and selected by John Letham for Mr. Brock. At Kansas City in October Overton Harris won the senior bull championship with the two-year-old Repeater, son of Disturber, and here was the beginning of another new and highly important bloodline in the western Herefords. Repeater at two years old was a bull of impressive bulk and substance, great rotundity of body, immense loin and heavy hindquarters. Still he was stoutly pushed upon this occasion by Giltner's Beau Columbus, a bull of somewhat similar conformation and very even from end to end. The aged bull class had been won by J. O. Bryant, Savannah, Mo., with the good bull Curtis, by Maynard, showing excellent character and particularly heavy quarters, besides being mellow in his flesh. Prime Lad 9th was now turned down to second. He had been on top for so long that he was of course about due to relinquish the crown. Mr. Tow of Iowa was now coming into the great public competitions as a factor to be reckoned with and received third here in the aged bull ring on Principal 6th, one of the low-legged kind, well spread.

Fortunately for the north Overton Harris had bought Gay Lad 6th before the bull had been started for the Taft ranch. At this show he was as fortunate with the McCray bull as he had been with Repeater, receiving first in the senior yearling bull class and subsequently the junior bull championship. McCray forged to the front in the get-of-sire class with a quartette by Perfection Fairfax. He also had first in aged cows and the senior female championship on the great cow Lady Fairfax 4th. At some of the state fairs VanNatta's Margaret had defeated her, and it was a question all through the season as to where the superior merit really reposed. In two-year-old heifers Luce & Moxley scored with Princess R 10th, daughter of Prince Rupert 8th, after a hard battle with McCray's Lady Fairfax 9th. It was in the senior yearling class, however, that a really sensational female developed in Dr. Logan's Scottish Lassie, by Young Beau Brummel. She was one of the ripest and richest specimens ever seen in the Kansas City showyard, and later in the week received not only the junior but the grand championship of the female classes. Another stunning presentation was that of the Harris Princess twins, daughters of Beau Donald 5th out of a Beau Brummel cow. These remarkable heifers bore away both the blue and red ribbons in a strong class.

At Chicago in December the Bryant bull was not shown, and Prime Lad 9th resumed his old-time position at the head of the senior bulls, McCray winning second on Principal 6th and Brock third on

Distinction by Disturber. Repeater repeated his Kansas City winning in the two-year-old bull class, as did also Gay Lad 6th in senior yearlings. When it came to the championships Mr. Harris had the rare honor of receiving the senior and grand championships on Repeater and the junior championship on Gay Lad 6th. John Letham, representing Mr. Brock, whose herd was now established at Lake Geneva, Wis., here made a phenomenal record for these extraordinary International competitions by winning first in junior yearling bulls on Discounter, first in junior yearling heifers on Disturber's Lassie 3d, first in senior heifer calves on Miss Annabel by Distinction, and first in junior heifer calves on Lady Viola by Disturber, besides the young herd prize. Mr. McCray again scored with Lady Fairfax 4th and won the get-of-sire competition with his Perfection Fairfaxes. Harris won the aged herd prize, and Scottish Lassie was junior and grand champion female.

Improvement Continues.—There were 1,203 head of pedigree Herefords put through the sale rings of 1911 at an average price of $160. This was an average advance of $16 per head over a like number during 1910, and indicated that the tide was still running slowly towards better things. That this was the case was all the more plain because of the fact that the offerings of the year were not notably attractive.

The ball was opened at a combination sale at Denver, Colo., on the 19th of January, when 50 head of bulls were sold at an average of $182. The best figure made was $475 for Mr. A. L. Weston's young bull Carlos 2d, taken by Mousel Bros. The Matador Company paid $400 for The Heir's March On. One of the largest indivdual buyers was M. K. Parsons of Denver. On Feb. 16 Mr. Hutcheon's herd was dispersed at Bolckow, Mo. At this sale the average of $326.50 on 13 bulls was encouraging, and the general average was $233.20. The bull calf Tempter 3d was taken by Makin Bros., formerly of Kansas, but now again breeding Herefords at Grandview, Mo., at $740. Mapleton 4th was bought by J. Secor of Iowa at $725. Handsome Lad, a double Beaumont, went to J. Wolf of Missouri at $500, and Beaumont himself, nearly eleven years old but looking very fit, was taken by H. D. Cornish of Missouri at $305. A. B. Cook, Helena, Mont., was a good bidder for females, his lot including the great producing matron Queen B, by Beau Brummel, from which Mr. Hutcheon had previously sold six calves for a total of $3,400. On Feb. 17 J. O. Bryant sold 81 head at Savannah, Mo., for an average of $141.70. A. B. Cook was also a bidder at this sale. The top price was $725 made for the Kansas City prize bull Curtis, purchased by Hann & Mayne of Iowa. This price was disappointing. Beaumont 2d sold for $700 to a Kansas steer feeder. On Feb. 18 J. A. Larson sold at Everest, Kans., 62 head for an average of $126.95. A. B. Cook was in attendance at this sale also and was a good bidder on the best animals offered, securing ten head. In this case the bulls again outsold the females, always a good indication of rising values.

Warren T. McCray sold at Kentland, Ind., on Feb. 22, 1911, 70 head for an average of $206. The prices were steady at around $200, with nothing going beyond $550, the price paid by J. F. Jones, Granville, O., for the cow Mercedes. The young show bull Beau Real 15th fetched $515, the top for bulls. J. P. Cudahy was a good bidder, securing Corrector Fairfax at $450. At Kansas City on March 7 and 8 59 bulls sold in a combination sale for an average of $174. The best figure reached here was $500 for the yearling bull Guy, consigned by J. A. Gibson. On March 10 W. J. Davis & Co., Jackson, Miss., held a sale at their farm which averaged $220, the bulls selling for $241. J. J. Early of Baring, Mo., reported in April the private sale of the five-year-old prize-winning bull Sunny U. J. to Williams Bros., Randolph, Wis., for $700. Gudgell & Simpson offered 62 head at Kansas City on April 28. The cattle had no great preparation and the general average was $141.30, 18 bulls averaging $172.20. The highest price paid was $330 for Beau Gaston by George J. Anstey of Iowa.

On Oct. 27 Cyrus A. Tow disposed of 49 head at an average of $181.90. The 15 bulls averaged $208.70. Mr. Tow during the year purchased the entire S. L. Brock herd, including some 75 head, thus acquiring possession of the highly prized blood of Disturber. Some of the get of this bull were included in this sale and they averaged about $110 above the general average. The top of this offering was $505, given by A. J. Minish for the yearling bull Dismora 2d. Thomas Mortimer topped the females by paying $375 for Lady Peel by Disturber. Mr. Cook of Montana got the show heifer Miss Laura B at $300. Mr. McCray was also a buyer at this sale. At Fowler, Ind., on Nov. 2 and 3 a disappointing sale was made by the executors of the estate of William S. VanNatta, at which 114 head averaged $184. Here, as at most other sales of the year, the bulls did better than the females, 24 head averaging $263 and the top price being $1,005 paid by O. Harris for Donald Lad 7th. Messrs. J. H. & J. W. VanNatta, relatives of the deceased who were breeding good cattle independently at Lafayette, Ind., paid the top price for a female—$505 for Prime Lady 2d. Mr. McCray was the most extensive purchaser, making investments which subsequently returned handsome profits. As the sale was to close the estate, no guarantee of any kind could be given, and this fact of course militated against high prices.*

---

*The well known herdsman, Tom Andrews, who had commenced his career with the first VanNatta show herd, assisted in this sale, and this therefore is a fitting opportunity to refer to his work in general. Tom has had a long and interesting identification with the fitting of show and sale cattle in leading western herds. Like most of the other English boys who came into our cattle business in the '80's, he was born in Herefordshire. This important event in his career occurred on Oct. 6, 1861, in the parish of Dorstone. His father was a farmer, who in 1869 leased a 450-acre tract at Winforton, where he remained for seven years. In 1876 Mr. Andrews Sr. leased a place of similar size at Weston Court, Pembridge. Reverses overtaking his father in 1880, Tom felt compelled to get out and hustle for himself, so in April, 1881, we find him approaching that Mecca of most of the Herefordshire lads in that day—Beecher, Ill.—in quest of work.

He engaged with T. L. Miller, and in the fall of 1882 went to Fowler & VanNatta to help George Waters in the fitting of the show cattle sent out successfully on the circuit that year. After this experience Mr. Andrews went to Earl & Stuart for twelve months, and from there to Kansas City with Frank Crane, who was at that

ARAMINTA 4TH, OF THE CARGILL & McMILLAN SHOW HERD.

DISTURBER 139989—IN BREEDING CONDITION—BRED BY JESSE ADAMS AND USED BY S. L. BROCK.

During the Kansas City show 50 head, mostly bulls, were sold at an average of $262. As far as bulls were concerned, this exceeded the average of prices at the sale of 1910 by $95. This was partly due to the presence upon this occasion of ranchmen buying for export to Hawaii. The highest price of the sale was $1,050 for the two-year-old show heifer Banza, contributed by R. H. Hazlett, El Dorado, Kans., and bought by J. P. Cudahy. Cudahy had in the meantime acquired possession of Dr. Logan's famous champion heifer Scottish Lassie. At this sale Laredo Boy was sold by his breeder, C. L. Browning of Laredo, Mo., to Carl Miller, Belvue, Kans., at $610. He had won a red ribbon in the show. On Nov. 9 Makin Bros. sold 63 head at an average of $146.30, the 12 bulls averaging $208.95. They were topped by Paragon 12th taken by John Gosling at $450. J. P. Cudahy bought Celandine 2d,

---

time maintaining a sale stable in charge of George Waters. While here Tom went to Quebec to take charge of 60 head of cattle then in quarantine that had been imported by George Leigh. These were intended for the Lee & Crane farm near Independence, Mo. From the quarantine station Andrews came to the Chicago Fat Stock Show with four steers, one of which, a purebred Hereford, won the two-year-old championship over all breeds.

This was perhaps the first job of fitting for which Tom had full personal credit, and he naturally felt proud of his achievement upon that occasion. After this he returned to the Crane farm at Independence. In the autumn of 1886 he went to England, and after his return was for about three years with Z. T. Kinsell of Mt. Ayr, Ia. In 1895 he was with the F. A. Nave show herd, and in 1896 was engaged by Mr. J. H. Miller, Peru, Ind., to take 23 head of cattle to Buenos Aires, South America. In 1902 he showed the F. L. Studebaker herd. Subsequently he fed for the Messrs. VanNatta of Lafayette for two years. In 1908 he revisited England and on his return assisted with the preparation of Frank Nave's cattle for the closing-out sale at Chicago. After this engagement he went to Michigan, showing a herd for Merritt Chandler, and in 1910 was with the Cyrus A. Tow cattle. Once again he went back to Herefordshire, and on returning to America was again with the herd of W. S. VanNatta & Son, staying with the firm until the closing-out sale after Mr. VanNatta's death. It thus appears that Andrews was helping with the first Herefords Mr. VanNatta ever exhibited, as well as with the last. In 1913 he was with the W. H. Curtice cattle in Kentucky, and at the present time (1914) is in the employ of the Messrs. Berry, Mt. Vernon, Ia.

the junior heifer show calf, for $500. It was announced at this sale that Claude Makin, who had been the guiding hand in the firm's Hereford breeding operations for thirty years, would continue in the business, retaining the bull Beau Paragon and a few females. Needless to say, this announcement was received with special satisfaction, because in the course of his long and honorable identification with the "white faces" he had been a consistent adherent of "quality" cattle and had gained the goodwill of all with whom he had come in contact. At the combination sale during the Chicago International 51 head made the gratifying average of $286.30. Twenty-one females averaged $344, this mark being reached largely through the sale of the junior champion heifer Perfection Lass to Mr. Cudahy at $2,000. C. M. Largert of Texas bought Mr. McCray's bull calf Victor Fairfax at $1,000, and William Andrews & Sons of Morse, Ia., gave the same amount for Jim Price's yearling Bonnie Brae 37th. George Leigh was a good buyer upon this occasion, taking a number of good bulls for the western trade.

Death of William S. VanNatta.—The veteran breeder, William S. VanNatta, died at Fowler, Ind., on May 26, 1911, at the mature age of eighty-one years. He was born in a log cabin on the prairies of Tippecanoe Co., Ind., on Sept. 27, 1830. These pages are so full of references to the great Herefords bred by Mr. VanNatta that it is unnecessary to enter into further details concerning his operations

THE LATE WILLIAM S. VANNATTA'S HOME AT FOWLER, IND.

in this connection. Suffice it to say that, everything considered, he has had no superior as yet in the ranks of American breeders of Hereford cattle. The dispersion of the old herd has already been referred to. In "The Breeder's Gazette" for June 7, 1911, the author of this volume endeavored to pay a suitable tribute to Mr. VanNatta's memory, from which we may here extract the following:

"Mr. VanNatta will live in American agricultural history as one who contributed heavily to the sum total of progress registered in his chosen field during the span of his activities. His career as one of the greatest cattle breeders this country has yet produced abounds in inspirations for those who would follow in his footsteps. But the one mountain peak standing out in bold relief in the outlines of his progress is unswerving tenacity of purpose. One can but admire the indomitable pluck, the infinite patience that characterized his unfaltering devotion to the cause he had espoused throughout all the alternating periods of adversity and depression. The lesson of his life is just this: It pays to stand by one's colors; it pays to keep one's head during 'booms,' and one's nerve during depressions; it pays to be ruggedly honest always. The fact may as well be admitted first as last that, as a people, we are not possessed in high degree of that quality of dogged persistence that finds probably its best national expression in the life of Gen. U. S. Grant. William S. VanNatta became convinced in the early '70's that the 'white faces' were the best beef cattle of the day, and in their cause he enlisted not for 'ninety days' but 'for the war'; and upon that line he fought it out to a finish that not only brought fame to himself but honor to the Hereford name."

The Judgments of 1911.—There was the usual preliminary skirmishing at the state fairs of 1911, with various changes in the rating at the hands of different judges. The Kansas City Royal this year found the new Secretary of the American Hereford Association, R. J. Kinzer, in charge of the Hereford exhibit as superintendent, Messrs. Shade and Hazlett being the other two members of the managing committee. The judges were Capt. Robson of Canada, S. W. Anderson of West Virginia and Edward J. Taylor of Michigan.

In the class of ten aged bulls Giltner Bros. were first on Beau Columbus, Makin Bros.' Paragon 12th, that had been very successful earlier in the season, being here turned down to fourth place. J. H. & J. W. VanNatta were second on Tippecanoe, an impressive bull forward, with drooping incurved horns and fine spread of back. Beau Columbus was universally admired for his good breed character, great evenness from end to end, strength of hind-quarter and fullness of twist. McCray came into third place with Fairfax 13th. In the two-year-old bulls Harris had an easy victory with Gay Lad 6th, that proved to be quite the sensation of the showyard year so far as Hereford bulls were concerned. Sired by Prime Lad 16th, son of old Prime Lad, and out of Sister Perfection, own sister to the champion bull Perfection, Gay Lad 6th was regarded as a more spectacular proposition than either of the champions whose names ornamented his ancestral tree. Fairfax 16th, shown here by Mr. Cudahy, was almost as

sensational. His beautiful front and wonderful finish all around rendered him an outstanding bull in any company. He was both senior and grand champion. Another remarkable youngster, Cudahy's Corrector Fairfax, son of Perfection Fairfax and Likely by Corrector, headed a class of seventeen senior yearlings and was subsequently made junior champion. He had a beautiful coat, grand front and quarters that left nothing to be desired—another one of the many top-notchers now coming so frequently from Mr. McCray's herd. Scottish Lassie was senior and grand champion cow and VanNatta's Donald Lass 9th, first-prize senior heifer calf, was junior female champion.

At Chicago under the judgment of Abram Renick and George Leigh Beau Columbus again headed the aged bulls, Paragon 12th, that had passed into the possession of W. A. Dallmeyer of Jefferson City, Mo., moving up to second place. Messrs. Robinson received third on Bonnie Brae 8th. Gay Lad 6th repeated his Kansas City victories, heading the two-year-old bulls and attaining the senior and grand championship. Corrector Fairfax duplicated his Kansas City victories.

In the female section Scottish Lassie was still the reigning favorite, again winning the senior and grand championship. The junior championship fell to Luce & Moxley on Perfection Lass, a heifer with a great top, beautifully balanced and of most attractive femininity.

Auction Sales of 1912.—During 1912 there were

957 Herefords disposed of at public sale at an average of $180.40. This was an advance of $20 per head over the previous year. During the Denver show in January 46 head fetched an average of $182.35. On Feb. 22 Mr. McCray sold 70 head for an average of $255, 20 bulls averaging $330 with none of them going higher than $680, which price was paid for Fairfax 13th by a cattle company operating in Old Mexico. On Feb. 29 at Grand Island, Neb., Mousel Bros. and G. G. Clement disposed of 44 head at an average of $201.80, on which occasion the young bull Beau Mousel was bought by Mr. DeWitt of Colorado Springs for $1,000. Mr. DeWitt had also been a good buyer at the McCray sale. On March 8 and 9 at Council Grove, Kans., Jones Bros., who were dissolving partnership, sold 158 head at an average of $128.70, bulls again outselling the females, the 22 males averaging $172.50. At a sale made by W. J. Davis & Co. of Mississippi on March 14 42 head averaged $163.50. A combination sale was held at Kansas City on March 26 and 27, but the offerings were not of the highest grade and the average on 116 head was but $116.35, with a top of $410 for a yearling bull consigned by the Funkhouser estate. On April 12 at Kingsley, Ia., Messrs. Edmunds, Shade & Co. sold 52 head at an average of $150, the best price being made by the yearling bull Beau Albany, sold to go to Canada at $500.

During the Kansas City show in October a lot of well conditioned cattle were offered, the 49 head averaging $303.90 and no exceptional prices being included. A top of $710 was paid by J. M. Curtice for Prince Rupert 39th, consigned by Luce & Moxley. Dr. Logan bought Beau Caldo, bred by Robert H. Hazlett, at $680. On Oct. 22 O. Harris & Sons made a good sale of 64 head which averaged $314. The two bulls Donald Lad 7th and Prizewinner made $1,000 each. Gay Lad 8th brought $1,025 and Gay Lad 2d $1,500, the latter price being paid by W. M. Braddock of Nebraska. The bull average here was $464. Five Gay Lad 6th bulls averaged $772. On Oct. 24 Gudgell & Simpson sold 50 head at Kansas City for an average of $177.60, the bulls averaging $254.30 and the top being $460 for Bright Lord. On the following day C. B. Smith sold 50 head at Fayette, Mo., for an average of $127.50. Nearly half of these went to Montana. Giltner Bros. sold 50 head at Shelbyville, Ky., on Oct. 30 at an average of $176.70, the bulls making $224. The yearling bull Beau Boston went to Texas at $500. At Chicago on Dec. 4 36 head averaged $325, the $1,000 notch being reached by the cow Prime Lady 2d, sold by J. H. & J. W. VanNatta to R. A. Thompson of Alberta, Canada. The highest price for a bull was $975 for Donald Rex by Mr. Zwick of Colorado. On Dec. 17 at Harlan, Ia., Hann & Mayne sold 59 head for an average of $152. The chief attraction at this sale was Beau Perfection 9th, taken by Mr. Cudahy at $1,000.

Fairfax 16th and Scottish Lassie.—Again passing by the ebb and flow of awards at the state fairs,

BEAU FAIRFAX 868360, BRED BY W. T. McCRAY.

THE CHAMPION FAIRFAX 16TH 816931, BRED BY W. T. McCRAY, SHOWN BY J. P. CUDAHY AND SOLD TO A. B. COOK.

we may summarize the show season of 1912 with the statement that the chief honors of the year, the Kansas City and Chicago championships, were reaped by Mr. Cudahy's Fairfax 16th 316931 and Scottish Lassie. The former had stood second to Gay Lad 6th in the two-year-old class at both the Royal and the International of 1911, while the Lassie had been senior and grand champion female. Fairfax 16th was bred by Mr. McCray. Sired by Perfection Fairfax he was out of Cherry Donald by Beau Donald 33d, the second dam being Mary's Cherry by Cherry Boy, son of old Fowler. He therefore represented a fine blending of the blood of the champions of other days, and in him surely all that was best in his ancestry lived again. At Kansas City under Thomas Mortimer Fairfax 16th was first in aged bulls, senior and grand champion bull and headed the first-prize herd. Scottish Lassie headed the aged cows and was again senior and grand champion female. O. Harris & Sons ran away with the junior championships on the wonderful senior bull calf Repeater 7th and the senior heifer calf Miss Gay Lad 6th. Mr. Tow had a remarkably fine lot of young cattle on exhibition, largely the get of Disturber, and gained high places in the prize-list.

Coming up to the International in December Mr. Cudahy was still unbeatable, Fairfax 16th again being first-prize aged bull, grand champion and the head of the first-prize herd. Mr. Tow stood second with his low-set masculine bull Standard, by Bonnie

Brae 8th. The Giltners were third on Britisher Jr., showing much of the scale of his famous sire, once champion in this same ring. Corrector Fairfax, the junior champion of 1911, led the two-year-olds and Mr. McCray was second with another bull of his own breeding—Byron Fairfax. In senior yearling bulls the VanNattas were first on Graceful Lad and Hazlett was second on Bonnie Lad 26th, by Bonnie Brae 8th. In junior yearling bulls Cudahy's Beau Fairfax was first, and Tow's Perfection Jr., by Perfection, was second, with Hazlett's Bocaldo third. In senior bull calves Messrs. Harris were easily first with Repeater 7th, also taking second on Gay Lad 9th. Luce & Moxley headed the junior bull calves with Prince Real by Beau Real 15th.

The place next to Scottish Lassie in the aged cow class was assigned to the VanNattas' Prime Lady 2d, McCray having third on Gay Lass 5th. In two-year-old heifers Cudahy headed the line with Perfection Lass, Tow's Disturber's Lassie 4th being at her side. In the championship rings the Kansas City verdicts were approved in the case of Fairfax 16th, Repeater 7th and Scottish Lassie, but the junior female championship was here sent to Miss Repeater 11th instead of to Miss Gay Lad 6th, the former having been awarded first prize in the junior heifer calf class. Miss Gay Lad 6th, the Kansas City champion, had been placed second in senior calves to Cudahy's superb plush-coated Pearl Donald by Beau Donald 40th.

Higher Levels Reached.—For the third consecutive year the Herefords sold at auction in 1913 realized the highest average attained by any of the beef breeds, a total of 1,311 head bringing $259.30 as against $220.35 for Shorthorns. It should be noted, however, that as a rule about twice as many Shorthorns as Herefords are sold publicly each year.

The year opened auspiciously in the west by a capital sale held during the Denver show on Jan. 23. The success was due in good part to the liberal buying of Mr. A. B. Cook of Helena, Mont., who took out the three-year-old bull Heir's March On 2d 359789 at $1,350. Bred by T. F. DeWitt of Colorado Springs, he had won the bull championship that week. He was sired by The Heir, he by Beau Donald 17th. Another youngster by this same bull was taken by D. Firm & Son, La Veta, Colo., at $1,250. Six other bulls by The Heir brought a total of $2,590. The highest-priced female was Defender's Lassie 2d, taken by Mr. Tow of Iowa at $600.

At Grand Island, Neb., on Feb. 14 and 15 another successful sale was held under the management of Mousel Bros.; 114 head sold for an average of $221, the 73 bulls averaging $241 and the 41 cows $186. Beau's Contender, consigned by Mousel Bros. and sired by Beau Mischief, topped the sale at $950, going to Wm. Hutcheon of Missouri. The bull calf Beau Lindell 2d was taken by Gudgell & Simpson at $925.

McCray's Big Average.—On Feb. 26 at Kentland, Ind., Mr. McCray had the satisfaction of selling 76 head for an average of $525, the females averaging

WARREN T. McCRAY.

$410 and the 27 bulls averaging $740. For several years Mr. McCray had been making Hereford showyard history, particularly with the get of Perfection Fairfax. Nine sons of this bull averaged $1,460 upon this occasion. Mr. Cudahy paid $2,500 for Crusader Fairfax. Thompson Bros., West Point, Neb., gave the same sum for Duke Real, out of a daughter of Perfection Fairfax. C. A. Tow paid $2,450 for Byron Fairfax, Dr. Logan $1,650 for Russell Fairfax, J. I. Moffat, Carroll, Manitoba, $1,575 for Perfect Fairfax, Charles Adams, Dickinson, N. D., $850 for Albany Fairfax and Watson & Puckett, Apple River, Ill., $725 for Prince Fairfax. Conrad Kohrs, the veteran Montana ranchman, purchased a bunch of good bulls at this sale. The day after this event Mr. McCray sold at private treaty Beau Perfection 11th to Downie & Sons of Alberta for $1,750, the prize-winning calf Real Fairfax to A. L. Weston, Golden, Colo., for $2,000 and a month-old calf by Perfection Fairfax—a full brother to Byron Fairfax—for $1,000.*

---

*Asked for a statement as to how he became interested in Hereford cattle Mr. McCray has supplied the following interesting story:

"I can hardly remember when I first became interested in cattle. My parents have told me that when I was a little boy in kilts I developed an early instinct for cattle. My favorite game was riding a stick horse and driving "cattleoes," as I called them at that time. My father handled a great many feeding cattle and I simply inherited my love for them. I can well remember when a small boy of going with my father to the home of Mr. W. S. VanNatta at the old Hickory Grove place to see his cattle, when he was in business with Mr. Fowler. The impression which those big broad-backed thick cows made at that time still remains with me. I was so enthusiastic about them that I earnestly pleaded with my father to buy two or three and take them home. He was not influenced to any appreciable extent by my pleadings, however, though I can well remember making a vow that I would some day have a herd of my own.

"After I grew to maturity and engaged in business for myself, and had acquired several acres of land, I concluded that it was about time for me to indulge in my desire to own some good

Other Mid-west Sales.—At Kansas City on March 4 and 5, a sale under the management of R. T. Thornton developed a good demand for bulls, 56 selling for an average of $274, the general average on males and females combined being $242. A steady range of values rather than exceptionally high prices prevailed. J. A. Shade sold 49 head

---

cattle. Accordingly, when VanNatta & Son advertised a sale of purebred cattle in the fall of 1905 I went over to Fowler. I intended to buy one or two head but when the sale was over I found myself the owner of five head of cows with three calves at foot. That winter the cattle were cared for by a farm hand. The next spring I realized I must have a bull, and inasmuch as I could hardly afford to keep a good bull for five cows I went to the sale of Mr. Huxley, at Bunker Hill, and purchased seven cows and a bull (Lord Improver). I had only ordinary farm help looking after the cattle, but by the fall following my purchase from Mr. Huxley I was forced to the conclusion that if I was going to handle cattle I must have a man who was experienced in the business. By this line of reasoning I also concluded that if I was going to have an experienced man at a higher cost I must have enough cattle to keep him employed. That fall Mr. Sotham was holding a series of sales at the cattle pavilion in Kankakee. I attended one of these and from the dispersion sale of Mr. Bullard I purchased eighteen head of splendid cows. By that time I was getting into the cattle game right. I then hired a first-class man in the spring of 1906.

"I watched the development of the herd that summer. My business instinct soon told me that while I might grow a few cattle for my own enjoyment if I was going to make any mark in the world I must become an exhibitor. I must get acquainted with cattlemen and have cattlemen acquainted with me. I also found that I must purchase and add to my herd the best bull I could find. That fall I visited the state fair at Indianapolis and watched the judging of the Hereford classes. After the ribbons had been tied which proclaimed Prime Lad 3d the grand champion bull and Prairie Queen the junior champion female of the show, I followed Frank VanNatta to the barns and asked, "Frank, what will buy Prime Lad 3d and Prairie Queen?" And he replied, "Thirty-three hundred dollars." This almost took my breath away but after I had sufficiently recovered I said, "Why, Frank, you don't mean it, do you?" "Yes," he said, "I mean it," and I could not take off a dollar. I finally secured an option on the two animals until the next week, as I wanted to think it over for a while. I reasoned it all out and eventually went over to Mr. VanNatta's farm and closed the deal. I had heard of a good cow out in Illinois and I sent my herdsman over to look at her with authority to buy. He found an outstanding animal in the cow Phoebe, which we developed and showed so successfully in the fall of 1907.

"That was my first season out and I only showed five head: Prime Lad 16th at the head of herd; Phoebe, the grand champion female; Prairie Queen, a two-year-old; Diana Fairfax, a yearling, and a calf by Lord Improver. With this aged herd I made my debut into fashionable Hereford circles. I mention this to show that with animals of proper merit one can compete with the most renowned exhibitors, even if he is a beginner and unknown. That

at Kingsley, Ia., on March 6 which averaged $294. At this sale 26 bulls averaged $308 and 23 females $283. The highest price here was $780 paid by Thomas Mortimer for the young bull Beau Matchless 382372. J. B. Ashley, Audubon, Ia., paid $725 for the bull Rex Dorian 323948. George J. Anstey,

---

year Phoebe was grand champion in almost every show and Prairie Queen was the reserve champion, and their places were accorded them without much contention. In the spring of 1907 Prime Lad 3d took sick and died after an illness of but a few days. This was a sad blow to my expectations, as I had paid $2,500 for him and had received but little benefit from his use, having but thirteen cows in calf to his service.

"I concluded that I could not afford to turn backward, as I had collected an unusual lot of females and had built barns and had arranged my pasture for a cattle business. So I commenced to look around for a suitable successor to the late head of my herd. I addressed fifteen letters to the prominent breeders of Hereford cattle, asking if they had for sale an outstanding two-year-old bull. I found them to be very scarce that year and after investigating the most interesting prospects I decided that Prime Lad 16th presented the greatest possibilities as a sire and show animal. I used him that season and the next fall. I was not sure of my judgment on him as a sire, so I looked around for the bull that sired calves which appeared to me to be the best that were being shown. After visiting a few shows I decided that the coming sire was Perfection Fairfax. I immediately opened negotiations with his owner, Mr. Huxley, and found that he could not be bought without my taking over the entire herd. This I did, counting Perfection Fairfax in the deal at $5,000. This was at a time when cattle were selling low. I afterwards bought the entire herd of Clem Graves, thereby securing Beau Real together with about twenty-five grand matrons of excellent quality and breeding. The well known principle of breeding, that the sire is half the herd, has been more than demonstrated in my experience.

"Prime Lad 3d died before he had an opportunity to contribute much to Hereford history. Prime Lad 16th also died early in life but in the two years that I used him proved his value as a sire and had he lived until old age his name would have gone down in history as one of the greatest of sires. He imparted to his offspring that size and quality, that level smoothness, that beautiful head and character, that boldness and style so much desired. His son, the grand champion Gay Lad 6th, shows the superlative character and quality he transmitted, and his daughters are uniformly large, smooth, well balanced cows.

"The record made by that grand champion and sire of grand champions—Perfection Fairfax—contributes a page of Hereford history that is truly remarkable, and today he stands the unchallenged king of Hereford sires. The uniformity of his descendants in all particulars, their excellent character, their general pleasing make-up, their ability to put on flesh with even smoothness, and their good bone and feet have distinguished them and stamped them as a remarkable family, clearly illustrating what influence a strong prepotent sire will have in a herd. Beau Real, a grandson of old Beau Donald, also had the faculty of imparting great uniformity to his offspring and his use added much strength to my herd. I have recently added the two-year-old Farmer to my list of herd bulls and I am expecting great results from his use."

Massena, Ia., gave $575 for Beau Dover. On March 18 Mr. Anstey sold at South Omaha 50 head for an average of $146. At a combination sale held at Omaha on April 16 under the management of W. N. Rogers 69 head sold for an average of $161, the 55 bulls making around $168. At East St. Louis on April 22 a combination sale under the management of Sydney B. Smith resulted in an average of $174 on 36 head of cattle. On April 29 Taylor & Welty sold at Wanatah, Ind., 51 head for an average of $182.

A Big Deal on the Range.—One of the most important transactions in recent years in the range country was the sale in May, 1913, by Montie Blevins of his herd at North Park, Colo., to go to Montana at $75 per head with all calves of the crop of 1913 counted. As there were between 700 and 800 head of cattle involved in this deal it was regarded as a notable event. The sale was all the more exceptional because of the fact that there were around 150 head of yearling heifers in the herd and no young bulls at all; the bull calves having already been contracted for. These were of course unregistered cattle, but Mr. Blevins had practically brought them up to a purebred basis.

Fall Sales of 1913.—The usual sale at Kansas City during the American Royal resulted in an average of $388 on the 46 head offered, 27 bulls averaging $474. They were a good lot and the demand was excellent. The top of this sale was $1,975 paid by O. Harris & Sons for the second-prize senior bull calf Vernet Prince 4th, that had been successfully exhibited by W. J. Davis & Son of Mississippi.* J. B. Burkett of Hereford, Tex., gave $750 for the yearling Proctor Onward. Spooner & Son, Mondamin, Ia., paid $700 for Perfect Donald. On the following day Gudgell & Simpson sold 46 head of females for an average of $232.

On Oct. 22 O. Harris & Son sold at Harris, Mo., 68 head for an average of $373, the 29 bulls averaging $518. This good sale resulted largely from the high quality displayed by the get of Gay Lad 6th and Repeater. Gay Lad 12th, by Gay Lad 6th out of a Beau Brummel dam, topped the sale at $2,600, being bought by Johnson Bros. of Colorado. Repeater 18th, just passed twelve months old, brought $2,100 from A. Christensen of Eagle, Colo. Gay Lad 9th was bought by E. H. Taylor, Jr., Frankfort, Ky., at $1,500. Gay Lad 13th and Gay Lad 15th brought $900 each, J. B. Gillette, Marfa, Tex., taking the former and J. E. Thompson, Martinsville, Ill., the latter. The top for females was $550 for Princess Repeater, also bought by Mr. Tay-

---

*The Messrs. Davis put a herd of purebred Hereford cattle on La Vernet Stock Farm in January, 1910, by the best Prime Lad and Anxiety cows they could get. They placed at the head of the herd the show and breeding bull Point Comfort 14th 337488, now five years old. The calves produced by him have fully met their expectations. They bought McCray Fairfax, a great son of Perfection Fairfax, to breed on the daughters of Point Comfort 14th, and the few calves to hand demonstrate that this is a good cross. They say:

"Hereford cattle excel all other beef breeds for this climate. They are great rustlers and great breeding cattle. They stand the long summers and fatten in the winters, go out March 1 on grass in good shape and are money-makers and soil-builders. We predict that in less than five years in this climate, where we can produce two crops on the same land, thereby making cheaper feeds and as good cattle, that the south will be able, with Hereford cattle, to make beef for the world."

OVERTON HARRIS.

lor. Mr. Christensen took Princess Repeater 3d at $510.

At the International sale at Chicago 46 head averaged $561, the 26 bulls exposed bringing an average of $626.70. The top at this sale was $2,200 for the senior bull calf Standard 11th, taken by A. B. Cook, Helena, Mont. James Chesney of Evanston, Wyo., gave $2,100 for the two-year-old bull Bonnie Lad 28th; C. G. Cochran & Son, Plainview, Kans., gave $1,130 for Prince Rupert 50th; Thomas Mortimer paid $785 for Royal Mail, and W. T. Jones of Texas took Diamond's Donald at $875. The highest-priced female was Defender's Lassie, taken by O. Harris & Son at $2,050. Mr. Taylor of Kentucky gave $1,250 for the two-year-old heifer Bonnie Lass. On Dec. 8 W. E. Hemenway & Son sold 33 head at Steward, Ill., for an average of $154, the top being $500 paid by Meier Bros., Bellvue, Ia., for Prime Star Grove.

William Andrews & Sons and James Price sold at Morse, Ia., on Dec. 17 17 bulls for an average of $321 and 21 females at an average of $269. Eleven head of polled Herefords sold at an average of $332, the entire sale averaging $314. Mr. Price's Bonnie Brae 69th, a double grandson of Bonnie Brae 3d, was bought by O. Harris & Son for $925. The Messrs. Andrews had been devoting their attention for some time to the development of the polled Hereford type and sold upon this occasion the two-year-old polled bull Prime Grove for $1,025, the buyer being Henry Smith of Nebraska. Another polled bull by the same sire went to Guy Jones of Missouri at $600.

Fairfax 16th and Beau Perfection 9th at $7,500.— In the month of August, 1913, Mr. McCray made one of the great private deals of Hereford history, buying the entire J. P. Cudahy herd at Belton, Mo., consisting of 437 head, and including the sensational winners at the big shows of 1911 and 1912—the bulls Fairfax 16th, Beau Donald 75th, Beau Fairfax, Corrector Fairfax, Crusader Fairfax and Beau Perfection 9th.

After this transaction was closed Mr. A. B. Cook of Helena, Mont., whose purchases of high-class cattle had for several years previous been one of the features of the trade, bought 32 head of the top Cudahy cattle for the sum of $18,000, including Fairfax 16th and Beau Perfection 9th at $7,500, the 30 females being figured at $350 each.

Many other illustrations of the keen demand for good Herefords could be cited from the records of private sales made during the year 1913. The activity in the trade in the west was specially marked, reflecting the steady trend towards higher values for beef cattle on the hoof at all leading markets. Scharbauer Bros. of Texas, who had for many years been breeding cattle of a character specially adapted for the range trade, reported that they had sold for shipment to the Dakotas in one lot thirteen carloads of bulls numbering 429 head, the trade involving between $30,000 and $40,000. During the fall of 1913 Mr. Richard Walsh, former manager of the

Adair ranch, visited the west and purchased 15 head of bulls for shipment to British South Africa.

A Champion From the South.—An interesting experiment in Hereford breeding in the lower Mississippi Valley has been going on for a number of years under the direction and ownership of W. J. Davis, Jackson, Miss. At the International show at Chicago in December, 1913, Mr. Davis had the satisfaction of gaining the senior and grand championship for bulls on Point Comfort 14th 337488, bred by Oscar L. Miles of Fort Smith, Ark. It is true that Mr. Harris did not exhibit upon this occasion, having dropped out of the race after the Kansas City Royal, where he had won both the junior and grand bull championships with Repeater 7th 386905. Nevertheless, the southern champion had to meet and defeat at Chicago such bulls as McCray's Corrector Fairfax, Luce & Moxley's junior champion Prince Real 396530, Beau Fairfax and Prince Rupert 50th.* Point Comfort 14th on the side of his sire ran through Patrolman 4th 133915, bred by Messrs. Ikard of Texas, to the Beau Brummel bull Patrolman of Gudgell & Simpson breeding. The dam of Patrolman 4th was Armour Poppy, bred by K. B.

---

*Mr. Moxley supplies these facts concerning the founding of the Luce & Moxley herd:

"When my father decided to give up active farming on account of his health, Mr. Luce, my brother-in-law, bought the farm, and we founded the firm of Luce & Moxley. He wanted to put the farm in bluegrass, and raise some kind of stock. We debated this question for some time, and then decided on Hereford cattle. Mr. Luce has spent most of his life in the cigar business, being a member of the firm of Powell & Smith Co. until he sold out to the American Cigar Co. and became vice-president of it. After about two years he resigned, and has since been interested in several companies in New York City. He tries to get to Kentucky once a year for about six weeks, as he loves the farm and his stock."

Armour from Beau Brummel Jr. The new champion's mother was Lady Christine, bred by S. L. Brock and sired by Disturber out of a daughter of Kansas Lad Jr. He therefore represented a rich combination of the best northern blood and was individually of a low-set, rugged type—one of the blockiest bulls seen since the days of Prime Lad 9th.

At this same show Perfection Lass was senior and grand champion female. In James Hendry's hands she had gone on satisfactorily since passing into Mr. McCray's possession, and was now a beautiful specimen of latter-day American Hereford breeding.* At

---

*The three Hendry brothers, of Scotch extraction, have had successful careers in this country. At one time they were all in the employ of Charles Gudgell. First George came over from England about 1892 and worked for Gudgell & Simpson continuously until 1909, when he left to go into business for himself. Early in 1914 he was engaged by J. M. Curtice to fit his herd for show. George was in full charge of Gudgell & Simpson's show herd and the breeding herd at Independence from 1898 to 1909, and Mr. Gudgell states that he was instrumental in producing and fitting some of the best show animals ever turned out from that establishment. Among other celebrities in his charge during that period might be mentioned Dandy Rex, Mischievous, Mischief Maker, Modesty, Bright Donald and Priscilla 5th. Mr. Gudgell in speaking of George Hendry's record with the Anxiety cattle remarks: "He was and is not only a successful feeder, but also has the ability to prepare animals for the showring without injuring their productiveness. He possesses all the faithful qualities of the best Scotch herdsmen, and at the same time is always kind and gentle with his charges, never impatient but even-tempered, and he seems to transmit this to the animals themselves."

James Hendry, now in charge of the Orchard Lake herd of Mr. McCray, has been with four leading establishments and has a record to his credit of which he may well be proud. He was first with Gudgell & Simpson, leaving them for a year or two to work for J. M. Curtice. He then returned for a few seasons to the Gudgell & Simpson cattle. When he first went to this firm he was on the Greenwood Farm. Druid was then a yearling, assisted in service by a brother to Don Carlos. After the World's Fair that bull and Beau Brummel were sent to the Greenwood Farm and Lamplighter was placed at the head of the Independence herd. Jim's second period of service with Gudgell & Simpson began in the fall of 1900, when he went to their sale barn at Independence. Among the good bulls there at that time was Beaumont. In 1901 he went to take charge of the Beau Donalds for Mr. W. H. Curtice. When Mr. Curtice hired Hendry he told him that he was particularly anxious to have him develop some good females if possible, for the reason that while the Beau Donald bulls were acquiring much reputation there was complaint that his heifers were not so good. Hendry certainly succeeded in

the Kansas City Royal Mr. Cook of Montana had exhibited a senior heifer calf named Joy, by Beau Carlos, that received, notwithstanding her youth, the female grand championship over Perfection Lass.

These shows are so fresh in the minds of contemporary breeders that we leave to some future historian the task of going into detail as to the truly extraordinary character of the exhibits of 1913 as a whole, and of the champion cattle in particular. They will not soon be forgotten by those who were so fortunate as to see them.

Opening Sales of 1914.—Interest in the sales for the new year centered as usual at the Denver show in January. Fifty-eight head were disposed of at this place on Jan. 21 at an average of $223, prices

---

demonstrating that the old bull's daughters could also give a good account of themselves, for the Belle Donalds 44th and 59th were winners at St. Louis as produce, and others such as Belles 28th, 56th and 60th were also shown with success. Speaking of the old bull's death and the subsequent purchase of Perfection Mr. Hendry says:

"When I came back from the fairs one fall Beau Donald was so lame he could not get around and he never recovered, dying that winter. We then tried to buy Perfection Fairfax, but failed. After this we went to Hoxie's and got 'Uncle Tom' Clark to close for us a deal for the purchase of Perfection. I believe today that the Perfection cross is one of the best in our modern Herefords. I left Kentucky in January, 1911, and old Perfection died in February. I then came to Mr. McCray's where Perfection Fairfax reigned supreme. There have been many good cattle sold from this herd in the four years that have since elapsed. I can but feel that I have had a little better chance than some of the other 'cattle boys', as not many of them have had such good material to work with as the get of Beau Brummel, Beau Donald, Perfection and Perfection Fairfax. I sincerely hope that Imp. Farmer's get will do as well or better, if that were possible."

The third Hendry brother was the last to come out from England, but he has been in the employ of Mr. Gudgell for the past eighteen years and during the last eight years has been in charge of the breeding ranch at Edmond, Kans. Prior to that time he was manager of the Gudgell farm in Anderson Co., Kans., for seven years. Mr. Gudgell says he is a splendid cattleman, but has devoted most of his time to the management of the farming operations, which include harvesting 1,000 acres of alfalfa and the production of over 1,000 acres of corn and small grains, with general supervision over the breeding herd of Herefords. This is William Hendry—good cattleman, excellent farmer and splendid handler of men.

Good work this, for one family!

being steadily good rather than sensational. On Jan. 28 Mousel Bros. sold at Cambridge, Neb., 57 head for an average of $348, the top being $1,025 for the young bull Mischief 40th taken by John McConnell, Somerset, Neb., who also bought the highest-priced female, the cow Germania 2d, at $710. Messrs. Williams & Lisle sold at Atlantic, Ia., in January 38 head for an average of $167, the best price being $575 for Beau Maid to G. W. Vinton of Exira, Ia. At Grand Island, Neb., Feb. 19-20, a combination sale resulted in an average of $193 on 114 cattle. The best figure reached was $510 paid by N. D. Meysenburg of Nebraska for the bull Freighter 14th. On Feb. 18 George J. Anstey made an average of $206 on 19 bulls and $201 on 41 females, M. A. Spooner & Son paying the top figure, $495, for Miss Albany 5th.

McCray's $604 Average.—Warren T. McCray made a sale at Kentland, Ind., on Feb. 25 at which the extraordinary average of $604 was made on 75 head. Fifty females averaged $577 and 25 bulls averaged $659. The wide distribution of the cattle and the steady range of values at a high level served to emphasize again the extraordinary success attending the handling of this herd. The champion Corrector Fairfax was bought by J. F. Gulick of Jasper, Mo., at $3,750. King Fairfax was taken by Ed. Kreisher, Mount Vernon, Ia., at $1,050. Don Fairfax, just turned twelve months, fetched $1,025 from L. O. Hill, Orange, Va., and Dale Fairfax brought $1,000 from A. E. Cook, Odebolt, Ia. The champion

show cow Perfection Lass, with heifer calf at side, brought $2,450, the successful bidder being L. A. Clifford, Oshawa, Ontario. The heifer Pearl Donald went to A. A. Berry & Son, Mt. Vernon, Ia., at $1,575. W. H. Hunter, Orangeville, Ontario, bought Nora Fairfax at $1,425, and E. H. Taylor of Frankfort, Ky., paid $1,350 for Teresa Donald.

Steady Bidding Continues.—On March 3 and 4 at a sale from the Funkhouser, Gabbert and other herds at Kansas City 114 head averaged $210. W. A. Dallmeyer, Jefferson City, Mo., received the highest price, $1,000, from Joseph Schmidt, Tipton, Kans., for the young bull Beau Dare. On March 6 J. B. Ashby, Audubon, Ia., sold at South Omaha, Neb., 59 head at an average of $210, the 30 bulls averaging $220. On March 5 J. A. Shade sold 71 head at Kingsley, Ia., for an average of $296, 32 bulls averaging $300, with a top of $1,500 paid by Wallis Huidekoper, Willis, Mont., for the bull Bright Lord by Beau Picture. The next best figure was $700 paid for Beau Shade by Henry Strampe of Paullina, Ia. O. S. Gibbons & Sons sold 50 head for an average of $267 at Atlantic, Ia., on March 11, the show cow Priscilla bringing $730 from Mousel Bros. On March 17 W. J. Davis & Co. sold 41 head at Jackson, Miss., for an average of $394.85, with a top of $1,300 for the cow Lady Druid 3d, taken by W. P. Connell, Baton Rouge, La. The young bull Vernet Prince 15th was bought for the Louisiana State University at $1,050. At this sale 7 head of young bulls sired by Point Comfort 14th, the International champion of 1913, averaged $521.45. The 14 bulls averaged $458.90.

On March 24 Ben Broughton sold at his Sunny Slope Farm near Lakeview, Ia., 52 head for an average of $256. The show bull General B, a son of Beaumont Jr., brought the best price, $950. He was taken by E. Gorman of Dougherty, Ia. On March 27 the Mossom Boyd Co. sold at Chicago a lot of polled Herefords, 19 bulls averaging $445 and 60 females $234, the entire lot averaging $280. The show bull Bullion 4th sold to the Renner Stock Farm, Hartford City, Ind., at $2,025. The bull Gemmation 2d sold to G. E. Pettigrew, Flandreau, S. D., at $1,350. The cattle were widely distributed, the most extensive buyers being the Beaver Lake Ranch Co. of Michigan. At Kansas City on March 31-April 1 113 head of cattle contributed by various western breeders averaged $171. J. W. Johnson of Childress, Tex., paid the top price of $625 for the two-year-old bull Woodrow Wilson, consigned by Messrs. Wadsworth of Missouri. At South Omaha, Neb., on April 8 at a combination sale 63 head averaged $200. The best price paid was $550 by J. W. VanNatta for the yearling bull Donald Fairfax.

The final chapter in the history of the old Brookmont herd was written at Odebolt, Ia., April 16-17, when A. E. Cook dispersed all that remained of this noted herd. The 188 head brought an average of $184, the 24 bulls averaging $287. The cattle were sold in their every-day working condition right from the fields, no attempt having been made at special fitting for this event. Much interest was shown in the disposition of the three stock bulls Generous, Dale Fairfax and Howard Fairfax. Nearly one-half of the cattle offered were the progeny of Generous, and although nine years old he was contended for by several discriminating buyers who appreciated the character of his get. He finally fell to the bidding of Cyrus A. Tow at $925. Dale Fairfax, by Perfection Fairfax, went to A. B. Tyler of Draper, S. D., at $1,110. The top price of the sale, $1,210, was paid by the Messrs. Hancock of Manilla, Ia., for Howard Fairfax.

Reporting from the W. H. Curtice herd at Eminence, Ky., early in May Manager Fraser stated that among their recent sales was that of Beau Perfection 23d to A. B. Cook of Montana for $3,000, together with 10 heifers for $2,500. Twenty bull calves were sold to Thomas Mortimer for $4,500. Beau Perfection 22d, a two-year-old son of old Perfection, and 5 yearling heifers were bought by Col. W. H. Roe of Shelbyville, Ky., for $1,750.* Mr. Fraser reported in all the sale of 56 head of cattle at an average of $306, adding that $5,000 had been refused for

---

*William Fraser, the present manager of the W. H. Curtice herd, was born in Aberdeenshire in 1886, and had his early training with cattle among the "doddies" of his native land. He was also employed at one time in connection with the management of a herd of blacks in Staffordshire, England. He next had to do with the handling of the Shorthorns of Sir R. P. Cooper of Shenstone Court, Staffordshire. While employed in these capacities he assisted in the fitting and exhibition of bullocks that were prize-winners at Birmingham and Smithfield.

Coming to America he was first engaged with the Carpenter & Ross herd. By this time Mr. Fraser had earned for himself a high place in the regard of those who appreciate good work in the training of cattle for exhibition, and he was chosen by Mr. Curtice to follow James Hendry in the handling of the famous Beau Donald Herefords.

CHAMPION HEIFER SCOTTISH LASSIE 305352, BRED BY DR. JAMES E. LOGAN, AND BULL CORRECTOR FAIRFAX 332053, BRED BY W. T. McCRAY.

PERFECTION LASS 342053, GRAND CHAMPION FEMALE AT THE 1918 INTERNATIONAL.

the show bull Beau Perfection 24th, and an offer of $1,000 declined for his yearling sister. A little later a sensational deal with Col. Taylor was reported.

At their public sale on Oct. 20, 1914, Harris & Sons realized an average of approximately $496 on 55 head. The 29 females disposed of on that occasion averaged $493, and the 26 bulls averaged $498. The highest price was $1,800, paid for the bull calf Repeater 38th, Fred Fleming, Dallas, Tex., being the purchaser. At $1,525 E. H. Taylor, Jr., Frankfort, Ky., secured the six-year-old cow Harris' Princess 81st by Beau Donald 5th. Gay Lad 25th, a yearling bull, made $1,500, William Henn, Denver, Colo., being the buyer. Repeater 19th, two years old, and Disturber's Lassie 5th, three years old, each sold for $1,250, the former going to S. B. Burnett, Fort Worth, Tex., and the latter to E. H. Taylor, Jr. Mr. Taylor also bought the seven-year-old cow Adeline by Prime Lad at $1,050, and the four-year-old cow Disturber's Lassie 4th at $1,000. Sixteen of the 55 head sold for $500 or more each.*

---

*Disturber was bought by Mr. Letham when a calf, and he won his way through the senior bull ring at Chicago at a weight of 1,245 pounds a few days before he was thirteen months old. As a yearling he was used heavily and showed in only half-fitted condition, winning third money. As a two-year-old his get were showing so well that again he was not fitted specially, retaining third place at Chicago. Both of these years he was the property of Mr. Brock. It was when Disturber was two years old that Letham showed him with his first calf. This was Distributor, the sire of Repeater, which defeated the St. Louis World's Fair champion Mapleton and the eastern champion Perfection Fairfax. Then came Distinction, own brother to Distributor. Disturber's heifers found favor perhaps more generally than his bulls during the first five years of his life, but after that there was probably about an even division of sentiment as to their relative excellence. During the year 1910 his get won more blue ribbons than any bull of any breed in the Chicago show, and that too on an exhibit of but thirteen head from the Brock herd.

When George P. Henry sold out his herd he had Kansas Lad Jr., Prime Lad and Disturber. The latter Letham took with him

Beau Perfection 24th Brings $12,000.—In the month of May, 1914, Mr. W. H. Curtice sold 20 head at private treaty to Col. E. H. Taylor, Frankfort, Ky., at the round price of $20,000. The deal included the sensational two-year-old bull Beau Perfection 24th, at $12,000, the highest figure as yet reached for a Hereford on either side of the Atlantic. The 19 females were taken at $400 per head. In the course of a letter announcing this remarkable transaction Mr. Curtice says: "This sale makes a total of 79 head of cattle disposed of since the Kansas City Show last year for a total of $41,135 cash without discount, freights or any extra sale expense whatsoever." Commenting further Mr. Curtice says: "I do not want the impression to go out that I am out of the Hereford breeding business as I still have 100 head of cattle, and expect to

---

to Missouri when he went to Mr. Brock. Speaking of the record made by the get of this bull, Mr. Letham says:

"He was the most consistent sire I ever handled. His get were not all show cattle, but everyone was a Disturber—perfectly marked, with good rich mossy coats and thick-fleshed always. In May, 1911, we sold Mr. Tow the entire Lake Geneva herd, including Disturber, Distributor and Standard. The old bull was then in his tenth year, but even so he carried more top meat than most show bulls and I still valued him at $3,500, a price which I had refused for him when he was three years old, at which time I had dared to ask $5,000. He made his best success on Kansas Lad Jr. and Prime Lad cows, of which we had only six. Without boasting, I believe that the young Disturber herd which Mr. Tow is showing this fall, thus far undefeated, and containing among the last of the old bull's get the junior champion bull Disturber Jr. and Disturber Lassie 12th, the junior champion female, and three others that are very uniform, is in my opinion the best I have ever seen. This I mean as an absolutely cold-blooded verdict. Disturber was grandsire to all the Repeaters as well as to Point Comfort 14th, the phenomenal Davis bull, Letham Fairfax in Mr. McCray's herd, and a lot of others on the road.

"Certainly the old bull's get are making much Hereford history, and I hope the old fellow is knee-deep in bluegrass and clover, if there are green pastures on the other shore, after twelve years of the best we could give him here. As a two-year-old at Chicago he weighed in at 2,140 pounds, and for seven straight years in breeding shape he stood between 2,250 and 2,300 pounds on grass alone."

breed Herefords as long as I live, both in Kentucky and in Alberta, Canada."

Lord Wilton was knocked down at the Stocktonbury sale for 3,800 guineas, or $19,000, but Mr. Vaughan, who was supposed to be bidding for America, failed to make good his offer and at a subsequent sale the famous bull went for $5,000. It should be stated, however, that Sir James Rankin's bid of 3,700 guineas at Stocktonbury was bona fide, although unfortunately, as it turned out, it was not accepted by the auctioneer.

The Show Herds of 1914.—The show season of 1914 opened impressively at the Forest City Fair, Cleveland, O., the last of August. The Kentucky herds were in prime form, and the old-time campaigners—Giltner Bros., W. H. Curtice and Luce & Moxley—were here joined by E. H. Taylor, Jr., who won grand championship on the $12,000 bull Beau Perfection 24th. From Indiana came the admirable herd of J. H. & J. W. VanNatta, and these fitted herds sounded a significant prophecy of the brilliancy of the fall campaign. The winners were almost without exception of the blood which has hitherto produced the prize-winners in these herds. Ohio and New York divided the Luce & Moxley and the Taylor herds the ensuing week, the Curtice contingent journeying intact to New York State Fair. Returning from the east the herd of Luce & Moxley encountered only local opposition at the Michigan State Fair while the cattle of Messrs Giltner, Taylor and VanNatta joined issue at the Indiana State

Fair with the herd of W. T. McCray, which had opened its campaign at Iowa. The Kentucky breeders concentrated their divided forces at their home fair at Louisville, and had the aid of the McCray cattle in presenting anew the merits of the breed in the Blue Grass State.

Meanwhile other spectacular exhibitions of the breed had been claiming public attention in the central west. Iowa summoned to its state fair no less than 15 herds of "white faces", from Iowa, Missouri, Indiana, Wisconsin and far-south Mississippi. Conspicuous on the prizelist were the cattle from the herds of O. Harris & Sons, W. T. McCray, Cyrus A. Tow, J. M. Curtice and W. J. Davis of Mississippi. It was clearly one of the bravest shows of the breed, emphasized by its setting at a fair where the Hereford has more than once overshadowed the other breeds in the uniformity of its excellence. Here again the names of the leading winners bring to mind bloodlines which have been most potent for years in the production of the ribbon-winning cattle at the western fairs. After Des Moines exhibits usually divide between the Minnesota and the Nebraska state fairs. J. M. Curtice, Cyrus A. Tow and A. A. Berry & Son journeyed north to Hamline, where they encountered three local herds not well equipped to meet such competition. Five herds left Des Moines for Lincoln, among them O. Harris & Sons, W. J. Davis & Son, and William Andrews & Sons, and met two Nebraska herds and one from Kansas. This fair provides grand championship competitions for all beef breeds, and the Harris champion Repeater 7th gained this honor for the breed.* The Harris heifer Miss Repeater 11th was second best female in this competition, while the calf herd from Mississippi won second in competition for such groups. The participation in the western campaign of the Davis cattle from Mississippi was among the notable features of the season, not only an evidence of the security of the foothold which the breed has obtained in the south, but an illustration of enterprise rarely equaled in our showyard annals.

Pushing resolutely the widely-planned campaign, the Harris and Curtice herds from Missouri, the Berry, Andrews and Tow herds from Iowa and the Green herd from Nebraska joined with five South Dakota herds in a sensational exhibit at Huron, the capital city. The Davis herd meanwhile had moved its colors to Topeka, along with the cattle of Biehl & Sidwell of Missouri and Thompson Bros. and O. E. Green of Nebraska, thus affording full classes at the Kansas fair. The exhibit carried to the western fairs brimmed with white-faced ripeness and quality.

---

*The Messrs. Harris are certainly doing their full share towards sustaining the cause of the Hereford in the United States at the present time. No other evidence is required to demonstrate the great enterprise they are displaying, and the liberal expenditure they are making in connection with the up-keep of their establishment, than the fact that there is at the present writing in service upon their Model Farms Repeater 289598 (an admirable photograph of which appears elsewhere in this volume), Gay Lad 6th, old Beau Donald 5th, Prince Perfection and Repeater 7th. Beau Donald 5th is still in service in his sixteenth year, and has sired cattle that have been sold by the Messrs. Harris for more than $100,000.

The herd of W. T. McCray coming up from Louisville, Ky., met the VanNatta cattle from its home state at Springfield, with J. E. Thompson, an Illinois breeder, supplying a few winners. Of all the state fairs this season the breed was least numerously represented at Illinois. On the succeeding week the Harris cattle from South Dakota, the McCray cattle from Indiana and the Curtice cattle from Missouri met at the Missouri State Fair, while the young cattle from the Kansas herd of R. H. Hazlett were winning most of the prizes at the Oklahoma State Fair in competition with the entries of Klaus Bros., from the same state, and several local exhibitors.*

---

*Mr. Hazlett has not been in the limelight as much as some of his contemporaries, but he has been honored with the presidency of the Hereford association, is one of the active managers of the Kansas City Royal Show, and is generally recognized as one of the ablest men now identified with the development of Hereford interests in the middle west. A statement, therefore, covering some of his personal experiences will undoubtedly be of much interest, and at our request he has prepared the following, which we have pleasure in inserting at this point:

"Really, my experience as a breeder has been so generally uneventful that I hardly know how to give you any very good idea of what I have done. I have never attempted much in the way of showing cattle. I first knew about Herefords when T. L. Miller and others of the early Hereford breeders were striving to obtain recognition by the management of the Illinois State Fair, Springfield, Ill., being my home at that time. My interest at that time, however, went no farther than just to feel that those men were hardly securing fair treatment and to be pleased when they obtained some recognition through the demonstrations at the fat stock show. In 1885 I moved away from Illinois to El Dorado, Kans., and had little experience and knew little of what was going on in connection with fairs for some years.

"I bought my first Herefords, an entire small herd, near this city in 1898. This herd had been kept on the farm where they were when I bought them for a good many years and I had frequently visited the place and admired the cattle. There were sixteen in the little herd when I bought it—two young bulls some eight to nine months old, and fourteen females, less than half of them being of breeding age, the others being yearlings and coming yearlings. Wild Beau, a full brother to Wild Tom, that first made Sunny Slope famous as a Hereford breeding establishment, was the sire of the young animals in this little herd. Wild Beau was by Beau Real and he by Anxiety 4th. One of these young bulls I kept and used in my herd for several years to some extent, on all the older cows except his dam. As I did not want to breed half sisters to this bull I secured a bull with a large per-

This bare summary of the battlelines of the 1914 show herds would be incomplete without the statement that from coast to coast and far into the south the banner of the breed was proudly carried. The Vermont and Virginia fairs staged exhibits that occasioned favorable comment and at some of the southern shows, notably at the Tri-State Fair at Memphis, the breed was surprisingly prominent.

centage of Anxiety 4th blood. The dam of this young bull, which had been named by the man from whom I bought him Major Beau Real, was by Stonemason by Beau Real, so that the first calf I used was a line-bred Anxiety 4th bull. This second bull was Bernadotte 2d, and he proved to be quite a good sire. I used these two bulls for several years, in the meantime having bought now and again a few females without much thought as to their breeding.

"Up to this time what I had done was without any real reason. It just happened that I had Anxiety 4th lines, as I really knew nothing of the science of breeding or for that matter of the prominent families among Herefords. Having daughters of these two bulls in my herd by this time and having the idea that is very prevalent—whether correct or not—that I ought to go outside for some fresh blood or a different line of breeding, I bought a Columbus bull, Dale Duplicate 2d. He was, I think, a little more than a half-brother to the champion show bull Dale, with whose history you are entirely familiar. I disposed of both the old bulls. I also bought another from a popular family at that time, by Improver out of a Corrector dam. Another one I tried was as close to The Grove 3d as I could find, having learned that The Grove 3d was considered by many a great sire. I used these bulls until their get were in the neighborhood of two years old and decided that I did not like the results I was getting. Whether this was the fault of the out-crossing or simply that it was not the proper nick with my cows, I am not prepared to say. I only cite the fact here that I did not get the results I hoped for and soon disposed of all three of these animals.

"In the meantime I had been trying to inform myself to as great an extent as possible in regard to the Herefords that seemed to give the best results from a breeder's point of view. This, together with the experience I had had, led me to return to the Anxiety 4th line of breeding through Beau Brummel bulls and cows. The next bull which I used was Beau Beauty, sired by Beau Brummel out of an Anxiety 4th-North Pole dam. I bought an entire herd of between fifty and sixty head in order to secure another son of Beau Brummel—Beau Brummel 10th, whose dam was much stronger in the blood of Anxiety 4th than the dam of Beau Beauty. Up to this time those two have been my chief stock bulls. Beau Brummel 10th is dead and I am using one of his sons, Beau Sturgis 2d, with good results. While Beau Beauty is living, I am using one of his sons also, Zelpho, on the daughters of Beau Brummel 10th, with very satisfactory results so far.

"I omitted one fact in connection with my herd bulls and that is this: I used a son of Bernadotte 2d of my own breeding for several years and still have a number of his daughters in my herd. This bull was Protocol 2d. He was a very large and very smooth bull, weighing at twenty-four months of age 2,000 pounds, and as a three-year-old 2,600 pounds. Protocol 2d is the only one

ROBT. H. HAZLETT.

In summary the Hereford show herds of 1914 have acquitted themselves brilliantly. More convincing testimony to the successful attainment of "white face" ideals in America's eminent nurseries of this breed could not be required. A season's exhibit

of these bulls that was ever shown at anything except a county fair and he but once at the American Royal, and as neither myself nor my herdsman had any sort of notion of fitting cattle for the showring our success was limited to getting inside the money.

"For the first few years after I began with the Herefords I did not cull my females at all, but kept them all for use in the herd, and I kept all the males intact to be sold as bulls. Of course, at this time I sold only to farmers and breeders of grade cattle and the importance of selection had not occurred to me so much up to this time. It was not many years, however, until I became very much interested in the matter of breeding and determined to give it as much thought and attention as was in my power. I decided to eliminate from my herd, through the stockyards, the unworthy males as steers, and the sub-standard females. I have never made much effort to sell females, as I have felt that in order to have a good herd I must keep the best for breeding purposes.

"I have made comparatively little reputation through the showring, though I have shown with credit the last two or three years at our principal state fairs in the west and at the American Royal. I bred and showed the heifer Banza, sired by Beau Beauty. She was the only one that ever beat the renowned Scottish Lassie until the present year. Banza, with her second calf at foot, was shown at the American Royal this year, winning first place in the new classification—"cows in milk or with calf at foot." I have shown quite a number of the get of Beau Beauty and Beau Brummel 10th at the western state fairs and have a good many firsts and championships to my credit in those shows.

"Speaking of this reminds me that I have omitted to mention another son of the old Beau Brummel which I bought about the time I was getting rid of those three above mentioned. This was Printer, of about the same breeding as Beau Beauty, largely Anxiety 4th and North Pole. I also have one of his sons in my herd and from his use have had more successful show cattle than from other bulls on the farm. His name is Caldo 2d. His dam was by Lucifer, a Beau Brummel-Anxiety 4th bull, bred by Steward & Hutcheon."

The Hazlett herd is in charge of William Condell, whose portrait appears elsewhere in connection with those of other prominent herd managers. Mr. Condell was born in 1882 at Lake Bluff, Ill., of Irish and Scotch descent. He was raised at Chicago Heights, Ill., both places being near the city of Chicago. When he was ten years of age his father bought a herd of Herefords, and since that time William has always been happiest when busy wtih the "white faces." On reaching his majority he determined to identify himself with some good herd and endeavor to render efficient service. On removing to Kansas in 1901 a connection with Mr. Hazlett was formed which has continued to the present time. All who are familiar with the management of the herd, with such cattle as have been exhibited from it, and all who have an acquaintance with Mr. Condell will find in the following sentence just about what they would expect from him, in answer to a query as to his methods: "Whatever success I have attained has been through trying to please my employer, and putting self into the work."

FARMER 426279, IMPORTED 1913 BY GEO. LEIGH AND OWNED BY W. T. McCRAY.

IMP. BRITISHER 145096, AS DRAWN BY THROOP—IMPORTED BY GEO. LEIGH, AND USED BY GILTNER BROS.

which sustains triumphantly the glorious traditions of the breed in its presentation of thickly-fleshed and perfectly fitted show ring cattle has again been placed to the credit of the master-builders of the breed in America.

CHAPTER XXIII.
IN FOREIGN FIELDS.

The question of the future meat supply of the world is one that is now receiving serious consideration. Whereas the United States but recently exported large numbers of live bullocks for slaughter at British ports, as well as great quantities of dressed beef, the passing of the open range in western America and the curtailment of beef cattle production and feeding by cornbelt farmers, due to the steadily advancing price of lards and grain, has within a remarkably short space of time converted us from an exporting to an importing nation. Beef and cattle shipments oversea are, for the present at least, at an end.* Our own ports have been opened to the free introduction of meats from other countries, and the first year's operation under this new dispensation has seen liberal shipments of frozen meats from Argentina to our Atlantic seaboard markets and the arrival of numerous cargoes from Australasia on our western coast.

Leading American packers are now operating their

---

*Owing to the abnormal situation developed by the great European war, in progress as this volume is written, our packers are selling big lots of canned, corned and pickled meats to the French and British governments for the maintenance of their embattled forces. But with the return of peace and the resumption of normal commercial relations this buying, on an extensive scale, will probably not continue.

own plants at Buenos Aires, and are also entering the Australasian field. They have been forced to do this or lose their Smithfield and other foreign business, built up in the past from American supplies. Not only have the packers entered these sub-equatorial markets, but large companies have been formed in recent years to engage heavily in cattle ranching in regions heretofore not stocked. Notable cases in point are the Brazilian Land, Cattle & Packing Co., of which Mr. Murdo Mackenzie, former manager of the Matador business, is the present executive head, and the British South African Chartered Co., which has engaged as manager Mr. Richard Walsh, long with the Adair ranch at Paloduro, Texas.

Exports to South America.—It was in January, 1911, that Mr. Mackenzie left the service of the Matador company and went to take charge of this great new venture in South America. Prior to his engagement the board of directors had purchased about 920 head of Herefords and 20 Shorthorns in the United States. All of these cattle except those coming from above the quarantine line were purchased from an infected district and were immune, and the cattle purchased from above the quarantine line were sent to the Texas Experiment Station, where they were held for six months and immunized. The cattle arrived in Brazil about the end of July and were placed in pastures in Parana. They stood the trip very well, only five head dying between

Galveston and the Brazilian ranch. On account of their being immune before shipment it was expected that little trouble would be experienced from tick fever, but in this the buyers were disappointed. So far as could be observed there was no difference between the cattle from above or below the quarantine line in the matter of susceptibility to the kind of tick fever prevalent in that country. It was found, however, that the Hereford cattle withstood the difficulties encountered and adapted themselves to all the conditions of that country much better than the Shorthorns. Such is Manager Mackenzie's testimony, and he adds:

"I consider that there is a great future for the cattle business in Brazil; all it requires is perseverance and push and the importation of the best breeds of cattle to make this country second to none in the cattle business. The climate is all that could be desired. Water is plentiful and well distributed over the range country, the annual rainfall being about 42 inches. There is practically no cold weather and grass will grow almost the year round. There is a great abundance of grass, but it is not so nutritious in some parts as it is in others, and in some parts not so nutritious as the grass you find in the range country of the United States."

Five Hundred Herefords to Brazil.—Through its representative, Alex. Mackenzie, son of Murdo Mackenzie, the Brazilian company purchased in Texas during the summer of 1914 500 head of Herefords, which were shipped for breeding purposes to the company's extensive ranches. This was a record shipment of American pedigree Herefords to a foreign country. It consisted chiefly of bulls about eighteen months old. It is understood that the total price paid was $65,000, or about $125 per head. The cattle were selected from herds below the quarantine line owned by J. W. & D. L. Knox of Jacksboro, R. H. McNatt of Fort Worth, M. W. Hovenkamp of Keller, W. N. Burns of Blanket, F. C. Vaden of Sherman, J. H. McCaskey of Decatur, J. P. Morris of Coleman, C. Sloan of Fort Worth, F. L. Smith of Graford, R. J. Johnson of Newcastle, J. T. Day of Rhome, J. O. Rhome of Kopperl, Ed Hayden of Moran and S. D. Penny of Watauga.

Although every animal was immune to Texas fever they were subsequently required to undergo immunization against another species of tick fever in Brazil, with the prospect of a considerable mortality.

The property of this company, consisting of some 10,000,000 acres of land, lies in the southern part of Brazil, the head offices being at Sao Paulo, which lies at an altitude of some 2,500 feet above sea level and is 60 miles inland. At last accounts it was estimated that the syndicate had acquired over 200,000 head of cattle, which number was likely to be increased to half a million. The entire country, however, is infested with ticks, and this complicates somewhat the problem of improving the native cattle with imported bulls. The Brazilian cow is a good-sized animal, much larger than the old-time Texan, and the entire country is covered with a wonderful growth of grass which is kept down by burning, the cattle following the prairie fires wherever they occur. When a strip of country is burned off the animals, attracted by the smoke, set out for it, and a few days after the fire has passed the entire country is green.

The fact that the Hereford has been chosen as the most likely type to successfully cope with the conditions there prevailing, is simply another tribute to its capacity to endure the hardship to which range cattle are usually subjected in all countries.

It may seem somewhat fanciful to speak of mentality as being a determining factor in the adaptability of a breed of cattle. However, there is no doubt that the peculiar mental qualities of the Herefords have contributed markedly to their success upon open ranges in general. In the mountainous regions of the western United States where mixed herds of other breeds were already in possession, when Herefords were introduced they very soon made themselves known by climbing to the highest slopes that carry grass. Ever afterwards, as long as the mixed herds persist, it may usually be noted that the cattle highest up on the grass-covered mountain-sides have white faces.

There is a sort of courage and resolution about the Hereford that makes him combat stormy weather away from the shelter of bank or tree or cliff and hunt for grass when cattle of more tender nature, developed under man's continuous and solicitous care, even though they may never have been fed, will be found waiting in the bottom of the canyon or in the shelter of a tree hoping that someone will come along that way. This fact as much as anything else has served to earn for the Herefords the admiration, and even affection, of their cow-boy caretakers.

Uruguay.—Here is another place where Herefords find high appreciation. Uruguay is a fine little country, in marked contrast to the flat and featureless Argentine plains being made up largely of rolling lands, in some parts almost hilly, though nowhere rising into real mountains. Rocks are often seen cropping out of the pastures or rising in good-sized cliff-like walls along the crests of the hills. Uruguay is a land of springs and many fine small streams, with also a few sizable rivers. We have nothing just like it in North America, although the high country in Texas somewhat resembles it. However Uruguay has a milder climate than Texas, with cooler summers and warmer winters. Uruguayan soils are good, but not so fat as those of the great plains of Buenos Aires. They support perennial grasses with fewer bur clovers and other legumes than are seen in Argentina. Alfalfa pastures are as yet infrequently seen in Uruguay.

Perhaps because of the more or less hilly nature of the country, perhaps because the pastures are less productive than those of Argentina, the Herefords are the most popular cattle of all breeds tried in Uruguay and are most frequently seen. They apparently make more fat on Uruguayan grasses than do the Shorthorns, the nearly universal cattle of Argentina. Uruguay has a fever line in the north, and above the line there are yet vast numbers of native Spanish cattle, greatly in need of improvement. Hereford blood is the kind most sought to effect this improvement. There is experienced the same difficulty that our own breeders have met in attempting to put northern cattle into southern pastures. The non-immune cattle quite often die when exposed to fever ticks. Wilson Bros., of Montevideo, who are large importers of cattle for breeding purposes, have expressed their opinion that northern Uruguay and Brazil could use many thousands of United States-bred Hereford bulls if they could be bought with any assurance of immunity from fever.

The truth is that our American breeders of both Herefords and Shorthorns have no adequate conception of the enterprise that has already been displayed by South American cattle-growers in the matter of elevating the standard of their cattle stocks, probably because nearly all of their buying has been done in Britain. Writing under date of Aug. 15, 1913, Mr. William Tudge stated that "Col. F. Braga, the leading Uruguyan breeder, has at the present time 800 head of pedigree Hereford cows and has just imported (on June 21) the most valuable lot of Hereford bulls, 27 in number, that ever left England at one time."

Argentina.—The chief cattle-rearing states of Argentina are Buenos Aires, Cordoba, Pampa Central, Santa Fe, Entre Rios and Corrientes. South of the Rio Negro few cattle are seen because of the aridity of the soil and the poverty of the grasses. Along the Andes, however, are fine rich pastures as far south as Santa Cruz. On these pastures are seen chiefly the native Spanish cattle, some of them of magnificent type being used largely for transport purposes, as it is a great sheep-growing country and the wool must be hauled a long way to market.

The province of Buenos Aires is chiefly low, black, fat land, and is devoted mainly to Shorthorn cattle, the few herds of Angus, Herefords or other breeds being quite inconspicuous amid such immense numbers of Shorthorns. In Cordoba more Herefords are seen, but even there Shorthorns largely preponderate. In Entre Rios the improved herds are chiefly Shorthorns, although along the northern edge will be found more of the Herefords and also many of the native Spanish long-horned cattle. Corrientes has a few estancias given over to cattle of good blood. Among these will be seen Angus, Herefords and Shorthorns, but in the main Corrientes is given over to the wild, unimproved Spanish native, living to be six years old before going to the salederos or salting works. These native cattle never reach the frigorificos because of their lack of quality. In northern Corrientes some cattle of Zebu or East Indian blood have come and are welcomed because of their tick-immunity.

Argentina is the only country beyond the seas that produces Indian corn in a large commercial way.

For many years past the proprietors of the great estancias have been buying Shorthorns in Great Britain, not only with great freedom but with a degree of enterprise unparalleled in the history of the British export business. The policy seems to have been to procure the best absolutely regardless of cost. For a long series of years buyers for the Argentine have been taking out the very tops of British herds.

It seems probable that the main reason why the Hereford has not as yet acquired such a dominating influence in Argentine cattle-ranching as in the United States is due to the fact that conditions throughout much of the interior of Argentine are not as forbidding as in the case of our own Rocky Mountain regions. This is simply another way of saying that the necessity for resorting to the peculiar qualities for which the Hereford is specially noted do not exist in Argentina to the same extent as with us. Where climatic conditions are favorable, and where food is abundant, it is not commonly claimed that the Hereford has any outstanding advantage over the Shorthorns. It is where the facing of grief has to be met that the Hereford practically gets away from all competition.

Foundations of Argentine Improvement.—In Volume 35 of the "Anales de la Sociedad Argentina" it is stated on the authority of Dr. Zeballos, a former Minister of Argentina to the United States whose acquaintance the author of this volume had the pleasure of forming many years ago, that the cattle trade of the River Plate had its remote foundation in the introduction in 1553 by the brothers Goes (Portuguese) of seven cows and a bull into Paraguay. The cattle were from Santa Catalina, Brazil. Space will not admit of our endeavoring to trace the gradual growth of the industry during the succeeding centuries of Argentine development. The point of real interest to us at this time is the fact that, so far as published herd book entries show, it was not until the year 1862 that the first introduction of the pure Hereford blood was recorded. In that year Don Leonardo Pereyra imported the bull Niagara, and in 1864 brought out the first two cows of the Hereford breed. In 1868 Mr. Juan Miller brought out the first Shorthorn bull, Tarquino, and several cows to his Nueva Caledonia ranch, thus founding the primal herd of that breed in Argentina. Without undertaking to present the details as to the subsequent importations, a general idea of the extent to which Herefords were introduced and bred during the years following this original importation may be gleaned from the statement that there were recorded up to the year 1907 in the first four volumes of the herd book established for the registration of pedigree Herefords 364 bulls and 649 cows. Prominent among those engaged in promoting the interest in the "white faces" in Argentina was Mr. Arthur Yeomans, of La Norumbega, Buenos Aires Province.

Shorthorn vs. Hereford.—The author put to a well informed and entirely disinterested authority in Argentina not long ago the query, "How do you account for the fact that the Shorthorn seems to be so much more popular in your country than the Hereford?" This is a point of so much interest that we can do no better than quote his reply verbatim:

"This question of yours, though natural and easily put, is the most intricate to be answered. It has agitated ourselves for more than two decenniums, and on their side hundreds of reasons have been adduced, all to no effect. Cool ciphers have shown the good qualities of the Hereford and its adaptability for an outdoor grazing life, dozens of times, and still the Shorthorn bears the palm a long way. Hereford enthusiasts have been dying away, without seeing their efforts crowned; large parts of considerable fortunes have for years been laid out with Herefords, at small returns, while the luckier rival was booming; and still at the slightest touch among the advocates of the Hereford the old fire of violently subdued enthusiasm breaks out again to the highest glow. As an observer, however, I should mention:

"First: That there is a great majority of Shorthorn breeders, and the largest extensions of the most fertile pasture land are devoted to Shorthorns, while the Herefords (owing to their hardiness and good feeding qualities under adverse conditions) are generally reared on poor pastures, consequently competing at disadvantage.

"Second: That the very pronounced hereditary power makes the people compare most commonly a first-cross Hereford with a fourth- or fifth-cross Shorthorn, to the great detriment of the former.

"Third: That a portion of the original native cattle, before and during the time of grading, possessed white heads (such animals being named 'pampa' out here), and that these 'pampas' are much disliked. And as the Hereford with his strong generic prepotence gave white heads even in the first cross, these two different classes of white heads were mixed up by an ignorant population."

Argentine Breeders Testify.—The fact that Argentina now looms so large in the matter of the world's beef supply, has led the author to endeavor to assemble the views of leading advocates of the Hereford in that country. The courtesy of our southern neighbors is proverbial. Some of the wealthiest and most deeply-engrossed of those who have stood by the Hereford cattle in the Argentine, in the face of many discouragements, have done us the honor to reply at length and in most interesting fashion to our inquiries as to the status of the "white faces" in the great South American republic. In view of the interest now attaching to the evolution of the cattle business in that country we feel that no more interesting contribution to contemporary cattle literature can be made than the submission herewith of liberal extracts from translations of their replies.

Cabana San Juan.—Reference has already been made to Don Leonardo Pereyra as the pioneer importer. The fact that his holdings of Herefords upon various estancias had extended up to 30,000 head at the time his statement was made lends weight to the following language:

"You ask me what reasons justify in the Argentine Republic the apparent supremacy of the Shorthorns as a race, the superiority of the Herefords being unquestionable as regards strength, hardiness, health, resistance and adaptability to all kinds of camps.

"I believe it is only a matter of personal preference. I acknowledge that such a motive has no weight as a commercial argument. It must also be borne in mind that the generality of our breeders have, for some time past, dedicated themselves to breeding Shorthorns, as these were more abundant than Herefords. The former spread thus easily over the country and today the owners of Shorthorns, although the origin of their cattle in this country may not be a pure one, have continued crossing with imported animals or with more or less pure bulls for such a long period that their herds are almost pur sang. Such breeders naturally are loath to give up the results of many years of assiduous work.

"I think, nay, I can assure, that there are Shorthorn breeders who are intimately convinced of the necessity of starting in the direction of a breed which, like the Hereford, offers them more endurance, is better adapted to all zones, shows greater resistance in times of drouth and during cold winters, than are displayed by the Shorthorns. Yet this conviction is an inward one; it is not openly avowed. Things will most likely continue in this state until the Hereford breeders, who on their side persist in their propaganda, do succeed in establishing their opinion, as I understand has happened in the United States. The Hereford breeders have kept up the struggle for a very long time; but they were few, compared with the number of their rivals, and the upshot so far favors the predominion of the Durhams. During recent years, however, a reaction has

BREEDING HERD ON THE PROPERTY OF DON LEONARDO PEREYRA, THE FIRST IMPORTER OF HEREFORDS INTO ARGENTINA.

begun, a reaction which is based or caused by the evidence gathered in bad times, when cattle are put to the proof by protracted drouths, intense cold, scarcity of pasture, etc.

"The San Juan Farm, established forty-five years ago, has always kept a valuable stock of pedigree sires. The development of both Hereford and Durham herds in our estancias has allowed me to compare the qualities of the two breeds, and I have reached the following conclusions: the Hereford cattle produce more and keep in better condition than the Durhams. I now possess from 25,000 to 30,000 head of Hereford cattle. The figures in my books speak eloquently in their favor, showing they give higher profits than an equal number of Durhams, although the latter have always grazed on a better camp (Tandil Leofu) which is situated in the same region, near the Tandil Mountains, where the Hereford cattle run (San Simon). I can therefore harbor no doubts. Facts have convinced me, and they certainly carry out my assertions."

We have the pleasure of submitting herewith several engravings which will demonstrate to the American reader that Herefords of the very first quality, equal, in fact, to the best of the breed in any country, have been utilized by Senor Pereyra in his extensive breeding operations. These illustrations have been prepared from a beautiful set of photographs sent to the author along with the manuscript from which the above statement is extracted. We have also been furnished with a detailed statement concerning the leading stock bulls used in this noted herd, the list including a large number of Royal English prize-winners taken to the Argentine at high prices. We regret that space will not permit of our setting forth the bloodlines of these imported cattle, but it must suffice to say that all of the great producing strains of the breed in Herefordshire were represented.

Las Hormigas.—Senor A. Ayerza, owner of Las Hormigas, established in Conchitas in the District of Quilmes, Province of Buenos Aires, in the year 1896, began with four pedigree cows imported from England, and with fifteen bred in Argentina, the produce of imported English cows. The first bull used was Eaton Defender, bred by Sir Joseph Pulley, Bart., of Lower Eaton, and was followed by the famous Red Cross, bred by Arkwright of Hampton Court and winner of many important prizes at English shows.

The natural increase from births in this establishment during a period of eleven years was 425 head. Comparing the "mestizacion" or effects of crossing with both "Durham" and Hereford bulls, Sr. Ayerza, after having practical experience with both breeds and having produced as many as 6,000 head of grades of both breeds decided to give up using "Durhams" and only employ Herefords. Referring to this he said:

"Above all I must tell you that the mestizacion in my establishment has been made principally with purebred pedigree bulls, but few of a 'mixed breed' having been used, but even those had been crossed at least four times. As to the cows, although they had some Hereford blood (only the color), yet on account of their poor development and form they were no better than the ordinary native cattle. With these 2,000 head as a base they gave, in the first instance, as a result steers which at three and a half years old were sold for the freezing establishment at the same price as Durham steers of the same age and fleshiness, but which had better blood on account of their mestizacion. I am certain that this result could not have been obtained with purebred Durham bulls and cows in the conditions of development and mestizacion to which I have just made reference.

"One point of great importance in the use of Hereford cattle is their rapid increase; for, comparing the annual increase of the Hereford cattle which I possess now with that of the Durhams which I had, I can affirm without fear of mistake that said increase is from 18 to 20 per cent more. I have also noted that in the case of this breed a breeder can with impunity, by means of crossing, produce an animal of pure blood without the least fear of losing in the smallest degree any of the strength and rusticity which belong to the breed, and which in my experience with the Durham cannot be obtained, because this animal once arrived at the grade of pure blood becomes exceedingly weak and unable to resist our system of rearing cattle in the open air, especially during the winters of the southern parts of Buenos Aires. If, for example, during the last winter I had in my establishment the Durham cows which I formerly had, I am quite sure that more than three-fourths of them would have died (as happened with my neighbors) whilst I only lost an insignificant part of my Herefords. You may be almost certain that a red native cow served by a pure Hereford bull will produce an animal with hair, short legs and somewhat of the roundness of body of the parent bull, which characteristics you will not find in the case of a Durham. I am fully persuaded that given the same conditions as to blood, pasture and care of the cows served by purebred bulls of either breed, the breeder of Herefords will obtain a larger number and a better type of animal in half the time.

"There are many factors to be considered to account for the superior popularity of the Durham:

"First: As there are few breeders of Herefords there are consequently few pure bulls to employ in the mestizacion, and this fact obliges many breeders to have their cows served by animals which have been only once crossed. These bulls with the facility with which they give to their produce the hair of the Hereford cause the offspring to have the name of Herefords without in reality possessing a drop of pure blood.

"Second: As everybody recognizes the strength and rusticity of this breed it becomes a reason why the breed is raised in those camps, where on account of their bad quality Durhams could not live, and therefore the want of grass, a bad climate, and bull of very little pure blood are the reasons why the produce show a want of development, causing one to believe on account of their color that they possess a high grade of mestizacion, when in reality they possess none.

"Third: A great deal is also to be attributed to the constant adverse efforts of the partisans of the Durhams against the Herefords. It is plain that the partisans of the Durham, being so numerous and powerful, those of the Hereford who are in a great minority are not listened to.

"For my part I can bring forward as a witness my commercial books to show that I sell annually to the freezing establishments for exportation to London from 10,000 to 13,000 steers, and I have never sold Durhams for a higher price, nor of less age, nor fatter than Herefords; with this peculiarity, that although both be fed on the same land the Herefords have fattened sooner, and I have never had a single animal of the Hereford breed rejected on account of a suspicion of tuberculosis, which disease is found largely developed in the Durham breed.

"One of the baseless reasons which the breeders of Durhams use in running down Herefords is the following which I have pleasure in giving you: In our great Palermo shows of 1890, 1895, 1896, 1897, 1898, 1899, 1900, 1901 and 1902 a grand special or champion prize was established to be awarded to the best bull for producing the best meat breed. During these nine distinct struggles the Durhams won five times and the Herefords four. I must tell you that in those different competitions the number of Durham bulls was four or six times more numerous than that of the Herefords, so that the triumph was greater for the latter. At the present day we cannot, unfortunately, compete in the same conditions, because a champion prize has been established for each breed, through the influence of the partisans of the Durhams.

"I have data given me by pedigree stock breeders of both breeds which says that given an equal number of cows and time, the Herefords have produced in the proportion of one and a half times more than the Durhams."

It should here be stated that the foregoing testimony, as well as that which follows it, was procured by the author of this volume several years ago, when this work was first projected. While general conditions in respect to the relative positions of the two breeds in Argentina have not materially changed since these interesting communications were originally written, it is but fair to state that several years have elapsed since they were placed in the writer's hands. It would appear therefore that Hereford breeders in Argentina find themselves up against about the same proposition that faced the early American breeders and importers in the United States—the Shorthorn power being entrenched at every point, making it an up-hill fight for the advocates of the "white faces."

San Gregorio.—Senor D. A. Villfane, proprietor of the San Gregorio estancia, substantiates what has already been said by his colleagues. His reply in part is as follows:

"One cause of the unpopularity of the Hereford is that he so easily imprints his type on the common classes, half-blood crosses are sent to market, and his premature product resulting from a hurried refinement served the Durham breeders to emphasize their anti-Hereford propaganda. As the Hereford type is easily imprinted, any product with a white face and horned was 'Hereford,' but what kind of Hereford? Of these I am no advocate, but I am a very great lover of the Hereford of quality.

"Again: In the shows the number of Durham animals exceeds the Herefords, and the public naturally rushes to wherever their attention is thus forcibly directed. The Durham undoubtedly will have its epoch; indeed it is enjoying it at the present moment, but then as it is only a fleeting custom (or shall I call it fashion?) it shall vanish like all other

STOCK BULL WONDERFUL, USED BY DON LEONARDO PEREYRA.

PRIZE BULL HOLMER 22229, BRED BY PETER COATS AND TAKEN TO THE ARGENTINE BY DON LEONARDO PEREYRA.

fancies and eventually the Hereford will again be in the ascendant."

Duggan Bros.—Mr. Edward C. Duggan of this firm says:

"The result of the Hereford crossing is undeniable to all those who have tested it. We have excellent herds resulting from crosses made with pure Hereford bulls and low-grade Durham cows, and we have also obtained in much less time excellent results with crossings made with pure Hereford bulls and Durham cows of high breeding. It frequently happens that many breeders in order to buy the Hereford cows separate from their herds all the inferior and useless cows, and placing these with Hereford bulls, wonder afterwards why they did not obtain a product of the 'cold storage' type and preach to the four winds that the crossing is not good. These gentlemen do not notice, or do not wish to recognize, that these same cows if mated with an excellent Durham bull would never give a superior product, but they expect the Hereford to do in one crossing that which they would not seek from the Durham in five.

"In a country like ours, which possesses every variety of climate, soil and pasture imaginable, it is a positive fact that in the cattle as well as the sheep there exists practically only one breed—in cattle the Durham and in sheep the Lincoln. It is somewhat difficult to determine to what can be attributed this strange anomaly. In the case of the Hereford we think it is mainly due to the slight knowledge of the breed. As the Hereford bull from the first crossing imprints his color on his offspring, it occurs that many persons think that every animal with a white face is a Hereford, although he has nothing else but the aforesaid characteristic. This lack of knowledge in connection with what we have written in the above paragraph about the little certainty prevailing among the majority of breeders proves that the crossing must be the chief cause of the small acceptability of this mixture compared with the Durham. We must here draw attention concerning the fact that all our neighbor live stock breeders, as well as ourselves, who have proved both species under similar conditions, are all adherents of the Hereford. We have herds of pedigree and numerous rounds of both crossings which enable us to speak of the matter with some authority.

"We have ascertained that the Hereford bull serves a greater number of cows than the Durham. Again, the Hereford cow gives a larger percentage of calves, there being comparatively few cases of sterile and tuberculosis cattle. Further, that the Hereford lives longer than the Durham, is more easily fattened and the 'cold storage' pay for good Hereford steers is a price equal to that offered for good Durhams. The fact of the Hereford being easier to fatten is of great importance, because on the same pasture you can place one-third more animals than you can of Durhams. Another strong feature for the Hereford is that it will fatten on the same good pasture even at the time when it is with calf, but the Durham, which is thin at this stage, will not improve while she is suckling and in the majority of cases will not improve without special care. Lastly, on pasture land where Durhams die the Hereford not only lives but keeps in a fair condition. As this last statement may seem exaggerated, we will add that on more than one occasion during bad winters we have had Durham and Hereford cattle on the same pasture separated by a fence, but we have been compelled to assist the Durhams, taking them to other pastures in better condition, while the Herefords remained to the end of the winter and exhibited a better condition."

At Esperanza.—Sr. Miguel G. Salas, one of the Argentine advocates of the Hereford, imported in the year 1882 fifty cows and one bull, and formed with these animals his first herd at Esperanza. This importation was followed periodically by others, which were used for renewing the blood, avoiding the necessity for close breeding. Of the bulls produced from the stock a number were prepared for sale when they were two years old, and the remainder utilized for improving the stock of "Creole" cattle, which he owned on different estates. Of the effects observed, it is said:

"The result of the crossing with the ordinary cattle was at first mediocre, but lately has been very satisfactory, the steers realizing prices which rival with the best obtained from among the Durhams. Notwithstanding all this, there are some objections raised on the part of market buyers for the Hereford cattle, and this opposition against such an excellent breed is chiefly based on the existence of two prominent features pertaining to this class of cattle, which have contributed to lessen their popularity here.

"First: The amazing facility with which the Hereford attains the coloring without more admixture than that resulting from a half-pure Hereford bull with an ordinary cow, which at once produces the characteristic color—red with a white face. But then an animal with so little strain of noble blood, although it has the color of the pure breed, naturally cannot have its other distinguishing features; and from these circumstances persons not very expert in the knowledge of the breed form wrong conclusions, considering the Hereford badly developed and difficult to fatten, etc.

"Second: It is a general opinion that as to rusticity this breed has no competitors, and in view of this fact the breeders allotted the poorest pasturage to the Hereford. In consequence of this the result was soon apparent, for the cattle thus treated—as was natural to suppose—did not attain to the size of their competitors, the Durhams, and from this circumstance arose the conclusion of the supposed inferiority of the Herefords, and this without once giving thought that the Durhams would have literally perished had they been grazed on similar pasture to that of the Hereford cattle. Now it is common knowledge that when placed on good pasture and being of a good cross the Herefords can compete favorably with the Durham or any other breed of beef, and if to this fact we add the dehorning, the Hereford can be converted into a polled steer, as beautiful and docile an animal as any Durham.

"The actual proprietor of Esperanza, Sr. Juan Cobo, had at about the same time established his Durham Stock Farm with the same number of animals, but he remains nowadays with only the Herefords, which he prefers to the Durhams."

La Estrella.—The author acknowledges with thanks the receipt of a long and particularly interesting statement from Dr. Emilio Frers on the general subject of the status of the Hereford in Argentina, including an even-tempered and scholarly analysis of the relative claims made for the two leading breeds in that country. Our only regret is that we have not space to publish this in full.

To condense it would only be to rob it of its vital interest.

Dr. Frers' experience began in the year 1882 when he took over the La Estrella establishment, at which time the cattle stock included a small herd of 50 or 60 Hereford-crossed Shorthorn cows and something over 3,000 head of "Creoles," showing some little blood. He decided to begin the work of improvement, and for this purpose selected the Hereford bull. He was fortunate enough to secure as his first purebred sire the imported bull Gordon, bred by Lewis Lloyd, that carried a double cross of Lord Wilton. He was procured through Mr. Yeomans, who stated at the time that "no Hereford bull of better blood had crossed the equator up to that date." His descendants at La Estrella certainly did high credit to his ancestry. During 1887 and 1889 several good lots of bulls were brought out from England. Since that date none but pedigree sires have been used in the herds. At the time this communication was written by Dr. Frers the herd numbered some 4,700 head of cattle, of which 119 head had pedigrees. Besides these there were between 400 and 500 cows which were already highly crossed, and which would be classed as of the pure blood. In fact, something over 2,000 of the cows were more than seven-eighths Hereford blood. Animals from this establishment have repeatedly been shown at the Argentine expositions with great success.

Dr. Frers has taken special pride in the Hereford steers he has shown, and these have not only been frequent winners, but have sold at fancy prices. Speaking of the value of the Hereford for the meat trade, he says:

"The Hereford steers are rapidly coming into favor, notwithstanding the prejudices of many breeders and exporters. Until a few years ago the average price obtainable for them was considerably less than that for the mixed-bred Durham. It was said that their net yield was less, but if so the reason was obvious. All white-faced animals produced by Creole cattle and those of the lower grades in general, were classed as Herefords, even though they possessed no other charactertistic than the red and white color. Those highly bred in Hereford blood were very few in numbers. Indeed the Durhams, with which they were compared, reached vastly superior refinement. It should be borne in mind that the proportion of the Hereford to the Durham is one to seven throughout the country. At the present time good Hereford steers command as high prices as any others at the public market, as well as for export. At recent cattle shows we have been triumphant. I have contributed three-quarters and seven-eighths blood Hereford bullocks that have secured on many different occasions the gold medal as the best lot of steers without distinction as to breed, in competition with the best Durham steers.

"In my opinion there appears to be a great future for the Hereford breed in this country, although I think it will never entirely dislodge the Durham. Indeed, I see no reason why it should do so. I think the Herefords equal them both as to value and economy of production. The Durham is somewhat more precocious. A Hereford steer will not give at three years the same weight and quantity of beef, but at four years this difference disappears. The Hereford is, on the other hand, more hardy. This is the outcome of a biological law. Precocity and rusticity are difficult to reconcile, and it is here the Hereford offers some advantages. He resists better the causes of general mortality, sickness, climatic variances, etc., in a higher degree, and this is the reason why where we have hard grasses, where the Durham cannot sustain himself and dies, the Hereford lives, sometimes weak perhaps, but still he survives. He fattens quicker when grazing, and preserves his condition better where hardships have to be met with, as is common in our country. The Durham is without doubt better suited for stall-feeding, but the Hereford has the advantage over him in grazing.''

Australasia.—Australia and New Zealand cut a large figure always in the world's supply of meats, more especially in the matter of mutton. In the production of Merino and cross-bred wools they hold a commanding position, and their exports of frozen mutton reach great totals. Cattle-growing is indeed subordinate to flock husbandry, and yet, as British colonies with good grazing available, they have naturally transplanted from the mother country the blood of the Shorthorn and Hereford in quantities that have resulted in the establishment and maintenance of many first-class herds. The bulls have made their impress upon the general cattle stocks. While the Shorthorn probably is to be found in larger number than the Hereford, the latter has met with special favor, as elsewhere, wherever harsh conditions have to be met, so that in those remote regions, as in lands lying nearer to our own boundaries, we find the "white face" as an important factor in the cattle business. Only lack of space precludes our going into details in this case as to their introduction and dissemination.

The Cape Colonies.—There are comparatively few Herefords in South Africa. A few have been imported into Cape Colony and Natal from time to time from England, but no herds of any size have been established. Mr. Walsh took out a small lot of Texas-bred "white faces" to Rhodesia in the fall of 1913, and expects to make further shipments.

One of the largest early importations of English Herefords into the colonies was that of the Transvaal Government in 1903. Then twenty-seven cows and heifers and four bulls, from good Herefordshire strains, were imported. Included amongst these was British Gold, by Gold Box (15339), presented to the Transvaal Government by the Earl of Coventry. This bull proved a useful and impressive sire. After this importation a few more cows and heifers were taken out for the Government and it was proposed to gradually increase the herd, which is located upon the Experimental Farm at Potchefstroom. The young bulls bred from imported females have been sold to farmers throughout the colony, and have been much sought after. Mr. Abe Bailey also established a herd in Cape Colony some years

A CROSS-BRED CALF—HEREFORD BULL ON AFRIKANDER COW.

A POPULAR CROSS IN CUBA—ZEBU BULL ON A GRADE HEREFORD COW.

ago. These were the first herds of importance in South Africa so far as we can learn.

Speaking of the general cattle-breeding situation in these regions, the Director of Agriculture for the Transvaal Government, writing to the author of this volume several years ago, said:

"Though a great portion of South Africa is by no means a dairying country, and so far very little has been done in the way of dairying, even in districts suitable for it, yet for some reason or other the first thought of the farmer when purchasing cattle is the amount of milk they will yield, the second consideration being the suitability of the steers for trek purposes. The carcass of the animal, the proportion of carcass to live weight, and early-maturing qualities have been greatly neglected. As a matter of fact, it would be far better economy on the part of many farmers if they were to go in for beef production pure and simple, and I have little doubt that before long they will do so. When they arrive at that stage Herefords should prove most useful.

"The native cattle—Afrikanders, as they are called—are very poor carcass animals and very poor milkers, though the little milk they do give is extremely rich. They are hardy and excellent for trek purposes. I enclose photographs of a bull and three young heifers.

"The favorite breeds of cattle, other than Afrikanders, in South Africa at present are Frieslands, Shorthorns, Devons, both North and South, and Ayrshires, but there are few herds of any size and merit of any of the breeds, and it would probably be hard to find any other part of the world in which cattle-breeding is, speaking generally, so backward as in South Africa.

"Until recently animals were allowed to run semi-wild upon the veld, and the only regard which the farmer had for his stock was that it should increase as rapidly as possible, and afford the minimum amount of trouble and expense.

"In many ways this is a trying country for stock, as in the winter, whilst the days are hot, the nights are bitterly cold. We are also bothered by many parasitic diseases caused by bacteria and protozoa, and conveyed by insects. The fact that South Africa once carried such an enormous herd of large game leads me to believe that when the various diseases have been overcome, and more sensible systems of management adopted, South Africa will be a good cattle country."

We reproduce herewith the photographs showing an Afrikander bull and females, kindly supplied by this correspondent, and in addition a plate showing the result of a cross of a Hereford bull upon a native African cow, from which it will be seen that the youngster, while nursing, is almost as large as his mother.

Mr. Walsh, who has selected something over 4,000,000 acres of land in Rhodesia for the British South Africa Co., believes that cattle-breeding can be successfully conducted in that region. It is not a well-watered country, that is, in the dry season, which is the winter. The summer or wet season is in November, December and January, during which time the weather is very hot, although not unbearable because the altitude is 1,500 to 5,000 feet above sea level. An abundance of water is to be had by digging or boring at a shallow depth. At the

AFRIKANDER COW AND HEIFERS.

AN AFRIKANDER BULL.

present time there are very few cattle in Rhodesia, probably not more than 6,000 head in the country, attributable to the rinderpest and South Coast fever which some years ago decimated the herds of the country to the extent of nearly 90 per cent. There is some difficulty in introducing better blood into Cape Colony on account of strict veterinary regulations, but it is believed that the next decade will nevertheless see a very large increase in cattle production in various parts.

CHAPTER XXIV.

PRACTICAL HERD MANAGEMENT.

This volume is designed purely as a record of accomplishments in the Hereford breeding field, and not as a treatise on feeding and general herd management. Nevertheless, it is certain to come into the hands of beginners, young and old, who may appreciate some practical suggestions on the handling of the Herefords, made by experienced men. A limited number of pages are therefore given herewith to a presentation of brief statements specially prepared for this purpose.

Hints from "Tom" Clark.—There is general recognition of the far-reaching influence of Thomas Clark upon the fortunes of the Hereford in the New World. During his long and active association with the work of breeding, feeding and showing the "white faces," he was ever in the front rank in point of actual accomplishments. He is now retired so far as enduring the heat and burden of the day is concerned, but as steward of the ring at the Chicago International he annually renews his youth by maintaining touch with those who are now bearing to still higher levels the standard of the breed which he did so much to uphold in bygone days.

Asked for a word as to the practical management of the breeding herd he replies:

"In regard to my method of handling a breeding herd, I shall first of all tell how I would handle the breeding bull. He should be kept in good breeding condition. I think some breeders keep their breeding bulls too thin. I believe a bull will sire calves with stronger constitutions and better flesh carriers if he is maintained in good flesh. In managing my breeding bulls I kept them away from the cattle as much as possible, giving them good roomy stalls with small yards adjoining so they could get all the exercise needed to keep them straight on their legs and active. I fed equal parts of ground corn and oats with a little bran and oilmeal added, and fed three times day about all they would clean up. But be sure that they clean it up at all times. Also feed good sweet hay; I prefer clover to any other kind.

"I managed my breeding cows as follows: I preferred to breed them so that they would have calves from Jan. 1 to April 1, except a few that I wanted to have calves for show purposes. Those I would breed to calve from Sept. 1 to Jan. 1. All cows that would not have calves until February and up to April I kept alone in a yard with an open shed well bedded, so that they could go in and lie down comfortably. If kept in that way they are more healthy and their calves will be stronger and more thrifty. Besides you save labor and feed. I believe that the less you move cows around while pregnant the better. Two or three weeks before they were due to calve I took them up and put them in loose boxstalls, and fed them liberally on ground corn and oats with a small allowance of oilmeal. After calving let the calves run with them for three or four weeks in the boxstalls so that they can suckle whenever they wish. But I would turn the cow out in the morning, and put her back in the stall at noon and feed her after the calf had nursed. Then I would turn her out again until evening, when I would put her back in the stall for the night with her calf. After the calf is four weeks old separate cow and calf, suckling the calf twice a day. Place shelled corn and oats in a trough where the calf can go and eat at will. You will be surprised how quick it will begin to eat.

"When grass came I turned cows and calves out together on pasture and let them run until flies got bad, and then took the calves up and fed as before, bringing the cows in mornings and evenings and letting the calves suck. I separated my bulls from the heifers, and fed the bulls all they would eat of ground corn and oats, equal parts, with a little oilmeal.

"I would not breed heifers until eighteen to twenty months old, so as to have them near three years old when dropping their first calves. I would breed them so as to have their first calves in the spring if possible, in order to get them quickly on grass, which will make them give more milk for the calves."

John Letham's Experience.—It is now near thirty years since the author first formed the acquaintance of a feeder contending for honors at the old Fat Stocks Shows in the Chicago Exposition Building on the Lake Front who impressed him as a man of exceptional capacity. During all these years this acquaintance has been continued, and with ever-increasing respect on our part for his judgment in all that pertains to sound methods of beef cattle management. We refer to John Letham. He has had a long and successful experience, and we doubt if a better or more practical statement touching the right handling of a breeding herd has ever been put on paper than that which he has prepared at the author's request, and is submitted herewith:

"The management of a breeding herd is not a very complex problem if you keep close to nature.

THE WOODS PRINCIPAL, CHAMPION BULLOCK INTERNATIONAL EXPOSITION 1901—Bred by Geo. P. Henry and fed by John Letham—Weighed 1,645 pounds as a yearling; sold at 50c per pound.

Abundant pasture and pure water easily reached in summer, well ventilated barns for the cows and young calves and good, dry, well bedded open sheds for the yearlings and two-year-olds in winter are all that are necessary for success. If these simple requirements were followed we would hear but little about abortion, tuberculosis, scours, foul-feet and many of the troubles that plague the caretakers and dishearten the owners. It is astonishing in going over the country how many cattle one finds which have insufficient pasture and filthy water, or only water at intervals. And yet they are expected to make good returns for their owners. In winter the conditions are deplorable even with men who mean well. Many of the costly bank barns are hotbeds of disease. Ventilation was never once considered by the architects and drainage was entirely forgotten, even where the cupola is a work of art resembling Joseph's coat of many colors. Go into such a barn at 5 a. m., where 50 to 100 cattle are housed. The hot, moist atmosphere meets you; it is past being unsanitary; it is impure, death-dealing to man and beast alike. At 8 a. m. these cattle are turned out into a yard resembling a hog wallow, there to stand in the storm or zero weather till 4 p. m. And breeders will talk about having bad luck! This is not an overdrawn picture. It is only too common and surely means the survival of the fittest in the end!

"When managing a breeding herd the bull is half the herd at all times, so we turn to him first. A paddock of 2 acres or less with a comfortable shed and boxstall in one end makes the ideal quarters for the herd bull. The shed should have a loft above to hold the hay and straw, so as to be handy in winter and to keep it cool in summer. Then you can have the breeding pit under cover in the shed. In many of the states no door is necessary. Let it open to the south and the bull will generally use good judgment. Should the young bull be lonesome turn a cow safe in calf with him for company. In this way you will conserve his virility and lengthen his life and usefulness. Feed him enough to keep him strong and vigorous all the time without loading him up with a lot of superfluous inside fat or outside tallow. Blood, bone and flesh are what you want in a herd bull. After getting his growth he should never vary 100 pounds summer or winter. This letting down and building up procedure is always disastrous. So far as my experience goes the single service gets as many calves as the double or triple service, provided the cows are in proper season and healthy. A radical change of pasture during a dry spell, ergot on the grass, changing to silage, heavy feeding of cottonseed meal, etc., have been the causes of charging up many a bull with unsatisfactory service.

"The pregnant cow should always be the herdsman's special care. She only drops one a year on an average. To save a good calf means to save a large part of the herdsman's salary, sometimes a year's salary. And right here is where you find the greatest difference in herdsmen. Keeping the cows bred up and saving the calves, far more than makes or loses the salaries of the best men. When the matron that is due has been on grass and raised naturally little need be done, the calves usually coming strong and healthy. The calves dropped on the green sod seldom get infected. Of course you have always the maggot, the screw worm or coyote to remember, depending on your location. The commonest evil is too much new milk at birth. Stale milk has killed many a calf. Therefore see to it the mother is properly stripped once daily even at pasture, and more especially should there be a retention of the placenta. It is astonishing how little the calves need to live on during the first week and how much damage can be done by too much, especially if the milk is stale or the mother at all feverish. In winter the calf cot is all-important. It should be cleaned and aired out every day and a little slacked lime sprinkled. It is always worth what it cost in the field. Do not wait till your calves get the scours, coughing, wheezing and running at the nose, then rush to town for disinfectants and diarrhoea medicine and have a general house-cleaning and a lot of sick calves. Prevention is always wise.

"The young calf is better beside the dam from 3 to 6 weeks of age. Then it can be put in the calf cot and nursed twice daily, 12 hours apart. This is especially good with a heifer's first calf. It develops her udder and makes her a better mother in the future. Supply the calf cot with the choicest morsel or hay and have shelled corn, oats, bran and a little oilcake in silage, so they can nibble at will. What good millers they are and how they enjoy doing their own grinding! But a word of caution about silage: Never let a young calf get frozen or musty silage. Alfalfa is rapidly replacing roots and silage, but I still believe good silage the best substitute for milk.

"The yearlings and two-year-olds in the open sheds need lots of roughness and should have some grain. Never let them stop growing a day if you expect to raise good young cows at the least possible expense. And remember that water is an all-important factor in winter as well as summer—not once every other day or a bellyful of ice water once a day. What a mint of money is lost in the cattle business in this country for want of water summer and winter!

"Alfalfa and silage are rapidly changing feeding conditions throughout this country but the general principles are still the same. It is still the good herd bull and the breeder who stays close to nature, watching the little details which the other breeder ignores, that forges ahead and gets the ripe persimmons. 'The eye of the master maketh his cattle fat and the righteous man is merciful to his beast.'"

Scale, Flesh and Fat.—Discussing the important subject of size and real flesh as against mere outside fat, the veteran English breeder, Mr. John Hill of Felhampton Court, in a letter written to the author some years ago commented upon type and the points to be observed in his judgment in selecting breeding animals in language which we deem worthy of preservation here:

"About the time of what may be called the 'Hereford boom' in the early '80's there were several popular sires which were especially adapted to get early-maturing cattle, and their progeny were unusually successful in the showring. Many breeders 'went mad' over these special strains and further set the seal on a type which had an extraordinary aptitude to fatten, put on flesh evenly, and mature early. Of course this is exactly what is wanted, but the greatest possible care is at the same time required to preserve scale and lean meat, and this was too often lost sight of both by breeders and by the judges in the showring.

"With reference to breeding for scale, it may be worth noticing that in old days when the breed was remarkable for this characteristic, the females were not usually of such dimensions as might have been expected that the dams of the large oxen would have been. But there was a peculiar look about them which can hardly be described, which experienced cattlemen can at once recognize. The words, 'she looks like a good breeder,' convey a particular meaning. Such cows are essentially feminine in their appearance, of moderate size, with well sprung ribs, roomy bodies, lengthy hind-quarters, often light in their fore-quarters, of clean-cut sweet-looking heads, with mild intelligent eyes. Usually she carries a good bag and is always a good handler. When looking for a suitable mate, at once discard any bull that has not got a good masculine head. I do not believe that an effeminate-looking bull, however good he may be in his quality and carcass, can ever make an impressive sire. Some few of them may get heifers, but never in my experience have I known them to get good bulls. A bull should carry himself majestically, and 'look a bull all over.'"

How the Harris Herd Is Handled.—The records of latter-day Hereford breeding in the United States present no instance of outstanding success more notable than that afforded by Overton Harris and his sons with their Model establishment at Harris, Mo. Asked for a brief resume of the methods employed in the handling of their cattle, Mr. Harris says:

"Our Hereford breeding herd since its establishment twenty years ago has been handled in as practical and economical a manner as we have known how to practice. Our pastures are more or less protected by timber, and many cows in our herd have never seen the inside of a barn or shed. Located in one of the best bluegrass regions of the world, the summer ration of our herd is bluegrass—plenty of it and nothing else.

"During the early winter stalk fields and bluegrass which has not been heavily grazed furnish an abundance of feed. Later, corn fodder, clover and timothy hay are supplied as required to keep the breeding herd in strong thrifty condition. In the late winter and early spring months cows that are heavy milkers, and we have many such in our herd, are given a little extra feed—anything in the form of corn chop or cottonseed. We find that the rugged hardy constitutions of the Herefords do not require

REPEATER 286598, BRED BY E. W. & A. M. HEATH—USED BY O. HARRIS & SONS.

that they be provided with expensive barns for winter shelter; in fact, we do not even find it necessary to provide sheds of any kind for them. We have never, even during the most severe winters known in this section of the country, found it necessary to give our older cattle protection, and we have never had losses from exposure. By this method our breeding herd is carried through the year at a minimum cost and maintains a very thrifty and healthy condition.

"Calves, except those intended for show purposes, are allowed to run with their dams on the pasture during the summer. When old enough to wean they are placed in a pasture or yard by themselves and are given a light grain ration during the winter, and except in the most severe weather are never housed at all. During the second summer bluegrass is their chief diet and they seldom if ever taste grain again. Our heifers are bred at from eighteen to twenty-four months of age and it is seldom that we find it necessary to give a heifer any assistance in the way of feed while nursing her first calf, except what she gathers herself. Young bulls generally require a light feed of grain once a day during their second summer to insure the best development and growth.

"Our herd and stock bulls, all of which have at some time been grand champions in the leading shows, are not kept in extremely high condition after we are through exhibiting them. They have the run of small grass paddocks. This gives them an opportunity for plenty of exercise and an abundance of fresh air and a sun shed, our aim being to keep them in strong vigorous condition. Beau Donald 5th, now sixteen years old, has been handled in this manner and is still active and doing good service.

"Our experience with close in-breeding has been

very limited, as we have never looked with a great deal of favor upon such a practice. It is our belief that it should only be undertaken in the hands of the most skillful breeders, and then only with such animals as are practically perfect in every respect. We have never kept an accurate account of the exact cost of maintaining our breeding herd, but we do know that our Herefords have made us plenty of money, as well as being the source of a great deal of pleasure and satisfaction.''

McCray's Methods.—Warren T. McCray, Kentland, Ind., has by common consent arrived at a position in the Hereford business attained by but few of his contemporaries. This is scarcely due to luck. Such success does not come by chance. In response to a request for a word as to the general line of treatment accorded his cattle he submits the following:

"Regarding the management and feeding of a herd I would suggest that one of the most necessary attributes of a successful cattle manager and feeder is a liberal endowment of good wholesome common-sense, or it might be more nearly correct to say good cattle-sense. There is a distinction between the two that comes naturally and they in whom the two are combined are the most successful cattlemen. I have never made any great discovery in the feeding and management of my herd. One's success or failure depends upon the care and watchfulness of the feeder. He must be regular, attentive and watchful. He must know the particular characteristics of each animal under his care and cater to them, so that the animal will at all times do its best.

"I am a great believer in the out-of-door life for the breeding herd. Nature has made laws which have never been improved upon by man. The only thing to watch is that the cattle have plenty of feed and water. Do not over-stock the pastures. Leave plenty of feed in them for fall and winter. Last winter I kept a bunch of dry cows on a good bluegrass pasture until the middle of January and they were as fat as one would wish when removed.

"We commence feeding our calves as soon as they are old enough to eat and keep this up until they are past the yearling stage. I have several small lots or grass paddocks and the young bulls are divided up and placed in these over night, but in the day time they are put in their stalls to protect them from the flies and heat. We commence to breed the heifers when they are from twenty to twenty-four months of age, and begin to use the bulls lightly when they are about fifteen months old. I have never practiced in-and-in-breeding to any great extent as I always considered it a dangerous proposition. However, I am now conducting some experiments by breeding some daughters of Perfection Fairfax to some of his sons which had dams with a decided out-cross and whose strong characteristics I want to maintain. I am hoping for satisfactory results but at this time I am not qualified to speak from experience on that subject.

"I have found a great deal of pleasure in the pursuit of cattle-breeding. The acquaintances and friendships formed among cattlemen are the most loyal and lasting that can be made. The business has also been most profitable, but aside from this there comes great pleasure and satisfaction in the feeling that one is really doing something in the world that is worth while, in trying to produce and improve an animal that contributes more to the support and material welfare of humanity than any other."

THE CUDAHY SHOW HERD, WITH FAIRFAX 16TH AT HEAD.

Photo by Hildebrand

Mr. Hazlett's Views.—Robert H. Hazlett, Eldorado, Kans., is known as one of the close students of the best contemporary methods and practices in the handling of Herefords for profit. Out of the fullness of years of successful practical work, in answer to our appeal for some hints from his book of experience with especial reference to his observations as to the effects of blood concentration, he writes:

"I shall state in the beginning that I believe in giving the young animals a chance. With this in view I try to grow them out as much as possible, keeping them at all times in good thrifty condition —not striving to make them fat, but on the contrary to produce real development by feeding for flesh, bone and size.

"Speaking of the heifers, specially: From the time they are weaned they are kept separate from all other cattle, in pasture in summer and in corrals, with open sheds for protection from storms, in winter. During the time they are not on pasture they are fed a grain ration consisting of a small percentage of corn, a larger percentage of barley usually, and a considerably larger percentage of oats. All this feed is ground and fed mixed with kafir corn and cane silage or cut cane fodder and alfalfa hay. In addition to this mixed feed they have as rough feed cut cane and alfalfa hay with occasionally a feed of prairie hay as a change of ration. They get practically no grain during the summer, although at times when the weather is dry and the grass not very nutritious it is necessary in order to keep them in condition to give those under one year old a light feed of grain once a day. The older ones have no grain in summer. It is my opinion that better breeding animals result from being handled in this way than if they are allowed to become stunted in any degree because of lack of proper nourishment during the developing period.

"After the breeding cows are two years old, with an occasional individual exception for a short time, they get no grain ration whatever, either summer or winter. They have the bluestem grass pasture for summer feeding and are wintered mainly on sorghum and alfalfa. Most of this feed is run through the cutter, and lately the sorghum is fed in the form of silage. Except those near calving and those having young calves, the cows are not put in barns at all, but for protection run to sheds open to the south during the winter months. Whether in pastures in summer or in lots in winter, I like to keep the aged cows, two-year-old heifers, yearling heifers and heifer calves separate.

"The bull calves are fed and handled in much the same way as the heifers, except that they are not allowed to run with the dams in pastures beyond the time they are around four to five months old. After that and until they are weaned they are kept in lots near the barn, the cows being brought in and the calves suckled twice a day. After being weaned they are not fed in open lots as are the heifers but are tied up and fed, each one by himself. We have what we call a 'bull barn' for these weanlings where they remain until sold. Around this barn are several lots in which these calves are kept, a few in each lot. They are in these open lots practically all the time, except when brought in for their feed of grain. I find that they do better when I feed them separately in this way than when I attempt to feed a number of them together. In the open pens they get the benefit of fresh air, sunshine and exercise.

UP-TO-DATE METHODS IN HEREFORDSHIRE.—MR. J. K. HYSLOP OF IVINGTON, LEOMINSTER, SPRAYING HIS CATTLE TO WARD OFF WARBLE FLIES, ETC.

"From the time our herd bulls are from twenty-four to thirty months old, being practically matured, they have a very light grain ration, usually once a day, with alfalfa hay, prairie hay and cane fodder. Each has a separate lot and an open shed, never being kept in barns. They are not kept fat, but in good thrifty condition.

"My first purchase of purebred Herefords was that of an entire herd consisting of only fourteen females, cows, heifers and heifer calves, and two bull calves. One of the cows was by Stonemason by Beau Real by Anxiety 4th. This cow was the dam of one of the bull calves, his sire being Wild Beau by Beau Real. Wild Beau was a full brother to the famous Wild Tom. All the younger females in the little herd were sired by Wild Beau. I retained this calf, out of the Stonemason cow, and used him on all the cows and heifers except his dam. I was without experience as a breeder at that time and knew practically nothing of different bloodlines or the different families, but the results of this very conservative close breeding were quite satisfactory and my young bulls found ready sale at fair prices from the beginning. I did not offer for sale any of the cows or heifers.

"To avoid breeding heifers to their own sire I soon bought another bull, a very good individual with a greater concentration of Anxiety 4th blood than anything in my herd at that time, and the results from his use proved very satisfactory. In fact, there is no doubt but that there was an improvement in the produce of my herd from his use. Later, when the heifers sired by this bull were coming of an age to be bred, having the prevailing idea that too close breeding was to be avoided I bought two young bulls of different breeding from my cows and from each other. One of them was a son of a famous show bull. Each was a good individual and of a family very popular at the time. These were used in the herd but with quite disappointing results, so much so that I disposed of both of them. Before they were sold, but after I had decided to dispose of them, I bought a third out-cross, a show bull with quite a record, a half-brother on his sire's side to a very famous prize-winner which was grand champion at the larger state fairs, the American Royal, and the International. This purchase was also an outstanding individual, but the results in my herd from his use were even more disappointing than those from the use of the two preceding him and he also was disposed of. I then decided to return to the Anxiety 4th breeding through sons of Beau Brummel and others tracing to Don Carlos, by Anxiety 4th, and have continued in the same line to the present time.

"From my experience with out-crosses I am compelled to believe that whatever success I may have attained as a breeder is due largely to the fact of persistent close-breeding. At least, so far the results of such close-breeding as has been practiced at Hazford Place have been very satisfactory. Undoubtedly some other elements have incidentally entered in, for instance, selection and environment. Both of these, however, are necessary to real successful constructive breeding in any herd.

"While it may possibly be true that the persistent promiscuous breeding of closely related animals, without discrimination or selection, may bring greater disaster than the persistent mating of promiscuously bred animals, yet from my observation and experience, if only worthy animals, those of good conformation, quality and breed character, are used for breeding purposes the ultimate results are bound to be better in the closely bred herd, produc-

LORD COVENTRY'S DOLLY MOUNT—ROYAL CHAMPION OF 1911.

SHOTOVER AND BULL CALF—JOHN TUDGE'S ROYAL CHAMPION OF 1904.

ing offspring more uniform in type and more uniformly equal or superior to either sire or dam. After all, this is only another way of saying that 'like begets like or the like of some ancestor.'

"If it is possible that close-breeding in itself will produce a weakness or defect in the offspring when both sire and dam are of outstanding merit, neither having this weakness or defect and both tracing back to the same ancestors of equal merit, there must be for such a result a demonstrable scientific reason. So far investigators and students of the science of breeding have not discovered any physiological or other scientific reason. If neither sire nor dam, however closely related, has a certain or particular defect or weakness and none of their ancestors has been affected with such defect or weakness their offspring will not have it as a result of this mating. The quite common notion that it is otherwise, and that defects and weaknesses are the necessary results of blood concentration, undoubtedly had its origin in sentiment and survives on 'common report,' scarcely anyone having attempted to solve the problem by persistent, patient, personal effort.

"Custom may make law, but tradition never established or created a physical or scientific fact, nor can folk-lore make or change a law of nature."

Tow's Practical Work.—Cyrus A. Tow has to his credit a marked success in breeding and development of "classy" Herefords in the recent past. In fact, he has proved quite a "disturber" in the calculations of his competitors at leading shows. He tells the story of the breeding of his cattle in simple language:

"It is about eight years ago since I became the owner of my first registered Hereford cow. It was bread and butter with me, so every cow had to make good. The cows of breeding age will care for themselves if given half a chance. We always aim to care well for the younger ones. Our cows are all run on grass in summer and those giving milk or heavy in calf are housed in winter. We feed nice clean oat straw in connection with silage for winter roughness. The young calves and yearlings are always run in paddocks around the barn where they get their feed. We always keep our calves separate from their dams. They are suckled in barns or yards. Our yearling heifers and short two-year-olds are left to run in open sheds in winter and are fed their silage and hay in racks. We breed everything at the halter and never breed a heifer younger than nineteen months. Our young bull calves are separated from the heifer calves at about four months old. We try and grow them all alike, giving them all an equally good chance to make good. Our herd bulls are housed in winter and grained daily the year around, except for only a few months in summer when grass is good their grain may be shut off. They have boxstalls to run in when in the barns.

"In regard to the 'doubling in' of blood in the breeding of cattle, I am very much opposed to the practice as a general proposition. I know that in some cases it has brought good ones, but we sometimes forget to mention the cases where it has failed. I may add that the cattle business has been a success with me thus far, and I believe that the beef cow is as sound an investment as a farmer can make."

Bluegrass Management.—Luce & Moxley, Shelbyville, Ky., of Prince Rupert fame, figure prominently in the showyard annals of recent years in all the great competitions. Their success in the bluegrass country has been pronounced. Mr. Moxley briefly outlines their herd management in the following terms:

"In this part of the state we usually have good bluegrass for grazing from May to middle of December. Our Herefords need no attention as far as feeding goes during this season. The calves that come during this season are left with their dams until bad weather comes, when they are taken to the barn and if old enough are weaned and fed a mixture of corn, oats, bran, and either cottonseed meal or oilmeal.

"The cows with calves at side are fed silage, cottonseed meal and a little hay. These cows are kept up at night and run to shock fodder in the daytime if the weather is not too bad. The dry cows are left out and run to a strawrick and shock fodder, unless we have a mean spell of weather, then they are fed a little hay. We aim to start our cows into the winter in the best shape possible. The cows that calve in the early spring are taken up a month before calving and fed the same as the cows with calves at side. By putting the younger calves in pens with older ones they soon begin to eat a little feed. We take the best of care of our calves until they are about fifteen months old. It is our experience that if we do this we have very few of the cheap kind. All of our breeding is done by halter. We have a small herd and find this plan quite satisfactory."

Fitting for Show.—The selection and making-up of cattle to be entered in the public competitions is not a topic that lends itself readily to treatment in cold type. No hard and fast rules can be given.

Especially is this true of any attempt at giving explicit directions governing the feeding of the animals intended for show or sale. And yet there is constant call for suggestions upon this subject from those who are without experience.

An Old-time Herdsman Speaks.—The name of Jim Powell is often mentioned in preceding chapters of this volume. None among the older generation of feeders is better qualified to talk upon this subject than he. We are glad to give space to a short discourse from him, prepared at our request:

"In starting to select a herd for showing I should get yearlings. In selecting these get smooth heifers with plenty of size, something that will make big cows. I would notice especially that they had good heads, with not too large horns, and that they were good in their heart-girths, and had good level backs with straight hind-quarters, and that they came down well in the round and that the tail was not set on too high. I would try to get them as near the same size and type as possible.

"Upon the selection of the bull a great deal depends, for he is more than half the herd. You want a bull to have a good bull's head, not feminine, wide between the horns, and with a short thick neck. He should be wide between the legs, with a good compact brisket. Be sure that he has good big bone, something that can carry weight. Another very important point to notice is that he is good in the heart-girth, and has a good level back and straight quarters. He should have a good thick mellow hide with good coat of hair, which denotes the good feeder.

"In feeding a herd I have found that corn and oats

IRON PRINCE (22250), BRED BY CAPT. HEYGATE.

GAINSBOROUGH (28303), BRED BY A. P. TURNER AND USED BY STEPHEN ROBINSON.

ground together in the proportion of two-thirds corn and one-third oats is a good winter feed. Alfalfa, timothy or clover make good roughness. In the summer I would reverse the ratio, making it one-third corn and two-thirds oats. In starting a young herd about 2 quarts of chops, 1 of bran and a half pound of cottonseed cake three times a day is a good feed. This can be increased to 4 quarts of chops as soon as the cattle get on their feed. However, different animals vary so much in the amount of feed they require, that it is impossible to give any stated quantity. Always be sure your feed-boxes are empty before feeding, as cleanliness is important. Never give an animal more than it will clean up. If any is left in the box, take it out before feeding again. If they do not seem anxious for their feed skip a meal. A little sulphur with their feed twice a week is good to give them an appetite and also to keep them healthy.

"Now, I think calves should have nothing but the cows' milk until they are three months old. Then commence to feed a little cottonseed cake, with oats and bran, as I think cake is the next thing to milk for young calves. When they are six months old, they should have about 2 quarts of oats and bran and a half pound of cake three times a day, besides the cows' milk. Then they will not lose their calf fat after weaning.

"In training young cattle much patience and time are required. In the first place tie them up in their stalls and handle them in there. Then commence leading them out. Have a buggy whip in your hand and make them stand when you want them to. A light cut on the nose will do this. Try to make them stand with their front feet well under them, not spread apart, and then they will show a good level back. Make them stand at ease. Do not try to do too much at once, as the calf may become sulky. They should have plenty of exercise.

"The young bull should have a boxstall, with a good yard to run out in, so as to keep him good on his feet. In the spring, when the grass is good, I would let the herd run out day and night, feeding them twice a day. When the weather gets warmer I would keep them in by day and turn them out by night.

"When the herd is brought back from the fairs great care must be taken to let them down easily. Feed twice a day, taking away about one-third of the feed. In a week or two, this may be cut down to one-half. But do not stop feeding, as you want to keep your herd in good thriving condition. In place of the corn chops, I would now feed about 2 pounds of cake with roughness, and this should carry a herd through."

Ed. Taylor's Views.—Edward J. Taylor has not only picked and trained champions, but is often called to pass upon the entries in the great showyards of these modern days. He has kindly set down his views on the preparation of cattle for exhibition purposes as herewith submitted:

"Much has been written from time to time regarding the fitting and handling of cattle in preparation for the showring, and few trainers agree as to the variety of feeds and methods pursued in the undertaking. Like the old lady who kissed her cow, it's everyone to their notion. There are no set rules to work by. Animals differ very materially in their dispositions, aptitude to fatten, etc. But there are a few fundamental principles which apply quite generally to the art.

"First of all, the man must be in love with his occupation and charges, and gain their confidence.

He should also be a good and impartial judge, and as quick to see the defects in his own cattle as in those of others. One who gets so enraptured with his own as to think they are the only pebbles on the beach is sure to be a bad loser, or as Burns puts it: If nature'd but the 'giftie gie us, to see oursels as ithers see us.' Eternal vigilance must be the fitter's watchword, and regularity and system cut no small figure in the game.

"Having selected the prospective winners and provided comfortable quarters, a light, well-ventilated basement, with well-bedded, roomy boxstalls, makes a nice place and he should be ready for business. In fly time the windows may be curtained through the heat of the day, and no unnecessary disturbance should be allowed around.

"I think that the calves are the easiest fitted members of the herd, and I shall touch on them first. Given plenty of milk and a mixture of grains, if they are of the right kind they will put on plenty of bloom. Calves seem to have a weakness for shelled corn, and this mixed with equal parts of ground oats and bran and kept where they can have access to it at will for the first three or four months of their lives will put them in shape so that one can select those most likely. Occasionally an outstander puts in an appearance, about which there is no question when only a day or two old, but this is the exception rather than the rule. When the calves are about four months old I think it better to feed the corn ground, and add a small amount of oilmeal, feeding what they will clean up at once. Absolute cleanliness of feed-boxes and stalls is essential with all, especially with calves.

"The yearlings receive similar treatment to the calves. Some prefer to continue with the milk portion of the ration while others think it time to call a halt when they graduate from the calf class. With a few exceptions I usually followed the latter course. It is when he gets to the two-year-old and older cattle that a fitter's discretion is put to the test, and right here many a one has met his Waterloo. The time has come now when corn must be handed out sparingly. More oats, a little barley if obtainable, succulent feeds such as roots, and anything that will aid them to hold their own without a tendency to obesity or patchiness should be used. Ground or cracked wheat is very helpful at this stage, but should be fed sparingly to start with. A double handful of bright clover hay, cut fine with a chaff cutter and mixed with each feed of grain, lightens the ration and makes it more easily assimilated in hot weather. I prefer to dampen the feed just enough to make it gritty, not sloppy.

"Exercise is a very important matter. A small pasture lot handy to the barn so that the females can run out nights through the hot weather, and suitable yards for the bulls are indispensable. All should be well halter-broken and taught to stand properly and show to best advantage. The feet should have close attention and be trimmed when necessary. The coat and skin must be kept clean by brushing and dampening occasionally; too frequent washing dries the hair and causes it to become thin. I never advocate rasping and paring the horns and polishing as we so often see them; it savors too much of the artificial. I think they look more natural with the rough shell and scratches taken out, and just wiped over with an oiled rag.

"There are many small details in connection with the business, and it is a difficult matter to touch on all, but I have tried to cover the ground in a general way. When a boy at home I have often heard my father say, 'One-half goes in at the mouth.' And while there is a good deal of truth in the assertion I think that is a pretty big percentage. As an illustration I recall a remark made by my old friend Robert Ewart, of Browndale fame, in regard to a cow I was showing at the Minnesota State Fair some years ago. In those days all breeds competed for championship-by-ages and she was a formidable candidate for that honor. After looking her over carefully he turned to me and said, 'If the old bull (meaning her sire) had done half as much for her as you have, she would be a world beater.'

"In preparing cattle for the auction ring the chief object is to have them in as good flesh and as presentable as possible. Fat hides a multiplicity of faults, and while many breeders claim to discriminate against highly finished animals for breeding I notice these are invariably the ones which command the highest prices. Whenever a sale falls flat and prices rule low the reporter's comments usually state the cattle were too thin to warrant the appreciation they deserved, or something to that effect. So there is little to be said other than to have each lot number in good consistent breeding condition (not overburdened, of course), well broken to the halter. Having been washed a day or two previous, and their coats dressed up for the occasion, they should present a good appearance and bring their full value. The vendor who values his reputation will of course guarantee all animals to be breeders."

"Jim" Hendry Heard.—The extraordinary success attained by the Beau Donald and Perfection Fairfax cattle in the hands of James Hendry is so universally recognized, that the matter herewith appended, setting forth a few of the ideas of that astute herdsman on the subject of training cattle for the

showring, will undoubtedly be read with interest by all who follow the fortunes of the breed in public competitions.

"I came to America in 1892. Before I came to this country my experience with cattle was very little, but what I had was with Angus, with Alex. Geddes, Blairmore, Glass. I started with the Herefords at Greenwood, Mo. I was but a helper then. The first summer I used to wish I had the cattle back in Scotland away from flies and heat. But as fall crept on and homesickness left me I began to see I was mistaken. The old imported cows came up in the fall so fat and sleek, and the nice curly calves by their sides made me think I was in the land of promise and made me forget heat and flies.

"I went to Kentucky in the fall of 1901 to take charge of the Beau Donalds and Mr. Curtice had a line-up of calves which was very easy to pick from. They were low-down, chunky fellows with coats of hair like velvet. I always try to get a calf with good back, not too long coupled, nice short head and good straight hind legs. I commence as soon as I think the dam is not giving enough milk and help out with a nurse cow—not too much to start with, one nurse cow between two calves until grass. Then if prospects are good for show calves, which can be told by this time, I give them a fresh nurse cow and dry up the dams. When about six weeks old I commence to put a little cracked corn and oats in a trough where they can learn to eat. In summer time I keep them in a dark, clean, cool stall away from flies, with plenty of water beside them. But in the winter months I turn them out in a dry lot with plenty of sunshine. But do not stand them in the mud a foot deep all day and expect them to do their best. The better care and feed you give them the first year of their life the better herd of cattle you have. It does not take so much when they are calves. Remember, it is not every calf that comes up to my estimation as a show calf by any means, because they have to have a good constitution and stand up to every meal and bawl for the nurse cow. Bull calves are harder to handle than heifers. They are more restless and sometimes you have to put them in single boxstalls or tie them up.

"And for fitting older cattle I might say a good deal. But of course there are differences in cattle. Some get too hard and the next too soft, so one has to gauge the different feeds on them. I do not believe in too much corn. More breeding cattle are ruined on corn than anything else. I generally mix my feed—bran, 100 pounds; corn, 150 pounds; oats, 150 pounds; cut hay, 30 pounds, and oilmeal, 10 pounds. Towards show time I add a little molasses or something sweet and it makes them eat a little more. And it helps their hair to grow. But remember, never feed them more than they will clean up at any time. Water is as important as feed, as they have to get plenty all the time. Keep your stalls clean and well disinfected all the time, because they love a good clean stall the same as we do a bed. In your spare time in winter get a good stiff brush and brush them, as it makes the hair soft and curly. Use it as much as possible against the hair. Some people try to say that too much brushing will take out the hair, but I have never found it that way, as the more brushing you give them the better fix you get their hair in. I do not mean to use a curry-comb but a good stiff brush. A curry-comb is liable to pull out the hair.

"As show time draws near wash them once a week. It helps them and also the calves. It breaks them so that when you wash them at the fair they are

ROB ROY, CHAMPION AT THE ROYAL SHOWS OF 1908-9.

CAMERONIAN, CHAMPION AT THE ROYAL OF 1906.

not so liable to get homesick as the calf generally does the first two weeks after you leave home. The best way to fit a herd is to stay right with it and raise your calves and keep showing them until they are mature cattle. Then you can see something you have done, and are not moving every year as some of the boys do. They cannot know in that length of time what success they are to have, as some years we have better prospects than others.

"In conclusion, I love to raise show cattle from babyhood up. And no one loves them more than I, but when the buyer comes along I am willing to sell and wish him success, and try and raise a better one."

CHAPTER XXV.

THE "ROUND-UP."

In the limited space now remaining it is impossible for us to take up in detail the winnings of Hereford bullocks at the International show since its establishment, and the achievements of the western-bred white-faced calves in cornbelt feedlots generally. A volume could be prepared on this one phase of the Hereford trade. Indeed, the value of well bred "white faces" in the baby beef business constitutes their one highest claim to the permanent consideration of the American public. They will get fat quickly if from well bred stock, and their record in the pens at the big Chicago show, as well as at the Kansas City, Denver, Fort Worth, St. Joseph, So. Omaha, Sioux City and So. St. Paul shows and markets, needs no detailed exploitation in this connection. It is a part of the current literature of the cattle business that is at all times accessible to the readers of the live stock press.

"Tom" Sotham was probably one of the first to exploit in a commercial way the transferring of range-bred Hereford calves direct to cornbelt farms. During the season of 1898 he purchased more than 2,000 calves in Texas to fill orders for customers in the middle west, paying $24 to $25 per head at that time at the point of shipment, charging his customers a commission of $1 a head for the service. He also assisted "Dick" Walsh in an auction sale of 400 JJ calves at Kansas City in November, 1898, at which a $37 average was made. The results of the feeding of these calves in good hands were so satisfactory that a large trade of this sort developed, which still stands as an important feature of the business of cattle feeding in the older states.

"Dan" Black of Lyndon, O., by virtue of his victories with Texas-bred calves at the International, contributed largely to the up-building of this trade. John G. Imboden, C. C. Judy and others have also been instrumental in bringing many feeders in direct contact with the southwestern producers. Mr. Imboden feeds regularly himself, and his intelligence and fairness are so generally recognized, and his experience as a butcher, exhibitor and judge at leading shows of the past twenty-five years has been so extended, that we have asked him to say a word at this point on this general subject.

Range-bred Calves in the Feedlot.—Mr. Imboden says:

"I have demonstrated the value of the range-bred Hereford calf and yearling for the cornbelt feedlot in a number of instances during the past twelve years. About the year 1900 I purchased in southern Texas at Beeville 50 Hereford steer calves. They reached my place about Dec. 20, when the thermometer was about 15° below zero. The change in altitude and temperature from southern Texas near the

gulf to Illinois did not unfavorably affect them. They made a satisfactory growth and gains from the beginning, and at the International the next year 15 head were first in class from the southern district. I have fed other Hereford calves from southern Texas, and they invariably did well.

"In 1906 I exhibited at the International 15 Hereford yearling steers averaging 1,174 pounds at $8.90 per cwt. They were first in their class from the southern district, and champion Hereford yearlings of the show. These steers were bred by Boog-Scott Bros. They were the top out of 50 calves that averaged 375 pounds Dec. 15, 1905.

"In November, 1904, '05, '06 and '07, I sold at my place several thousand southern calves and yearlings. They came covered with the Texas fever ticks, were put in the barn with native cattle, and when sold were shipped as far east as New York, and south to Virginia. I never heard of a case of Texas fever that developed from exposure to these tick-infested calves and yearlings that left Texas for the north after Nov. 1. They invariably fed well.

"For a number of years I fed Hereford calves and yearlings from the noted '6666' herd, bred and owned by Mr. S. B. Burnett. These yearling and two-year-old steers have been exhibited at the Chicago International, and have made quite a record. A number of times the two-year-olds were the champion Herefords of the show, and at the International of 1913 the '6666' Hereford yearlings that I fed were the champion Hereford yearlings of the show. They averaged 1,100 pounds and sold at $10.45.

"All the Herefords I have sold I have handled in very much the same way; I have had them about

Photos by Hildebrand

TWO ENDS AND A MIDDLE—PRIZE LOADS AT THE INTERNATIONAL LIVE STOCK EXPOSITION.

11½ months on feed and about 10 months on full feed. The calves and yearlings have made about the same average gains—from 600 to 700 pounds. I feed principally corn and cob meal and linseed or cottonseed meal. I have fed and exhibited and won first in class with Hereford calves bred in Wyoming and Colorado. I think the heaviest load of Hereford yearling steers ever shown at the International were Colorado-breds that I fed in 1907. These steers had a foundation of Shorthorn in their breeding and averaged 1,270 pounds.

"I find the Wyoming- and Colorado-bred calf has more bone and scale than the Texas-bred calf, but the growth and development of the southern calf coming to the higher altitude of the cornbelt is more noticeable than those of the Wyoming or Colorado calf coming to a lower altitude. I am now feeding 20 Matador yearlings for the International of 1914. These steers were champion Hereford yearling feeders at last International, their weight then being 818 pounds. They have been on full fed since March 10; they averaged on May 1, 1,180 pounds.

"The average feeder of course is not interested in the production of show steers, but with the increasing demand for lighter cuts of prime beef, the prevailing high prices of all feeding cattle, and the high average cost of all feed products that enter into beef production, the successful feeder of the future must consider early-maturity, economy of gain and value of product, and where these are considered the Hereford calf or yearling, whether range-bred or farm-raised, for the cornbelt feedlot has no superior.

"With the present-day demand for lighter cuts of prime beef and the increased advance of the initial cost of all our feeding cattle, and a high average cost of all feed products that enter into the production of beef, the successful feeder of the future must seriously consider the question of early-maturity, cost of production and value of product produced. When these questions are considered, the favor in which the Hereford calf, whether range-bred or farm-raised, is held by the cornbelt feeder is merited, from the fact that for early-maturity, economy of production and value of product the Hereford has no superior."

The Polled Herefords.—That hornless cattle are popular among feeders goes without saying. The polled characteristic is certainly not without its decided advantages. Indeed, this fact has had much to do with the success met with by the Aberdeen-Angus and Galloways in this country. Dehorned Herefords are common at the stockyard markets and in our great fat cattle shows.

At the time Gov. Simpson made his selections of pedigree cattle in Herefordshire for importation to the States, he was also buying Aberdeen-Angus. In fact, Messrs. Gudgell & Simpson originally had one of the best collections of black polls in the west. Gov. Simpson at that time endeavored to locate a naturally polled white-faced bull somewhere in Herefordshire for importation for experimental purposes, but in this was not successful. Had he succeeded, the polled Hereford might have become a feature of American cattle-breeding at a much earlier date.

In the year 1901 Mr. Warren Gammon, Des Moines, Ia., circularized the members of the American Hereford Cattle Breeders' Association, asking if any hornless "freaks" had ever appeared in their respective herds. As a result of this correspondence 14 head of registered Herefords that had never developed horns were brought to light. They were the property of reputable breeders, so that there could be no question as to their pure descent. Ten of these were females, and four were bulls. Mr. Gammon bought all of the latter and seven of the cows, and began mating them, with the result that practically all the calves dropped were polled; and it is stated that these polled bulls when mated with horned Hereford cows gave 50 to 75 per cent of hornless calves. This was the beginning of the Polled Herefords of the present time.

A national organization was formed about 1907, with a membership of five. In 1913 this had grown to 296, and the herd book now maintained by the organization shows a total registry of over 4,000 head. Mr. J. E. Green, Muncie, Ind., has been the President of this association from the beginning. Up to 1911 the founder of the type, Mr. Warren Gammon, served as Secretary, but he has now been succeeded by his son Mr. B. O. Gammon, who estimates that there are at this writing between 5,000 and 6,000 head of these cattle in the country. The movement corresponds identically with that which resulted in the foundation and formation of the Polled Durham association, whose members are handling hornless cattle of the pure Shorthorn or Durham blood, both cases representing an effort to dispense with the horned characteristic of each breed without resort to artificial means.

Recent Importations.— In a previous chapter there was presented a tabulation covering the early importations from England, and extending down through the entire period of active buying on the other side by American breeders. After the $100 registration fee on imported cattle was put in effect importations practically ceased for a time. This so-called "tax" was repealed by the association in 1891. From the accompanying supplementary

IMPORTATIONS OF HEREFORD CATTLE TO THE UNITED STATES FROM ENGLAND SINCE 1893.

Date	Importer	Address	Number
1893	H. H. Clough	Elyria, O.	3
1893	Gudgell & Simpson	Independence, Mo.	1
1898	C. S. Cross	Emporia, Kans.	43
1898	K. B. Armour	Kansas City, Mo.	85
1898	Shadeland Stock Co.	Lafayette, Ind.	1
1899	K. B. Armour	Kansas City, Mo.	152
1899	W. G. Busk	Coleman, Tex.	17
1899	Geo. Leigh	Aurora, Ill.	70
1899	A. J. Libby & Son	Oakland, Me.	1
1899	John Sparks	Reno, Nev.	8
1899	T. F. B. Sotham	Chillicothe, Mo.	1
1900	F. A. Nave	Attica, Ind.	30
1900	John N. Taylor	Huntsville, Mo.	6
1900	A. E. Reynolds	Denver, Colo.	38
1901	K. B. Armour	Kansas City, Mo.	219
1901	W. G. Busk	Coleman, Tex.	11
1901	C. A. Jamison	Peoria, Ill.	1
1901	Geo. Leigh	Aurora, Ill.	25
1901	T. E. Miller	Oak Park, Ill.	2
1901	F. A. Nave	Attica, Ind.	7
1901	T. H. Pugh	Carthage, Mo.	6
1902	Geo. Leigh	Aurora, Ill.	107
1902	W. B. Tudge	Craven Arms, Salop, England	13
1903	Charles W. Armour	Kansas City, Mo.	112
1904	A. R. Firkins	Worcester, England	1
1913	Geo. Leigh	Aurora, Ill.	50
Total			1,010

tabulation it will be observed that during the decade ending with 1913 but one animal was imported. More recently there have been indications of a possible revival of this business, although not on any ex-

PRIZE-WINNING RANGE-BRED HEREFORDS AT THE INTERNATIONAL SHOW.
Photographs by Hildebrand

tended scale. The importation made by Mr. Leigh in the summer of 1913 included the good show bull Farmer, purchased and now owned by Mr. McCray. Mr. Leigh undertook an additional importation during 1914, but the embargo laid on exports by Great Britain on account of the European war interfered with purchases and shipments. A special license for the exportation of pedigree animals has to be obtained from the Privy Council pending the termination of hostilities.

Distribution of the Herefords.—Believing that an approximate general idea of the distribution of pedigree Herefords in different parts of the United States would be of interest, the author requested Secretary Kinzer of the Hereford association to prepare some figures covering this point. These are submitted herewith. The figures are based on

APPROXIMATE NUMBER AND PERCENTAGES OF REGISTERED HEREFORDS IN THE VARIOUS STATES.

State	Per cent	Number	State	Per cent	Number
Iowa	16.5	19,800	Arizona	1.	1,200
Texas	14.	16,800	New Mexico	0.75	900
Missouri	11.5	14,800	Oregon	0.75	900
Kansas	10.	12,000	California	0.75	900
Illinois	9.	10,800	Wisconsin	0.75	900
Nebraska	6.	7,200	Tennessee	0.5	600
South Dakota	5.	6,000	Utah	0.5	600
Indiana	3.	3,600	Maine	0.5	600
Wyoming	2.5	3,000	Virginia	0.5	600
Minnesota	2.2	2,640	Michigan	0.5	600
Colorado	2.2	2,640	Idaho	0.5	600
West Virginia	2.	2,400	Washington	0.25	250
Montana	1.5	1,800	Arkansas	0.25	250
Kentucky	1.2	1,400	Mississippi	0.25	250
North Dakota	1.	1,200	Nevada	0.25	250
Oklahoma	1.	1,200	North Carolina	0.25	250
Ohio	1.	1,200			

an estimate of a total of 120,000 head of registered cattle now living, and include all states in which there are 250 head or more.

Some Interesting Tabulations.—It had been the purpose of the writer to undertake at this point a detailed analysis of the various bloodlines and combinations entering into the production of the greatest Hereford sires and show bulls of these latter days, but again we are faced with the fact that such endeavor must be deferred from an absolute lack of space. All that has preceded leads up logically to such a procedure, by way of drawing conclusions based upon the lessons of the sales, shows, breeding operations and importations herein recorded, but this must now be reserved for another occasion or for such students as may see fit to work them out from the mass of facts contained in the foregoing chapters. There are several other matters still to be touched upon, and we have either to end this part of our study here or face the necessity of beginning another volume. This one is already too fat. We shall, therefore, content ourselves with merely inserting at this point a few tabulations that suggest themselves as of special interest at this time in connection with contemporary sale and showyard events.

PERFECTION 92891.

```
              ┌ Columbus   ┌ Earl of Shadeland 41st 33378 ┌ Garfield 7015
              │ 51875      │                              └ Gertrude Wilton 19017
Dale 66481..  │            └ Pet 36054..................  ┌ Prince Edward 7001
              │                                           └ Jessie 4th 10907
              │ Rose Blos- ┌ Peerless Wilton 12774......  ┌ Garfield 7015
              └ som 39225  │                              └ Peerless 10902
                           └ Blossom 12866..............  ┌ Auctioneer 9572
                                                          └ Blowdy 12867

              ┌ Hoosier Tom┌ Anxiety 2d 4580............. ┌ Anxiety 2238
              │ 7732       │                              └ Alice 4858
Melley May    │            └ Isabel 4577................. ┌ Corsair 4581
  41752       │                                           └ Juliet 4578
              │ Rosebud 6690┌ President 2058.............  ┌ Chieftain 2059
              └            │                              └ Leonora 2060
                           └ Carabassett Rose 2120......  ┌ Kennebec Hero 2100
                                                          └ Necklace 6th 2105
```

THE "ROUND-UP"

FAIRFAX 16th 310931.

- Perfection Fairfax 179767
  - Perfection 92891
    - Dale 66481
      - Columbus 51875
      - Rose Blossom 39225
    - Melley May 41752
      - Hoosier Tom 7732
      - Rosebud 6606
  - Imp. Berna 138482
    - Fairfax 84159
      - Salisbury 84174
      - Decima 84153
    - Belle 138483
      - Leander 83622
      - Lavender 111575
- Cherry Donald 189271
  - Beau Donald 33d 109867
    - Beau Donald 58996
      - Beau Brummel 51817
      - Donna 33735
    - Sir Carroll's Earl Grove Maid 56110
      - Sir Carroll 2d 40067
      - Earl's Grove Maid 46193
  - Mary's Cherry 52077
    - Cherry Boy 26495
      - Fowler 12899
      - Cherry Pie 2d 17849
    - Lady Mary 4th 36936
      - Beau Monde 9903
      - Lady Mary 2d 24493

REPEATER 7th 386905.

- Repeater 289596
  - Distributor 176433
    - Disturber 139989
      - Beau Donald 3d 86140
      - Columbia 76779
    - Elfin Lass 106907
      - Kansas Lad Jr. 75104
      - Shadeland Elfin 51367
  - Mina 184985
    - Missouri Chief 2d 104368
      - Hesiod 17th 56467
      - Missouri 41560
    - Evelyn 126208
      - Rajah 91721
      - Pretty Face 88462
- Harris Princess 31st 266423
  - Beau Donald 5th 86142
    - Beau Donald 58996
      - Beau Brummel 51817
      - Donna 33735
    - Sophia 56115
      - Sir Carroll 2d 40067
      - Earl's Lilian 2d 46194
  - Lottie Macon 139290
    - Sir Macon 63693
      - Valentine 46544
      - Portrait 12245
    - Queen Bess 92820
      - Norwood Chief 70814
      - Gipsy Maid 4th 60573

BEAU PERFECTION 24th 394173.

- Perfection 92891
  - Dale 66481
    - Columbus 51875
      - Earl of Shadeland 41st 33378
        - Garfield 7015
        - Gertrude Wilton 19017
      - Pet 36054
        - Prince Edward 7001
        - Jessie 4th 10907
    - Rose Blossom 39225
      - Peerless Wilton 12774
        - Garfield 7015
        - Peerless 10902
      - Blossom 12866
        - Auctioneer 9572
        - Blowdy 12367
  - Melley May 41752
    - Hoosier Tom 7732
      - Anxiety 2d 4580
        - Anxiety 2238
        - Alice 4658
      - Isabel 4577
        - Corsair 4581
        - Juliet 4578
    - Rosebud 6606
      - President 2058
        - Chieftain 2059
        - Leonora 2060
      - Carrabassett Rose 2120
        - Kennebec Hero 2100
        - Necklace 6th 2105
- Belle Donald 11th 287191
  - Beau Donald 44th 109865
    - Beau Donald 58996
      - Beau Brummel 51817
        - Don Carlos 33734
        - Belle 24629
      - Donna 33735
        - Anxiety 4th 9904
        - Dowager 6th 6932
    - Cinderella 61048
      - Sir Carroll 2d 40067
        - Earl Shadeland 9th 16900
        - Elena 5th 27141
      - Wilton's Grove Maid 2d 51158
        - Earl 2d of Pine P'rk 41061
        - Wilton's Gr've Maid 33276
  - Belle Donald 76th 187362
    - Beau Donald 58996
      - Beau Brummel 51817
        - Don Carlos 33734
        - Belle 24629
      - Donna 33735
        - Anxiety 4th 9904
        - Dowager 6th 6932
    - Minnie H. 61053
      - Sir Carroll 2d 40067
        - Earl Shadeland 9th 16900
        - Elena 5th 27141
      - Lily Princess of Pine Park 46199
        - Earl of Pine Park 41060
        - Lily Princess 26729

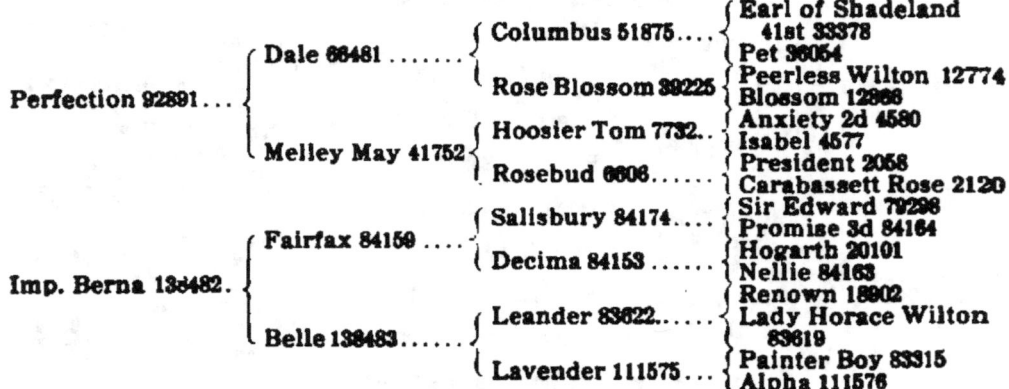

About Beau Donald.— The history of cattle-breeding abounds in surprising developments in connection with the careers of various celebrities. A number of these tales have already been related, but none is of deeper interest than the story of old Beau Donald's early history and subsequent extraordinary success as a sire. Some details are therefore submitted.

Mr. Charles Gudgell says:

"Beau Donald had no nurse cow nor special fitting as a calf, but was treated in exactly the same way as our other calves at the time. As a calf he was remarked for his generally sappy appearance and make-up, and was naturally thick-fleshed without necessarily being fat. He was built close to the ground and had a hide on him as the saying is, 'like a bull pup's'. At a little past a year old he gave promise of having a decidedly drooping horn, a characteristic which along with natural thick flesh he has transmitted to his descendants in a remarkable degree.

"He came from one of our very best families. His sire and dam, grandsires and grandams were all extra good. Beau Brummel's record as a breeder is well known, but when it is noted that Donna 33735, the dam of Beau Donald, is out of Dowager 6th 6932, a cow that was also the dam of Don Carlos and Don Quixote, it can readily be seen that he was no accident, but came by his good qualities honestly. Dowager 6th was bred by Mr. T. Lewis, Woodhouse, the well known English breeder from whom we secured her, and was one of the best cows in all our importations. She was a very smooth, medium-sized cow and had a decided droop to her horns (a feature we desired to cultivate) and transmitted the same characteristic to all her produce. Donna 33735, mother of Beau Donald, also had a pronouncedly drooping horn. She was a straight-lined, low-down cow of the breedy type and above the average size and weight. She died at the age of seventeen years."*

There was nothing special to be observed in the make-up of the calf in his earlier days at the side of his dam, but as he came along into bullhood he was picked up by H. B. Watts, a man who had ever a

---

*The somewhat unsatisfactory portrait of Mr. W. H. Curtice appearing in this volume shows him mounted on his favorite saddle horse, Champagne, so called because of his peculiar color. Like all Kentuckians, Mr. Curtice has a fondness for a good saddle horse, and in speaking of Champagne he states that he can go a running walk at a 9-mile-an-hour gait "without shaking the rider in the least." Moreover, this horse is a successful weight-carrier, having won several prizes over large fields in the blue-grass shows with Mr. Curtice up.

W. H. CURTICE AND HIS FAVORITE MOUNT—CHAMPAGNE.

BEAU DONALD—FROM THE DRAWING BY CECIL PALMER.

keen mind and eye. While it was therefore not mere chance that guided him in this selection, it is doubtful if his vision was prophetic enough accurately to forecast the youngster's brilliant future. Certainly Gudgell & Simpson would not have let him go had they been able to read the stars aright. They knew he was one of the best yearlings they had ever bred and really intended to reserve him for their own use, but the trade was then passing through a period of acute depression. Judge Watts apparently caught Gov. Simpson napping one day and bought the youngster for $125! Watts' own story of Beau Donald's "discovery" and purchase as told to the writer is substantially as follows:

"Some time in April, 1914, I went to Independence and told Gov. Simpson I had come to select a yearling bull to head my herd, and asked him what he had on hand. He replied, 'Fifty-six as good yearling Hereford bulls as anybody in America, but they are so d—m cheap and low that I propose to make steers of them rather than sell at prevailing prices.' We then got in his old spring wagon and drove out to the farm. Going into the pasture where the bulls were grazing, he said, 'Now, Watts, there's the lot. Pick your bull, and I'll see if you know a good one.'

"The Governor had certainly stated facts when he said they were a good lot of yearlings; indeed, they were exceptionally good. After looking over the lot for fully an hour I finally selected two bulls, and asked the Governor to show me their dams and give me the sires of the two before I would determine which calf I would take. We drove over into another pasture where the cows were quartered. He drove close to a massive cow and said: 'This is Donna, the dam of the smaller bull of the two you picked. She is by Anxiety 4th and out of Dowager 6th, which makes her a full sister to Don Carlos.' I said, 'Governor, we won't look for the larger bull's dam; I want the small calf. What is your price on her?' He replied: 'Watts, confound your little picture! You've picked the best bull in the bunch, a bull that I thought of keeping to breed from ourselves. However, I like you and you can take the calf at $125. I told him I thought the figure a little high, as low as cattle were selling at that time, that I had thought when I left Fayette that $100 would buy the top. He said: 'I can't take it, that is one of the best calves I ever bred. I will take $100 for the other calf you picked.' 'No,' I said, 'I want this calf.' He replied, 'Well, if you promise not to tell what you gave for him you can take him, and if he don't make one of the most impressive sires in the country, send him back, get your money, and Simpson will say he has no cattle sense.'

"Beau Donald was shipped to Fayette the following week, and it was then that I first learned his name, as I had neglected to ask it of Gov. Simpson. A week or so after getting him rested I led him to the courthouse yard in Fayette so that my cattle friends could see him. All thought him a wonder. From the time I bought him until sold to Mr. Curtice I never saw or knew him to attempt a vicious act.

"Beau Donald was intensely bred in the renowned Anxiety 4th blood, and came naturally to transmit so unerringly to his offspring his marvelously straight lines, strong front and superb quarters, together with all his elegant style, finish and symmetry. As a yearling he was bred to eight or ten cows, getting all with calf. He was kept in good thrifty growing condition until two years old, and then my son Will (now my partner) concluded to develop him. He fed well from the start, and in October, 1895, at thirty-one months old, he weighed 2,200 pounds. In fact, at that time and for two years afterwards he was a great show bull. On Aug. 12, 1897, I received a wire from Secretary Thomas at Kansas City, asking me what amount would buy Beau Donald. Not caring to sell him at any price, I concluded to put the price so high no one would think of paying it, cattle at that time being extremely low. I priced him at $1,000.

"The next day Mr. Curtice came to my farm, and told me he had asked Mr. Thomas to see the bull. My son brought him to the lawn, and I don't think the old fellow ever did or ever could make the show he did that afternoon. Curtice looked him over and said, 'Turn him out, I never saw a bull I'd pay $1,000 for.' I replied, 'Very well, I'm glad to hear you say that; and rather than sell you Beau I'll give you a check for $100 to release me from my proposition.' He had me take him to the train and just as the whistle sounded for the station Curtice said, 'Put Beau in the barn for a few days, and then ship him to my address in time for him to reach the fair at Shelbyville on Aug. 24.'

"I thought so much of Beau Donald that for fear he would get hurt in transit I got in the car and rode through to Kentucky with him, feeding, watering and keeping the bedding under him. As I frequently said to Mr. Curtice, 'I am the architect of your fortune.'"

This interesting narration may be supplemented by a brief statement from Hendry, Mr. Curtice's former manager, who says:

"When a calf Beau Donald was recognized by Mr. Simpson and herdsman George Shand as a good mellow-hided, short-legged calf. Shand, my old Scotch friend, when showing him to any one, always caught hold of his hide to reveal the looseness of it. I can remember seeing him first on a visit to Independence from the Greenwood Farm in the fall of 1893, when he was still nursing his dam. He was weaned when about seven months old. The following summer he was with a bunch of bulls Mr. Simpson used his knife upon, passing by Beau Donald.

"The bull landed at Shelbyville the week of the fair there. Mr. Curtice started to take him home, but being advised by some of his friends to show him had his old darkie, Joe, take him to the fair grounds, where he received the blue ribbon next day and sweepstakes over all breeds, although some of the Shorthorn men told Mr. Curtice he was no breeding bull, but simply a mess of beef. These same Shorthorn men say today, however, that he was one of the greatest breeding bulls that ever came into Kentucky. The following year Mr. Curtice showed him at Lexington, Ky., Shelbyville, Ky., Columbus, O., and Wheeling, W. Va., winning first place with him at the places named. In 1899 he showed him on the same circuit with about the same results. He then took him the same fall to the Kansas City show and sale with a bunch of his calves, which averaged $629 a head, and the following year did about the same, always bringing in a ribbon from the big shows with the old bull. After that Mr. Curtice decided to keep him at home and show his get.

"Beau Donald was never used very hard, having about forty cows a season. He always ran in the pasture with the younger bulls of which he was king. This reminds me of an anecdote of the old bull. On bringing him home his son, Prince Rupert

79589, from the show circuit we turned him into the pasture with his sire from whom he had been separated for three months. I accompanied them to the pasture, being a little afraid that the young bull would be too much for his dad; but to my surprise it went the other way. The old bull rolled the Prince over, and held him to the ground, I scarcely being able to pull him off by the ring, and from that day Prince Rupert never ventured within twenty-five yards of him.

"Beau Donald's disposition was good. Any child could handle him by the ring. He was always ready for the feed bucket, which is also a trait of his descendants."

A Story of Perfection Fairfax.—Mr. A. C. Huxley bought the imported cow Berna in calf to Perfection at the Hoxie sale at Thornton, Ill., May 14, 1903, for $365. She was due to calve in October and as the cow was a good individual and a heavy milker and as Perfection had been many times champion Mr. Huxley had a right to expect something particularly choice. The calf arrived on the first of October, and was about as disappointing a prospect as could be imagined. In the first place, he was small and so crooked that herdsman Willard Pierce used to put one hand on one of the calf's knees, take hold of the foot and push on the knee and pull on the foot until the tendons fairly cracked, all in an effort to straighten the legs of the future champion. This was done twice a day for two weeks, and in the course of about six weeks his legs were about normal.

This, however, was not the only difficulty with

PERFECTION FAIRFAX. Photo by Hildebrand

PERFECTION AND BEAU DONALD 86th. Photo by Hildebrand

the youngster. He was a disgusting light yellow-red in color. Now this pale red has from time almost immemorial been tabooed by Hereford breeders. While Pierce was still operating on the legs of the "future great" Mr. Huxley had a visit from Mr. E. W. Furbush, a piano manufacturer of Boston, Mass., who had a farm at Freedom, N. H. He was a reader of "The Breeder's Gazette" and became an admirer of good Herefords. He arranged a visit to Mr. Huxley's farm, desiring to secure a cow with a bull calf at foot. The proprietor offered to trade imp. Berna and calf for a $500 piano, but the visitor replied that he "would not have a cow on his farm that would not produce a better calf than that." Mr. Furbush was not alone in having a poor opinion of the calf at this time.

Huxley was expecting to show a full herd at the fairs of 1904 and had two other bull calves that were receiving extra attention with a view toward exhibition. Meantime Perfection Fairfax was running around with the other calves on the place quite neglected. As time passed, however, Huxley and Pierce engaged in arguments as to which of the two bull calves in preparation was the better, and finally Pierce remarked, "If you will get another nurse cow, I'll just take the yellow calf and beat all the others." The nurse was finally procured and under the stimulus of this additional milk the young bull began spreading out into fine form. Best of all, in the spring as he was shedding his coat, lo and behold, he was a fine rich red!

This was of course a very welcome surprise to the owner. From that time there was never much question as to his destiny.

At the fall fairs of 1904, Perfection Fairfax was probably the heaviest calf in his class. At the International his official weight was 1,220 pounds. This is certainly remarkable in view of the late start he had. At the Kansas City Royal of 1905, just as Mr. Huxley was leading Perfection Fairfax out of the ring with the purple badge that meant the junior championship of the show, friend Furbush came up and congratulated him upon the winning, and his surprise can be better imagined than described when told that this was the calf that he had said was too mean to have around his New England farm. Mr. Huxley was offered $2,000 for Perfection Fairfax as a calf and $3,000 as a yearling; he declined both propositions. Out of a possible thirty-one first prizes shown for as a calf, yearling and aged bull he won thirty firsts and one second. As a two-year-old he was not shown. Mr. Huxley having disposed of Beau Donald 33d, his chief stock bull at that time, could not spare Perfection Fairfax from the breeding herd. Beau Donald 33d, it will be remembered, was also a most successful stock bull, the progenitor of the Lady Fairfax line, two members of which, Lady Fairfax 4th and Lady Fairfax 9th, made enviable records.

The later career of Perfection Fairfax is too familiar to call for extended comment in this connection. Messrs. Harris, Curtice and McCray were all interested in trying to secure him after Mr. Huxley decided to close out his herd, but McCray was the successful bidder. Mr. Huxley closed out his cattle, not because of any lack of interest in them, but in order to fulfill obligations elsewhere and in another field which he could not justly ignore. He is now located in Minnesota and has recently informed the writer that he expects to re-enter the business in the near future. Meantime, commenting upon the record of Perfection Fairfax, Mr. Huxley says:

"Several of the old breeders used to tell me that Perfection Fairfax was not the type to prove a great sire, a fact which goes to show how risky is the occupation of the prophet in such matters. I think the first calf that he sired was Diana Fairfax, which Mr. McCray showed, and she proved a winner. I do not believe there has been a bull of the breed that has been a winner himself and afterwards sired so many good bulls and heifers as has Perfection Fairfax. I claim that the reason why he is entitled to be called the greatest bull of the breed, living or dead, is because he sired both good bulls and good heifers all of his own stamp, so that almost any one can distinguish them from the get of other sires. I regard it as particularly fortunate for the breed that he fell into the hands of Mr. McCray, because in that herd he has been given opportunity to be mated with a grand lot of cows. I look back with pride on Perfection Fairfax, and have vivid recollections of the various hard-fought battles we had before reaching the top."

His Trainer Testifies.—Willard Pierce takes a keen enjoyment in the progress of the Perfection Fairfax stock. Here is his own account of the old bull's early career:

"Mr. A. C. Huxley bought at the Hoxie spring sale of 1903 three cows—imp. Berna, Dauntless and Miss Peerless. All were bred to Perfection. Dauntless and Peerless dropped heifer calves. Berna dropped a bull calf in October, 1903, which was named Perfection Fairfax. He did not look like much of a calf, as he was rather slim and thin. But in February he seemed to be developing into something better than we had expected, so we decided to put him on a nurse cow. He kept on doing well. His stablemate, Beau Dale 2d by Beau Donald 33d, was much the better calf to start with. But time went on and by May there was not much difference in them. By fall Perfection Fairfax had the best of him. We always raised a small patch of beets. Of these Perfection Fairfax was very fond, and I gave him plenty of them at all times. When he was shown at Chicago as a calf he was weaned. Not many of the show calves are weaned at that time.

"The first show we made was at Columbus, O., where he stood first and was junior champion. He stood that way all around the eastern circuit until Chicago, where he was second.

"The bull wintered well as a yearling, and I could fairly see him grow and widen out. He never had a sick day during the four years I fed him. His constitution was wonderful, and I believe that has much to do with making his offspring all good feeders and good show cattle. I believe constitution is the greatest thing to consider when we come to feed cattle for the showring. One thing more about Perfection Fairfax was that he was very quiet—what I would call a lazy bull. He would never fret about anything.

"When he was a yearling we went to some county fairs where he was champion, and then to Columbus, where he was first and junior champion. As a two-year-old Perfection Fairfax was not shown. We used him as our herd bull, but never let him down from show shape. As a three-year-old we showed him and came near cleaning up the most of the boys. They realized he was 'some show bull' at three years old, and by that time we knew he was 'some breeder'! But there is one thing very funny to me about Perfection Fairfax, and to think of it

Wm Condell   Willard Pierce   Andrew Meikle

often makes me smile. When Mr. Huxley and I showed him lots of the breeders would grant that he was a good show bull, but usually added that he would be no breeder. I meet these same breeders around the fairs now, and I often say in a joking way: 'Perfection Fairfax is a good show bull, but he'll make no breeder.' I usually spring this when I see his get head towards their stalls with blue ribbons tied to them, and sometimes the red, at our very biggest fairs.

"When Perfection Fairfax was a four-year-old

we never intended to show him. But then Mr. McCray bought out Mr. Huxley, and decided to show him instead of Prime Lad 16th, though he was never fed to be shown that fall. It is surprising to see him so fresh at eleven years old. And I doubt very much if any of the bulls which showed with him nine or ten years ago can come up and beat him now in the showring. I do know that they have a hard time to beat his get, although it was this same Perfection Fairfax which was 'a show bull, but would never make much of a breeder.'"

The Dam of Dale.—Too late for incorporation in our text at the proper point we received the following interesting recital from Clem Graves, the breeder of Dale, as to the mother of that famous bull:

"In appearance Rose Blossom was like the Garfields. Her color a rich red, neither light nor dark, her form thick-fleshed, evenly balanced and set on short legs, her face broad, with full eyes and the short 'Berkshire' nose, the identifying feature of the Garfields. She was sold to Mr. Harness, Galveston, Ind., when a yearling and her calves, Little Phil and Hopeful, were dropped at his farm. I purchased her in the spring of 1892, paying $75 for her. This was a time when the breeders were overstocked on account of the depression in the cattle trade in 1891 and 1892, when I saw a 1,200-pound grade Shorthorn cow sell for $14 and good yearlings sell for $8 a head.

"I showed Rose Blossom in 1896 at Toledo, and the leading county fairs in Ohio and Indiana. At that time many of the fairs did not classify the beef breeds, Shorthorns, Angus and Herefords all showing together. She was defeated but one time and then by a Hereford cow of great scale and even finish. Vivien, the first calf Rose Blossom brought me, I sold when six years old for $500. This cow was the mother of two heifers. One was Armel, bought by A. P. Nave for $220 in my first public sale at the Palmer House Stables at Chicago in 1897. He sold her a short time afterward to K. B. Armour for $1,000 and she was one of the attractive members of that celebrated herd. The other heifer, Viola by Columbus, I listed in the sale at the Kansas City Royal in 1899, where she was purchased by Col. Slaughter and Thomas Mortimer for $1,250, selling next to Armour Rose, which brought the highest price in that very spirited sale.

"You will note from the full list of the progeny of this famous cow, herewith appended, that three years elapse between the birth of Vivien and Dale. This is a long period for a young cow to pass not breeding and I was almost at the point of disposing of her, all in ignorance that the story of Dale, Perfection, Perfection Fairfax, on down to Joan Fairfax, was already written 'among the things that are and the things that shall be hereafter.'

"The story of Dale is now a matter of common knowledge. There are two points in his showyard career that I wish to emphasize: in 1897 he was the first Hereford to win the grand championship, all breeds competing, at the Ohio and Indiana state fairs; and after five years of showing when he was matched against Perfection, Christopher, Dandy Rex and many other famous bulls he won the purple ribbon in his final appearance in the showring at the International in 1901. Dale rests in the burying ground of the little Christian church on Jesse Adams' farm, his grave cared for by Amy Adams.

"Mr. Adams purchased Columbia from me for $1,100 and Columbia 2d for $1,325 in a breeders' sale in Kansas City in January, 1901. He bred Columbia to Beau Donald 3d, bringing Disturber. This was the initial trial of the Columbus-Beau Donald cross. This same line of breeding is now in practice in the herd of Mr. Curtice of Kentucky. I sold Rose Blossom in 1902 for $500, to the Wabash Cattle Co. Her record follows:

"1890—April 21—bull—Little Phil 41937 by Earl of Shadeland 41st 33398.

"1891—April 18—cow—Hopeful 46919 by Earl Wilton 47th 46333.

"1892—Aug. 26—cow—Vivien 51183 by Earl of Shadeland 33398. This heifer became the dam of Armel and Viola, sold to G. McWilliams in 1898 for $500.

"1895—Sept. 15—bull—Dale 66481 by Columbus 51875. Sold to F. A. Nave, Nov. 5, 1897, for $1,100. Bought in Mr. Nave's dispersion for $7,500. Sold in 1901 to Wabash Cattle Co. for $8,000.

"1896—Sept. 18—bull—Earl Wilton 69585 by Columbus 51875. Sold to A Wolcott, Concord, Mich., in 1896. No record of price, I think $200.

"1897—Oct. 4—cow—Columbia 76779 by Columbus 51875. Sold to J. C. Adams, 1901, for $1,000. The dam of Disturber.

"1898—Aug. 16—cow—Columbia 2d 86594 by Columbus 51875. Sold to J. C. Adams, 1901, for $1,325.

"1899—Aug. 21—cow—Rosamond 100492 by Cherry Ben 56767. Sold to Wabash Cattle Co., 1902, for $300.

"1900—Dec. 25—bull—Dara 117715 by Imp. Freedom 76005. Sold to Wabash Cattle Co. for $135.

"1901—Nov. 13—cow—Rose Blossom's Princess 135358 by Le Roy 70778. Owned by S. H. Godman."

Harking Back to the Range.—As this chapter is called a "Round-Up" we feel warranted in introducing at this point certain facts concerning the operations of another one of the ranking western outfits of the days when the range cattle business was getting squarely upon its feet. We refer to the famous old-time firm of Lee & Reynolds. Their operations were not only extensive, but they always advocated the use of well bred bulls. The author had prepared the following statement as to their work for inclusion in a preceding chapter, but the copy was inadvertently mislaid until it was too late to incorporate it, in this edition, in its proper place. It is deemed of sufficient historical

THREE-YEAR-OLD HEIFER MAY MORN, A ROYAL WINNER IN 1913.

MOTHER AND SON.

importance, however, to be given space at this point, rather than be omitted entirely.

This firm started business at Camp Supply in the Indian Territory in the fall of 1869 as post traders at that military camp. Their consignment of six or seven cars of merchandise from New York was the first through freight which crossed the bridge at Kansas City, shipment having been timed to reach there just at the opening of the bridge, and was sent through from New York to Hayes City without breaking bulk.* The firm continued business at Camp Supply and at other points established and owned by them in the Territory, and at Fort Elliott in the Panhandle of Texas—there being four or five of these stations—until the year 1881, when the co-partnership was dissolved and the property divided. During this time in the Territory there were two Indian wars, one being the final round-up of the Cheyennes, Arapahoes, Kiowas and Comanches, upon their reservations at El Reno and Fort Sill, where the red men and their descendants have ever since remained peaceably occupied and intent upon the struggle for existence.

Lee & Reynolds commenced establishing a herd of cattle about the year 1876, placing in service purebred Shorthorn bulls purchased of Frederick William Stone of Guelph, Ontario. This first herd

---

*Mr. A. E. Reynolds, following up this shipment, in company with one man similarly mounted, rode a mule from Hayes City to Camp Supply, 180 miles, sleeping on the ground en route, with the saddle for a pillow and the blankets used under the saddle as his bedding; and as there had been a blizzard across the country at that time, he has a very vivid recollection that it was rather a painful cold journey, and somewhat limited as to commissary.

was sold in 1880 to Mr. Bud Driskell, and about that time Mr. Lee purchased for the firm a tract of land on the Canadian River at or near the New Mexico line from Messrs. Gunter & Munson, and contracted for a herd of cattle from Reynolds & Mathews of Albany, Tex. These cattle were the beginning and basis of the herd carrying the LE brand which afterwards became well known. In the dissolution of the firm and division of the property, this herd and the lands were taken by Mr. Reynolds, and the Reynolds Land & Cattle Co., was formed to operate at that point in the Panhandle. Other lands were accumulated to a total of over 250,000 acres; the herd was enlarged to the capacity of the ranch, or about 12,000 head, and was bred up by the use of purebred bulls exclusively, and by care and attention the quality of the herd was improved until it was probably excelled by none. The first bulls used were purebred Shorthorns, mostly from Mr. Stone's herd, and these were followed by purebred Herefords from the same source. Others were obtained from Messrs. Gudgell & Simpson, and firms of like prominence.

This company operated until 1902, when the lands were sold through their manager to the Prairie Cattle Company of Edinburgh, Scotland, and the cattle to Mr. J. J. Hagerman, and moved to a point near Roswell, N. M. All of the cattle were so disposed of, excepting a few hundred purebred Herefords which were brought to some lands owned by Mr. Reynolds near La Junta, Colo. This herd was run there a few years, when the better portion of it was sent to the mountains near Delta, Colo., where they are now a part of the herd of the E. J. M. Cattle Company at that place, in which Mr. Reynolds is still interested.

During the operations in the Panhandle of Texas the firm bought at one shipment fifty registered bulls from the principal Hereford herds of England. These were turned upon the range. In reply to an inquiry from the author as to the results of the use of the Shorthorn and Hereford bulls upon the LE herd, Mr. Reynolds says:

"The cattle bearing the LE brand in the Panhandle were originally Texas cows from the vicinity of Albany, Tex. The first cross on these cows was made with Shorthorn bulls. We used these bulls about two or three years in the herd before they were displaced to any great extent with Herefords, so that we might say the basis of the herd was a cross of Shorthorn blood on the Texas cattle, followed by the Herefords on these cows. There can be no question but what we got good results from this method.

"I noticed evidence in the herd for many years of the original use of the Shorthorn bulls, even after we had been using exclusively Herefords in the herd. I think the chief evidence of the breeding back to the Shorthorn blood, was shown in the diminished size of the horns of our white-faced cattle, occasional Shorthorn markings on the bodies, such as roans, and occasionally a red spot on the face or nose, which clearly showed the Shorthorn strain. We changed to the Herefords on the theory that they were the more hardy animals for range use, as well as being in themselves as good if not better than the Shorthorns. I certainly should at this time prefer to take my chances with the Herefords if I were going into the cattle-raising business, chiefly on account of the probability of their greater endurance."

Westward Ho!—With the passing of the open range the establishment of real "quality" herds in the newer west bids fair to result in a material extension of the field of pedigree Hereford breeding in the United States. The character of the herds that have been founded in recent years in the Rocky Mountain region, as illustrated by the exhibits at the Denver show, indicate clearly that the production of top cattle of this favorite western type is likely to become a large and important industry in connection with the further evolution of the cattle trade of the mountain and inter-mountain states.

Typical of this new condition is the case presented by the persistent purchase of cattle of the highest class by Mr. A. B. Cook. He is engaged in concrete construction work in a large way in Canada and our own Northwest, but has lived in Montana since the early '80's and knowing what has been accomplished in the past by the Herefords on the open ranges of that region, he finally decided to assemble a herd of the best registered "white faces" obtainable, in order to see what Montana climate, Montana bluegrass and alfalfa hay, Montana oats and barley and Montana sugar beets, coupled with proper care and attention, would do for the highest type of modern American-bred Hereford. His great enterprise in securing valuable material for this purpose is deserving of all praise, and at our request he furnishes the following statement as to his operations:

"Look back a few years and remember the range cattle shipped from Montana—cattle three to four years old that had never even tasted hay until they were en route to the eastern market. Could you beat them any place on earth, raised under similar conditions?

"My first effort in connection with the establishment of a registered herd was to buy the best foundation stock procurable. I visited most of the dispersion sales for several years, buying only cows that were outstanding, and all as near one type as possible. First I would select from pedigrees, going through the catalogs thoroughly and selecting only cows whose breeding suited me. I then passed on them as individuals. If the cow came up to my standard I invariably bought her in the ring, if not, no matter how cheap she went, I would not buy. It was quality I had to have. Often I would only find one cow that suited me at a sale, other times two or three, and sometimes a carload. I had my standard and the cows I bought must come up to it.

"My greatest problem was the selection of herd bulls. I was fortunate in my first selection—Beau Carlos 248915, one of the greatest of the breed, sire of the grand champion Joy and other winners. Up to this time I have never sold but one of his heifers and I have since tried to re-purchase her. The selection of the balance of my herd bulls caused me considerable trouble. I traveled thousands of miles, inspecting many animals. The right breeding I wanted first, then conformation and type. This combination was hard to find, but I finally succeeded, far beyond my expectations, in securing Fairfax 16th 316931, undefeated grand champion bull of America during 1912, Beau Perfection 9th 368012, Beau Perfection 23d 394172, Premier 2d 311882, and Standard 11th 411222.

"We try to keep our breeding cows in good thriving condition, breeding them to the bull with which we think they will nick best. We are not always successful in this, but the percentage is good and we keep trying until we do hit it. We try to keep our calves growing. The calf, once stunted, never regains what it has lost and never makes as good an animal as if it had been kept growing and developing steadily.

"No enterprise that I have ever been connected with has given me greater pleasure than the raising of purebred cattle. When it comes to real enjoyment here is where I get mine."

A Word About Herdsmen.—The author cannot bring this volume to a close without a word on the subject of the men who have the actual care of herds. Upon them rests to a marked degree the responsibility for success or failure. We have taken pleasure all the way through this narration in referring from time to time to various individuals who have contributed largely toward the accomplishments of the cattle of their employers. Unfortunately in a volume such as this it is quite as impracticable to mention by name all those in this profession who are really entitled to this recognition as it has been to make a record of the operations of all owners and breeders of good "white faces."

A number of portraits of some of the better known

herdsmen are presented. Others would have been shown had the author succeeded in obtaining the necessary photographs. In several notable instances most capable men have shrunk from this publicity, even though it was unsought on their part. It will of course be understood that the portraits shown have been made from photographs supplied at the special request of the author. A long period of observation of the work of men of this type on both sides the water has convinced us of the justice of generous recognition of their efforts. While as a rule they receive full consideration at the hands of their employers and of cattle breeders in general, the author has such a keen appreciation of the importance of the part they play in the work that he believes frequent note should be made of their valuable services.*

An Involuntary Tribute.—One more little story and we close. As is commonly known, most of the

---

*We are reminded at this point that we have not yet supplied certain details concerning the work of George Mason, another of the "old guard" of good herdsmen. He was born in Aberdeen, Scotland, Sept. 2, 1852, worked with Angus herds in Scotland for about ten years and came to America in 1882, bringing over an importation of Angus cattle for Geary Bros. in Canada. He spent three years with the herd of Mossom Boyd and came to the United States in 1889 to the Hereford herd of C. H. Elmendorf, Kearney, Neb., where he remained six years.

At the Chicago World's Fair the show herd in George's charge included Earl of Shadeland 30th, Lily, Lady Daylight and Lady Laurel. In the fall of 1895 he went to the herd of W. S. VanNatta and in the fall of 1897 to C. S. Cross, Emporia, Kans., thence to John Hooker, New London, O., for about eighteen months. After that he spent several years with Angus and Shorthorn herds and later a year with the Hereford herd of John E. Painter, Roggen, Colo. He now has charge of the Herefords on Highland Ranch, owned by K. H. Zwick, near Pyramid, Colo.

Fred Corkins, who tended Dale and other celebrities for Jesse Adams, ought not to be forgotten, and among contemporary workers we should not omit to mention William Burlton, now with Mr. Tow; "Andy" Meikle, who went from Cudahy to Harris, and "Bob" Johnson, who has made up many Hereford as well as black polled champions.

cattle buyers for the packing houses at the stock yards have a special fancy for Aberdeen-Angus bullocks. While good bidders for prime Herefords, they sometimes like to find a little fault. In this connection the following incident is of interest as illustrating this inclination on the part of some of those who ride the alleys at the yards in quest of good steers. It happened at the International some years ago at the time "Dan" Black exhibited his best load of JJ cattle. The Krambeck blacks had been given the grand championship over them, which Hereford men generally characterized as a mistake, pointing out that Irwin Bros. bought Mr. Black's Herefords but only took the ribs and loins of the Angus. The latter were killed at Swift's and the former at Armour's. "Billy" Kay, a son of old Scotia, was at that time with Swift's; in fact, he had been with the company some twenty years. He fell into a warm dispute with John Gosling as to the relative merits of the breeds and being somewhat "put to it" for a rejoinder to some of the Hereford arguments advanced, finally blurted out:

"All yer Herefords are guid for is to mak' money for the farr-mers"—with a fine Scotch burr on the latter word. Some time afterwards "Billy", who owned land in Oklahoma, decided to buy some Hereford bulls for his own use.

As to Fashions.—Fortunately there is little occasion for warning the friends of the Hereford against the pitfalls of fads and fashions in dealing with the pedigrees of their breeding animals. In the first place they have always utilized the tabulated pedigree—a form of presenting bloodlines that sets forth the facts in their proper relation. Their colleagues in the Shorthorn and Aberdeen-Angus world have not always been so wise. By the method of pedigree-printing long in vogue among the latter one of the maternal lines was paraded in such way as to build up a more or less mythical basis of family or tribal prestige which, however convenient it may have been, was wholly misleading and tended to exaggerate out of all reason the importance of a certain fractional part of the real pedigree. Some remote female ancestress gave name, and in many cases supplied almost the sole measure of value to great groups of cattle, to the exclusion of the immediate ancestors on both sides of the house.

In so far as they use "family" nomenclature at all, Hereford breeders commonly group their breeding animals under heads that convey a definite meaning. They have their Wiltons, Anxieties, Grove 3ds, Garfields, Beau Donalds, Perfection Fairfaxes, Disturbers, Repeaters, etc., thus laying stress where it of right belongs, on the great producing bulls; and as a rule they have followed a great bull's blood only so far as it appeared to make good in his descendants.

The record price for a Hereford bull has already been noted. The American top for a Shorthorn bull was higher—$17,900 for the 14th Duke of Thornedale—but in his case the bidding was influ-

SAILOR KING. CHAMPION OVER ALL BREEDS—PRINCE OF WALES PRIZE—AT THE ROYAL OF 1913.

MARINER. CHAMPION AT THE ROYAL DUBLIN SHOW OF 1911.

enced to a very large degree by the mere fact that he was of the so-called pure Duchess line. Unlike Beau Perfection 24th, it is more than doubtful if His Grace of Thorndale could ever have gained a prize on his own merits in any great showyard. However, the breeders of Shorthorns long ago got over that folly and are today doing a good business on the sound basis of inherited individual worth.

It is sometimes well to recall the mistakes of others, because weeds grow in every garden and as Mr. John W. Cruickshank of Aberdeenshire once wrote to the author, "the weeds produced even by the good sorts should be carefully avoided." Popularity, however attained, breeds the tendency to save and exalt all the material that emanates from a given source. There is temptation at times to throw the mantle of a great and deserved blood popularity over the defects of animals that by the rigid tenets of the Hereford faith should be discarded. We counsel all true friends of the "white face" to stand steadfastly against any such practice. One of the fundamental elements in Hereford success has been freedom from fads and fancies. Their breeders have never been dragged as slaves at the wheel of an indefensible fashion or pedigree speculation. They have the blessed privilege of buying and using any bull, out of any herd, in England or America, so long as it comes up to herd book requirements, and there are none to say "thou shalt not!" In this liberty lies the seed of all true progress, all lasting success.

Modern English Herefords.—It will be noted that our detailed narration of the Herefordshire side of this story ended with the general suspension of importations about 1890. Writing as we are for American readers and limited as we must be in the matter of space, it has been found quite impracticable to refer specially to what has been doing in more recent years in the old home of the breed. The truth is that our own breeders have felt for a long time past that they had actually passed their colleagues across the sea in the matter of the elevation of the Hereford standard of merit.

And yet the old blood is still doing its wondrous work over there in the beautiful Severn vale. Handed down from father to son, as in the days of old, prized and preserved as the proudest possession a Herefordshire farmer can boast, undisturbed by the ceaseless rise and fall of prices, unmoved by any extraneous influence whatsoever, the descendants of the Hereford fathers, staunch defenders of their heritage, are still producing cattle the equal of any that have hitherto been seen in their native pastures or in the great forum of the Royal showyard. This is clearly indicated by the illustrations of recent prize-winners reproduced in these pages.

And in acknowledging our debt to these steadfast men of Hereford, notwithstanding the claim that we have evolved here a more uniform type of cattle, let us not be hasty in declaring our independence. Great as have been our results in the blending and doubling of the Anxiety, Garfield and other bloods of the old importations, we do not have to go back far to meet the March Ons in the hands of VanNatta, Funkhouser and their contemporaries; and — lest we forget — Kirk Armour brought out from England in comparatively recent

AVONDALE, HIS MAJESTY THE KING'S ROYAL WINNER OF 1914.

years the mother of Perfection Fairfax! So let us, while rejoicing in the marvelous character of our modern American-bred Hereford, not forget that the highest development in the future, as in the past, is most likely to attend a resort to the best material the entire breed affords, regardless of international boundaries or blue water.

Complete liberty of selection, limited only by the

entire Hereford body, both at home and abroad, is the surest way of retaining present excellence and insuring future progress. Proud as we may well be of what we have already accomplished, let us not wrap ourselves up in the mantle of infallibility or self-sufficiency but rather keep minds and hearts open for the reception of all that is of proved goodness wherever and whenever it may be found.

Conclusion.—And so our story ends—leaving off, as it began, with a reference to the old home across the sea. It seems but yesterday that we wandered first among the green fields and apple blossoms of Hereford, but this attempt at following the fortunes of the white-faced cattle has surely led us far and held us long. Our survey of their origin, development and wide distribution up to date, incomplete and fragmentary as it is, must now be brought to a conclusion. Not even all the "high spots" have been touched. Much that should by right have found a place in a volume of this character remains untouched. There is therefore only this to be said:

Nothing that any man has ever done to further the cause of the Hereford cattle, whether it be set forth in written page or not, can ever be really lost. The Herefords of today are the sum total of all the effort that has been put upon them from the days of Ben Tomkins and his predecessors up to now. The mark of every owner, for good or for evil, is on the breed as it stands. Whether public acknowledgement be made or not, the impress of even the least of those who have assumed the responsibility of handling these good cattle has been left upon them. The Herefords of today are what they are by reason only of the skill, or lack of skill, of those who have received them as an inheritance from the generations gone before. And if we may judge of the work of American Hereford breeders as a whole by the type as it exists in our western states as we bring these notes to a conclusion, we may conscientiously enter up the verdict: "Well done, good and faithful servants."

The history of live stock husbandry affords no account of stewardship more honestly, more faithfully fulfilled. American breeders in particular, prone as they are to be influenced by the coming and going of remunerative prices, and not specially inclined to travel in the footsteps of their fathers in any calling, have certainly in this case kept the faith, conserving loyally the material handed down by those who have gone before. Mindful of the debt they owe both to the past and the future, they have not only upheld the highest standards of the olden days, but possibly have set the mark of finish, breed character, quality and prepotency at levels never heretofore attained.

But, after all, the most appealing note developed by this narration is that which records the courage and the bravery of the Hereford wherever there has been peril to be faced on cattle ranges!

Throughout all the world, wherever, in order to improve upon a native stock, there has been a forlorn hope to be led—whether on the ice-bound banks of the Saskatchewan, in the desolation of dusty deserts, under equatorial suns, in Australian bush or upon African veld—there has the lion-hearted Hereford practically walked alone!

Over-lord of the grazing world! Pathfinder of the sands and snows! Filler of feedlots and generous provider of prime beef! From the grass-roots of the plains and prairies that have known the touch of thy hardy hoof more gold has sprung than has ever yet been wrested from all the rocks and rivers of thy vast western kingdom!

THE END.

THE COMING OF THE CATTLE.

[PUBLISHER'S NOTE.—While the author of this volume was writing the concluding paragraphs of the chapter entitled "The Long Trail", he began, purely for his own mental diversion, a Hiawathan imitation having as its motive the dramatic phases of the expulsion and virtual extinction of the Indian tribes, the buffalo and the countless other "children" of the western wilderness that followed the general occupation of the ranges. The first crushing reverses suffered by the cattle, the lessons learned by the pioneer cattlemen through over-stocking and general lack of foresight, and above all the re-formation of their lines under the leadership of the hardy Hereford, supplied the material for its elaboration.

The real story is, of course, already concluded, and the author naturally feels that none but himself can have any special interest in "The Coming of the Cattle". The publishers do not assume to pass upon the presence or absence of literary merit in the composition. It was, as above indicated, not intended for publication. They do believe nevertheless that as an unconventional portrayal of the Hereford's greatest achievement up to date, the adherents of the great grazing breed may possibly wish to possess the fanciful picture painted.]

Ever as the evening shadows
Deepen o'er the plains and prairies,
Ever as the darkness gathers
'Round the foot-hills and the mountains,
In the fire-light there are phantoms,
In the pine-trees mystic murmurs,
Spirit voices calling ever
From the land beyond the sun-set.

There is moon-light on the mesa,
Stars are shining o'er the sages,
And the night-wind from the desert
Bears upon its wings the wailing
Of the red men in their lodges,
Of the dwellers in the cañons,
Of the children of the vegas,
Of the bison on the meadows,
Of the grizzlies in the gulches,
Of the wolves upon the barrens;
And forever in the gloaming
As the Great Bear watches o'er them
Can be heard their plaintive story
Of the peace upon the ranges,
Of the fatness of the grazing,
Of the plenty in the valleys,
Of the shelter in the forest
In the days before the coming
Of the pale-face and the cattle.

Countless moons had passed above them,
Nature's creatures of the dry-lands,
And their comrades of the high-lands.
Generations came and vanished;
Still there came naught to appal them.

Feared they not the fangs of winter,
Nor the flaming breath of summer,
For the North-wind was their keeper
And the South a loving mother;
And the wandering breezes told not,
And the rippling rivers sang not
Of the evil days impending.
But the thunder clouds were hanging
Heavy o'er the hapless races.
Moons of plenty shine not always,
Bluest skies at last are blackened,
Lightnings hover in the sunshine,
Longest trails must have an ending.
And there came the day of waking.

Signs portentous in the heavens,
Fires by night and clouds at noon-day,

Copyright photo by Erwin E. Smith

"Fires by night and clouds at noon-day."

Told of trampling hosts advancing,
From the distant Rio Grandé.

Hoofs were heard along the Brazos,
Horns were tossing on the Pecos!
From the far-off Southern pastures,
From the waters of the Concho,
From the grassy realms of Texas,
Day by day in countless numbers
Pressed the cattle to the conquest.
Northward, Westward, ever Northward,
Toward the sunny plains of Kansas,
Toward the walls of Colorado.

Night by night their bed-grounds found them
Nearer still and always nearer
To the nameless unknown perils
Of the Northland they had entered
On the trails that led not backward.

Not the pangs of thirst nor hunger,
Not the Northern storm-clouds' warning,
Not the stampede in the darkness,
Not the seas of fire that threatened
On the wind-swept blazing prairies
Stayed them in their great migration
As they journeyed ever onward
Toward the sand hills of Nebraska,
Toward the Bad Lands of Dakota,
Northward, Westward, ever Northward.

And the Chinook came to cheer them.
Higher still and ever higher
Newer pastures bloomed and beckoned.
Where the Yellowstone was flowing,
Where the wide Missouri wandered,
Where Montana's peaks were gleaming,
Where the Big Horn dreamed of battle,
Where Wyoming's highest ranges
Led up to the lofty passes,
To the parting of the waters,
Came the cow-men and their cattle,
Came the bronco and the buster,
Came the camp-fire and the cabin,
Came the round-up and the branding.

Where the silent snowy summits
Guard the Colorado's sources,
Where the darkly-frowning forests
Hide the Rio Grandé's fountains,
Lo, the west wind came a-sighing,

"Came the cow-men and their cattle."

"Came the bronco and the buster."

Came a-telling of the coming
Of the cattle to the empire
That belonged to Montezuma
In the days before the Spaniards.
Told of hoof-prints of the Longhorn
And of lowing herds a-basking
In the sunshine everlasting,
Where the antelope and bison
And the cliff-men of the cañons
Had for ages all unbroken
Roamed and reared their happy children.

Vainly had the dread Mojavé,
Vainly had the high Sierra
Stayed the coming of the cattle
On the trail of Coronado;
For they failed not in their daring
'Til beyond the burning desert
Far beyond the jagged sky-line
In a flowery land and fruitful
Billows beating on the sand-dunes,
Thundering on the rocky headlands,
Marked the ending of the grazing.

From their ancient haunts the hunted
Creatures that the wild had nurtured,
Driven from their lands and waters,
Now in sullen stealth retreated
To their secret rocks of refuge,
Calling on their sleeping war-gods:
Prayed that elemental furies
Might be loosed upon the ranges.

And the strangers all unconscious
That the earth would soon be shaking
With the anger of the heavens
Went their way in peace and feared not.

As the eagle from his eyrie
Hurls himself upon his quarry,
As the arrow from the cord flies,
As the lion on his prey springs,
As a wounded herd bull charging,
So the wilderness revolted;
So did Manitou awaken,
Swift to punish and to chasten.

Through the North-land Arctic demons
Rode the frozen ice-bound ranges;
Through the Southland fiery dragons
Scourged the earth with blazing horrors.
Then the drifting to the death-traps!
Hopeless struggling of the helpless!
Herds a-wreck from drouth and famine!
Bleaching bones to tell the story!

As the spear by shield is shattered,
As the shore turns back the waters,
As the rock resists the torrent,
So the wild enforced her mandates,
Claimed her tribute of the reckless,
Taught the lesson of the ages.
Nature brooks not mad defiance!

But the earth renewed its fruitage.
Sunbeams dancing on the ranges,
Waters from the purple mountains,
Soft airs from the Western ocean,
Called the grasses from their slumbers,
Clothed again the world with verdure.
And again the herds were gathered,
Not with folly in the councils,
Not with blind chiefs in the saddles.
Children scorched by fire have wisdom.

On the trails that led not backward
Once again the cattle entered;
Once again the herds were scattered
Far and wide across the pastures;
At their head a pale-faced stranger
Staunch of limb and lion-hearted,
From beyond the deep sea waters,
From the distant shores of England.
His the heritage of ages
From the hills of grim Glamorgan;
His the power that was descended
Through the Hereford generations,
From the wearing of the burdens
Of the yoke of heavy hauling,
From a life of toil and travail
In the service of his masters.

Proud the bearing of this chieftain
As he armed them for the battle;
Wrapped them in red robes of courage,
Bound them by the ties of kindred

As of tribes by blood united;
Filled them with his dauntless spirit,
Taught them how to meet privations,
Taught them how to face the northers,
Winter's stress and summer's terrors;

Copyright photo by Erwin E. Smith

"At their head a pale-faced stranger."

"Taught them how to face privations."

Fought their fight through many perils,
Led them bravely through all dangers,
Grasped dominion of the ranges,
Held them in secured possession,
Brought the cattle to their kingdom.

As the leaves fall in October,
As the stream dies in the quicksands,
As the snow melts in the sun rays,
So the children of the open,
Of the mountain, plain and valley,

"Brought the cattle to their kingdom."

Copyright photo by Erwin E. Smith

"Fled before the conquering cattle."

Fled before the rail and rifle,
Fled before the conquering cattle,
Farther still and ever farther
To the bosom of the river
That is bearing them forever
Through the land of the Hereafter.

www.ingramcontent.com/pod-product-compliance
Lightning Source LLC
Chambersburg PA
CBHW062346220526
45472CB00008B/1718